Advanced VLSI Design
and Testability Issues

Advanced VLSI Design and Testability Issues

Edited by
Suman Lata Tripathi, Sobhit Saxena, and
Sushanta Kumar Mohapatra

CRC Press
Taylor & Francis Group
Boca Raton London New York

CRC Press is an imprint of the
Taylor & Francis Group, an **Informa** business

First edition published 2020
by CRC Press
6000 Broken Sound Parkway NW, Suite 300, Boca Raton, FL 33487-2742

and by CRC Press
2 Park Square, Milton Park, Abingdon, Oxon, OX14 4RN

© 2021 Taylor & Francis Group, LLC

CRC Press is an imprint of Taylor & Francis Group, LLC

ISBN: 978-0-367-49282-3 (hbk)
ISBN: 978-0-367-53836-1 (pbk)
ISBN: 978-1-003-08343-6 (ebk)

Typeset in Times
by codeMantra

Contents

Preface to the First Edition ix
Editors... xiii
Contributors xv

Chapter 1   Digital Design with Programmable Logic Devices 1

                            M. Panigrahy, S. Jena, and R. L. Pradhan

Chapter 2   Review of Digital Electronics Design 17

                            Reena Chandel, Dushyant Kumar Singh, and P. Raja

Chapter 3   Verilog HDL for Digital and Analog Design 39

                            Ananya Dastidar

Chapter 4   Introduction to Hardware Description Languages 67

                            Shasanka Sekhar Rout and Salony Mahapatro

Chapter 5   Introduction to Hardware Description Languages (HDLs)...............93

                            P. Raja, Dushyant Kumar Singh, and Himani Jerath

Chapter 6   Emerging Trends in Nanoscale Semiconductor Devices 111

                            B. Vandana, B. S. Patro, J. K. Das, Sushanta Kumar Mohapatra, and Suman Lata Tripathi

Chapter 7   Design Challenges and Solutions in CMOS-Based FET 129

                            Madhusmita Mishra and Abhishek Kumar

Chapter 8   Analytical Design of FET-Based Biosensors 147

                            Khuraijam Nelson Singh and Pranab Kishore Dutta

Chapter 9   Low-Power FET-Based Biosensors 169

                            Prasantha R. Mudimela and Rekha Chaudhary

Chapter 10 Nanowire Array–Based Gate-All-Around MOSFET for
Next-Generation Memory Devices 189

Krutideepa Bhol, Biswajit Jena, and Umakanta Nanda

Chapter 11 Design of 7T SRAM Cell Using FinFET Technology 203

T. Santosh Kumar and Suman Lata Tripathi

Chapter 12 Performance Analysis of AlGaN/GaN Heterostructure
Field-Effect Transistor (HFET) 215

Yogesh Kumar Verma and Santosh Kumar Gupta

Chapter 13 Synthesis of Polymer-Based Composites for Application in
Field-Effect Transistors 225

Amit Sachdeva and Pramod K. Singh

Chapter 14 Power Efficiency Analysis of Low-Power Circuit Design
Techniques in 90-nm CMOS Technology 233

*Yelithoti Sravana Kumar, Tapaswini Samant, and
Swati Swayamsiddha*

Chapter 15 Macromodeling and Synthesis of Analog Circuits 251

B. S. Patro and Sushanta Kumar Mandal

Chapter 16 Performance-Linked Phase-Locked Loop Architectures: Recent
Developments 271

*Umakanta Nanda, Debiprasad Priyabrata Acharya,
Prakash Kumar Rout, Debasish Nayak, and Biswajit Jena*

Chapter 17 Review of Analog-to-Digital and Digital-to-Analog Converters
for A Smart Antenna Application 291

B. S. Patro, A. Senapati, and T. Pradhan

Chapter 18 Active Inductor–Based VCO for Wireless Communication 311

*Aditya Kumar Hota, Shasanka Sekhar Rout, and
Kabiraj Sethi*

Chapter 19 Fault Simulation Algorithms: Verilog Implementation....................325

Sobhit Saxena

Chapter 20 Hardware Protection through Logic Obfuscation............................339

Jyotirmoy Pathak and Suman Lata Tripathi

Index...351

Preface to the First Edition

The objective of this edition is to provide a broad view of VLSI design in a concise way for fast and easy understanding. This book provides information regarding almost all the aspects to make it highly beneficial for all the students, researchers, and teachers of this field. Fundamental principles of advanced VLSI design and testability issues are discussed herein in a clear and detailed manner with an explanatory diagram wherever necessary. All the chapters are illustrated in simple language to facilitate readability of the chapters.

CHAPTER ORGANIZATION

This book is organized in 20 chapters.

Chapter 1 covers various aspects of digital design with programmable devices, e.g., programmable logic arrays, complex programmable logic device, and field-programmable gate array (FPGA). Moreover, this chapter emphasizes on uses of FPGA on various image-processing and biomedical applications.

Chapter 2 mainly focuses on the review of the basic concepts of digital design, i.e., number system, logic families, combinational and sequential circuits, registers, counters, and memories.

Chapter 3 deals with the basics of Verilog HDL and its special features toward implementation of digital design as well as analog design. It starts with a background on the emergence and importance of hardware description languages (HDLs) with a brief introduction to VHDL and Verilog HDL.

Chapter 4 focuses on the basic understanding of digital design in perspective of HDLs and its structure, which will enlighten the orientation of its usage with the various scopes of HDL. Next queued up some advanced topics such as high-level synthesis, HDL tool suites, and HDL-based design flows. Depending upon the basic ideology of HDLs, the applications and various types of HDLs for analog circuits, digital logic, and printed circuit boards (PCBs) are elaborated.

Chapter 5 introduces basics of various HDLs available for analog circuit design, digital circuit design, and PCB design. The last two sections of the chapter are devoted to the most popular HDLs for digital design, i.e., VHDL and Verilog.

Chapter 6 describes a general overview of the various technologies, materials, and architectures; researchers are concentrating to continue the technology beyond Moore's law with low power consumption.

Chapter 7 presents the ebb and flow of current research related to the replacement of silicon by new material or using new geometry for silicon-based FET (field-effect transistor) design, considering various design parameters while providing some desired solutions.

Chapter 8 begins with the introduction of the biosensor, followed by its classification. Then a detailed study of the FET-based biosensor which includes the detection principle and its types is presented. The detection principle is based on the variation of electrical characteristics of the device upon the interaction of biomolecules and the biosensor since every biomolecule has distinct dielectric and charge values.

Chapter 9 describes the study of biomolecular interactions with Bio-FETs (FET-based biosensors) that have been developed so far. MOSFET (metal oxide semiconductor FET)-based biosensors are used to find DNA–protein interactions, whereas enzyme FETs are glucose-sensitive sensors. FET-based biosensors can be used in clinical applications such as cardiovascular diseases, cancers, diabetes, and HIV diagnosis. This chapter begins with the introduction of FET and biosensors, followed by types of Bio-FETs, working principle, and applications.

Chapter 10 deals with the arrangement of nanowires to form the channel that may play a vital role during device performance metrics calculation. Distribution of nanowires in the different geometrical arrangements may exhibit better performance compared with random arrangements of the nanowires. With efficient driving capability, the proposed device can be a fruitful member of GAA (gate-all-around) family for next-generation memory applications.

Chapter 11 presents different 6T, 7T, 8T, and 12T SRAM (static random access memory) cell architectures based on CMOS (complementary metal oxide semiconductor) technology. Different leakage power reduction techniques for SRAM cell design have been discussed. The chapter also describes SRAM cell design with advanced FET such as FinFET and compares the same with existing CMOS-based SRAM cell.

Chapter 12 deals with the performance analysis of AlGaN/GaN heterostructure field-effect transistor with respect to different spacing between source–gate and gate–drain. It has been revealed that the drain current reduces significantly when the spacing between source and gate is reduced.

Chapter 13 describes the composite materials having properties different from their constituent elements that are used in a large number of applications such as capacitors, supercapacitors, electrolytes, FETs, electrochromic applications, and so on. Organic FET is a three-terminal device consisting of the drain, source, and gate.

Chapter 14 presents two adiabatic logic to reduce the leakage power consumption; these are positive-feedback adiabatic logic and two-phase adiabatic static clocked logic. Low-power and high-speed digital circuits are elementary for any digital circuit design. In this chapter, different design techniques are introduced, such as low-risk conventional technique, CMOS, transmission gate, fate diffusion input, pass-transistor logic, and dual pass-transistor logic, and their comparison analysis is discussed on the basis of power, delay and area (number of transistors).

Chapter 15 mainly focuses on a review of various types of macromodeling techniques used for analog integrated circuits (ICs). So various macromodels such as symbolic models, polynomial models, and neural models, and so on have been analyzed based on the literature. These macromodels are used in place of SPICE simulation in the circuit synthesis flow.

Chapter 16 provides a concise survey of the latest phase-locked loop (PLL) techniques where proficiency is achieved by deploying different methods or architectural modifications for different building blocks and the performance of the PLLs is comprehensively presented.

Chapter 17 deals with the digital beamforming that has many of the advantages over its analog counterpart. In most cases, less power is needed to perform the beam steering of the phased array antenna. This chapter will focus on the main components involved in the implementation of digital beamforming such as analog-to-digital converters, digital down converters, and so on.

Chapter 18 describes a cross-coupled pair of transistors that are used to generate negative resistance. Again, active inductors (AIs) are also used in the circuit to make the voltage-controlled oscillator (VCO) more tunable over a larger bandwidth as compared with the passive inductor–based VCO, where AIs are there for coarse tuning and variable capacitors are for fine-tuning of the frequency of oscillation.

Chapter 19 describes the basics of simulation and its categorization into logic and fault simulation. The importance of fault simulation algorithms before actual testing on hardware is explained. The evolution of algorithms for reducing test time from the simplest serial to differential fault simulation algorithm is described by explaining each algorithm. The concept of fault simulation is further practically implemented using Verilog language in which parallel and concurrent algorithms are implemented and simulation results in terms of waveforms are explained for better understanding.

Chapter 20 deals with hardware obfuscation that basically means hiding the IC's structure and function, which makes it much more difficult to reverse-engineer by the adversaries. Thereby, obfuscation provides a means of making the circuit structurally and functionally difficult to comprehend, which increases the cost and time required to reverse-engineer it, providing security.

MATLAB® is a registered trademark of The MathWorks, Inc. For product information, please contact:

The MathWorks, Inc.
3 Apple Hill Drive
Natick, MA 01760-2098 USA
Tel: 508-647-7000
Fax: 508-647-7001
E-mail: info@mathworks.com
Web: www.mathworks.com

Editors

Dr. Suman Lata Tripathi is associated with Lovely Professional University as a professor with more than 17 years of experience in academics. She has published more than 35 research papers in referred journals and conference. She has organized several workshops, summer internships, and expert lectures for students. She has worked as session chair, a conference steering committee member, an editorial board member, and a reviewer in international/national IEEE journals and conferences. She has been nominated for "Research Excellence Award" in 2019 at Lovely Professional University. She had received best paper at IEEE ICICS-2018. Her area of expertise includes microelectronics device modeling and characterization, low-power VLSI circuit design, VLSI design of testing, and advanced FET design for Internet-of-things (IOT) and biomedical applications, etc.

Dr. Sobhit Saxena has a vast teaching experience of more than 10 years in various colleges and universities. He has designed a new hybrid system of Li-ion battery and supercapacitor for energy storage application. He worked as an SEM (scanning electron microscopy) operator for 4 years against MHRD fellowship. His area of expertise includes nanomaterial synthesis and characterization, electrochemical analysis, and modeling and simulation of CNT-based interconnects for VLSI circuits.

Dr. Sushanta Kumar Mohapatra is working as an assistant professor in School of Electronics Engineering, Kalinga Institute of Industrial Technology, Bhubaneswar. He has more than 18 years of teaching experience. He has extensively contributed as author and coauthor in several national and international journals and conferences of repute. He also received the "Best Paper Award" at International Conference on Microelectronics, Communication and Computation, San Diego, United States. He has been a part of committee member of various international conferences, an editorial board member, and a reviewer of many international journals. He is a life member of ISTE, IETE, CSI, and OITS and senior member of IEEE. His research interests include modeling and simulation of nanoscale devices and their application in IoT such as energy-efficient wireless sensor networking, ad hoc networks, cellular communications, metamaterial absorbers in THz application, UWB-MIMO antenna, reconfigurable antenna, and performance enhancement for high frequency.

Contributors

Debiprasad Priyabrata Acharya
Department of Electronics and
    Communication Engineering
NIT Rourkela
Rourkela, India

Krutideepa Bhol
School of Electronics Engineering
VIT-AP University
Amaravati, India

Reena Chandel
Department of Sciences
Dr. Y.S Parmar University of
    Horticulture and Forestry
Solan India

Rekha Chaudhary
School of Electronics and Electrical
    Engineering
Lovely Professional University
Jalandhar, India

Ananya Dastidar
Electronics and Communication
    Engineering
College of Engineering and Technology
Bhubaneswar, India

J. K. Das
School of Electronics Engineering
Kalinga Institute of Industrial
    Technology, KIIT Deemed to be
    University
Bhubaneswar, India,

Pranab Kishore Dutta
Department of Electronics and
    Communication Engineering
North Eastern Regional Institute of
    Science and Technology
Itanagar, India

Santosh Kumar Gupta
Department of Electronics and
    Communication Engineering
Motilal Nehru National Institute of
    Technology Allahabad
Prayagraj, India

Aditya Kumar Hota
Department of Electronics and
    Tele-communication Engineering
Veer Surendra Sai University of
    Technology
Burla, India

Biswajit Jena
Department of Electronic and
    Communication Engineering
Koneru Lakshmaiah Education
    Foundation
Guntur, India

S. Jena
School of Electrical Engineering
Kalinga Institute of Industrial
    Technology, KIIT Deemed to be
    University
Bhubaneswar, India

Himani Jerath
School of Electronics and Electrical
    Engineering
Lovely Professional University
Jalandhar, India

Abhishek Kumar
Department of Electronics and
    Communication Engineering
IIT Jodhpur
Karwar, India

T. Santosh Kumar
School of Electronics and Electrical
    Engineering
Lovely Professional University
Jalandhar, India

Yelithoti Sravana Kumar
School of Electronics Engineering
Kalinga Institute of Industrial
    Technology, KIIT Deemed to be
    University
Bhubaneswar, India

Salony Mahapatro
Digital Design Engineer
Asiczen Technologies
Bhubaneswar, India

Sushanta Kumar Mandal
School of Electronics Engineering
Kalinga Institute of Industrial
    Technology, KIIT Deemed to be
    University
Bhubaneswar, India

Madhusmita Mishra
Department of Electronics &
    Communication Engineering
NIT Rourkela
Rourkela, India

Sushanta Kumar Mohapatra
School of Electronics Engineering
Kalinga Institute of Industrial
    Technology, KIIT Deemed to be
    University
Bhubaneswar, India

Prasantha R. Mudimela
School of Electronics and Electrical
    Engineering
Lovely Professional University
Jalandhar, India

Umakanta Nanda
School of Electronics Engineering
VIT-AP University
Amaravati, India

Debasish Nayak
Department of EIE
Silicon Institute of Technology
Bhubaneswar, India

M. Panigrahy
Kalinga Institute of Industrial
    Technology, KIIT Deemed to be
    University
Bhubaneswar, India

Jyotirmoy Pathak
School of Electronics and Electrical
    Engineering
Lovely Professional University
Jalandhar, India

B. S. Patro
School of Electronics Engineering
Kalinga Institute of Industrial
    Technology, KIIT Deemed to be
    University
Bhubaneswar, India

R. L. Pradhan
HAL, SLARDC
Radar, Hyderabad

T. Pradhan
Kalinga Institute of Industrial
    Technology, KIIT Deemed to be
    University
Bhubaneswar, India

P. Raja
School of Electronics and Electrical
    Engineering
Lovely Professional University
Jalandhar, India

Prakash Kumar Rout
Department of EIE
Silicon Institute of Technology
Bhubaneswar, India

Shasanka Sekhar Rout
Department of Electronics and
    Communication Engineering
GIET University
Gunupur, India

Tapaswini Samant
School of Electronics Engineering
Kalinga Institute of Industrial
    Technology, KIIT Deemed to be
    University
Bhubaneswar, India

Amit Sachdeva
School of Electronics and Electrical
    Engineering
Lovely Professional University
Jalandhar, India

Sobhit Saxena
School of Electronics and Electrical
    Engineering
Lovely Professional University
Jalandhar, India

A. Senapati
School of Electronics Engineering
Kalinga Institute of Industrial
    Technology, KIIT Deemed to be
    University
Bhubaneswar, India

Kabiraj Sethi
Electronics and Tele-communication
    Engineering
Veer Surendra Sai University of
    Technology
Burla, India

Dushyant Kumar Singh
School of Electronics and Electrical
    Engineering
Lovely Professional University
Jalandhar, India

Khuraijam Nelson Singh
Department of Electronics and
    Communication Engineering
North Eastern Regional Institute of
    Science and Technology
Itanagar, India

Pramod K. Singh
Materials Research Laboratory
Sharda University
Greater Noida, India

Swati Swayamsiddha
School of Electronics Engineering
Kalinga Institute of Industrial
    Technology, KIIT Deemed to be
    University
Bhubaneswar, India

Suman Lata Tripathi
School of Electronics and Electrical
    Engineering
Lovely Professional University
Jalandhar, India

B. Vandana
Department of Electronics and
    Communication Engineering
KG Reddy College of Engineering and
    Technology
Hyderabad, India

Yogesh Kumar Verma
School of Electronics and Electrical
    Engineering
Lovely Professional University
Jalandhar, India

1 Digital Design with Programmable Logic Devices

M. Panigrahy and S. Jena
Kalinga Institute of Industrial Technology

R. L. Pradhan
HAL, SLARDC

CONTENTS

1.1 Introduction 1
1.2 Factory-Programmable Devices 2
1.3 Read-Only Memory 2
1.4 Programmable Read-Only Memory 4
1.5 Erasable Programmable Read-Only Memory 4
1.6 Electrically Erasable Programmable Read-Only Memory....................5
1.7 Field-Programmable Devices 5
1.8 Programmable Array Logic...6
1.9 Programmable Logic Array...7
1.10 Generic Array Logic Devices...8
1.11 Complex Programmable Array Logic 8
1.12 Field-Programmable Gate Array 10
    1.12.1 Internal Architecture of Field-Programmable Gate Array.............10
        1.12.1.1 Configurable Logic Blocks 10
        1.12.1.2 I/O Blocks 12
        1.12.1.3 Programmable Interconnects 12
    1.12.2 Design Flow of Field-Programmable Gate Array 12
    1.12.3 Applications of Field-Programmable Gate Array in
        Medical Imaging..14
1.13 Summary 14
Bibliography 15

1.1 INTRODUCTION

Programmable logic devices (PLDs) are integrated circuits, used for implementing multiple functions that can be modified by programming and can be reconfigured by the programmer according to the need of end user. These devices offer high performance and flexibility to any digital system.

1

Digital designs mainly consist of logic functions made up of logic gates, flip-flops, integrated circuits (ICs), and their combinations. Usually, this association involves long interconnections, consumes large board space, and requires enough design time for proper insertion and testing. Additionally, power consumption is one of the major concerns in these designs. Apart from functional specifications such as speed, power, and size, features such as programming yield, production complexity, reprogramming, and debugging after manufacturing and software support are also important concerns while evaluating chip performance. PLDs overcome these difficulties and make the system more reliable and allow further modification at later instant with reduced debugging effort. Desirable features such as reduced chip size, faster execution, and low power consumption achieved by PLD attract users for several applications. Available PLDs are based on either bipolar or complementary metal oxide semiconductor (CMOS) technology.

Programmable devices are broadly classified into two types, i.e., factory programmable and field programmable, depending upon the flexibility of programmability. One-time programmable devices that can only be programmed by the manufacturer at the factory and cannot be updated by the user at later instants are called factory-programmable devices. After programming. i.e., after the metallization mask step, these devices are tested at high temperature for stability and usually available at wide frequency range. Examples of factory-programmable devices are read-only memory (ROM) and mask-programmable gate array (MPGA). On the contrary, field-programmable devices can be programmed a number of times and have much programming flexibility with the user without involving semiconductor fab. These devices require specialized equipment and software for programming. These devices have comparatively small frequency range and are used for a wide variety of applications. Nowadays, PLDs are also available with security features and include security bits in its standard. Figure 1.1 shows the hierarchical structure of PLDs.

1.2   FACTORY-PROGRAMMABLE DEVICES

This section provides different programming implementation approaches and their internal architectures.

1.3   READ-ONLY MEMORY

ROM is a factory-programmable memory circuit that is programmed in the IC manufacturing lab. Mask programming is intended for such high-density devices where the configuration cannot be altered once being manufactured. Information can be extracted from any random memory location. Access time of all locations is the same.

These devices store binary data in an array format. Binary bits in 1's and 0's form combine together to form a word that represents numbers, alphanumeric characters, and so on. Eight bits are grouped to form a *byte*. Memory size of ROM is generally specified by the number of words that it stores along with its word length. Each word is accessed from a memory location that is addressed by address lines. For example, an 8K byte memory stores 8K *words* of 8-bit *length*. To access these 8K (i.e., $2^3 \times 2^{10}$) words, 13-bit address line is required.

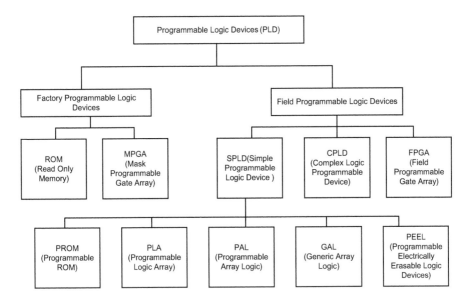

FIGURE 1.1 PLD hierarchical structure.

A decoder followed by a memory array constitutes the basic ROM structure as shown in Figure 1.2. Decoder generates 2^n combinations with n-bit input. Each of these combinations represents the address for each word stored in the memory array. The memory array architecture is represented by a 2D array where each row and column represents a word and a bit line. Usually, a number of words present in the memory are much larger compared with the bit length. This leads to a tall structure that needs long vertical interconnections and creates problem in chip floor planning. To overcome this difficulty, larger memory is subdivided into multiple smaller

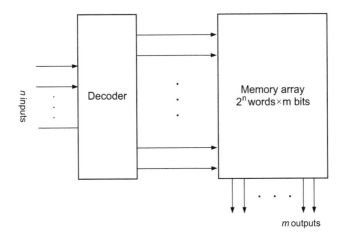

FIGURE 1.2 Basic ROM structure. ROM, read-only memory.

memories. To have low power dissipation and better control over these memory organizations, row and column decoders are included in the memory architecture.

As the memory size is fixed by the manufacturer and cannot be updated by the user, these devices are mainly used for high-density well-verified data such as firmware updates.

1.4   PROGRAMMABLE READ-ONLY MEMORY

Programmable read-only memory (PROM) is a type of memory that stores data permanently like ROM. However, programming capability of this device is enhanced as the data can be changed after fabrication by blowing on chip fuses. A special device called *PROM programmer* can be used to provide a sufficiently high-voltage pulse to store electronic charges. This type of devices are usually considered as nonvolatile memory and programmed only once after manufacturing.

Internal architecture of PROM contains a fixed AND plane and programmable OR plane. Fixed AND plane provides all possible product terms and is created by nonfusible hard-wired link. As n input variables are available in their normal and complementary form, the AND array provides 2^n min-terms.

1.5   ERASABLE PROGRAMMABLE READ-ONLY MEMORY

Erasable programmable read-only memory (EPROM) is advancement over PROM devices and was developed by Dov Frohman in 1971 at Intel. Programming in this device is achieved by passing ultraviolet (UV) rays over the PROM chip. Initially, existing data in the memory are erased by passing UV light through the transparent quartz crystal opening on the top of EPROM. The UV light cleans the data on the chip for reprogramming. However, this process of erasing can be performed a limited number of times, as high-intensity UV ray may damage the SiO_2 layer. This action may cause reliability issues in the memory chip. To program EPROM, one needs to remove the chip from the board, and erasing can be performed selectively.

Floating gate transistors form an array structure to store data in an EPROM. In floating gate metal oxide semiconductor, actual gate remains isolated, forming a floating DC node. Inputs are provided to the secondary gates, which are connected capacitively to the floating node. Resistive materials surrounding the floating gate retain the charge for longer instance and can be erased by applying high voltage. Modification in this memory is possible in two phases, namely, tunneling and hot carrier injection by means of two transistors, namely, storage transistor and access transistor.

Main advantages of EPROM are as follows:

- Programming flexibility
- Low-cost logic implementation
- Nonvolatile memory
- Better testing and debugging facility

However, EPROM devices are less cost-effective compared with PROM. Use of UV ray for erasing requires special tool and programming skill. Again, reprogramming process increases the design time and slows down the overall system. EPROM is used as bootstrap loader for storing computer bios and has also been used for micro-controller for making video games.

1.6 ELECTRICALLY ERASABLE PROGRAMMABLE READ-ONLY MEMORY

Electrically erasable programmable read-only memory (EEPROM or E^2ROM) is a nonvolatile ROM chip used for storing small amount of data. Data stored in this type of memory are permanent until the user modifies its content. EEPROM works on the same basic principle of an EPROM device. However, this type of memory can be erased and reprogrammed by using electrical signal. Content of this chip can be modified byte per byte without removing it from the board. The erasing and rewrit-ing data into the memory chip is a time-consuming process that operates at a slower rate compared with random access memory (RAM). Hence, usually the stored data in an EEPROM are extracted at the start-up phase for any system.

The first EEPROM was developed by Intel. At the initial phase of development of EEROM, an external high-voltage source was necessary for erasing operation. Later on, requirement of external voltage source for EEPROM was avoided by integrating the memory chip with inbuilt voltage source.

Operation of this memory device is dependent on the electrical interface. Accordingly, EEPROMs are operated with serial or parallel interface. Examples of serial interface protocols are serial peripheral interface, MICROWIRE, I2C, and so on. Serial protocol transfers data bit by bit in serial manner with fewer pins and is interfaced to slower devices such as EEPROM, microcontrollers, and analog-to-digital converter (ADC) and digital-to-analog converter (DAC) circuits.

Parallel EEPROMs are usually available with industrial and military grade and used in most of the avionics and telecommunication applications. These devices inherit comparatively lager pin count and are operated at a faster rate. A parallel EEPROM uses an 8-bit bus for data transfer and a wide address bus to access the whole memory. Examples of parallel EEPROMs for military- and industrial-grade applica-tions are AT28C256 and AT28HC64BF, respectively, from Microchip Technology. Nowadays, EEPROM becomes an important part in most of the advanced microcon-trollers such as Atmega328 for storing small chunks of data.

1.7 FIELD-PROGRAMMABLE DEVICES

Field programmable devices enable programmers to modify the logic and reprogram after being manufactured. Programming flexibility depends upon the technology and available resources. Depending upon device density, field-programmable devices can be simple or complex. Programmable logic arrays (PLAs), programmable array logic (PAL), and generic array logic (GAL) are the subsets of simple programmable logic devices (SPLDs). Similarly, complex programmable logic device (CPLD) and field-programmable gate array (FPGA) deal with complex logic functions and include enhanced feature for their application in several domains.

1.8 PROGRAMMABLE ARRAY LOGIC

PAL is also a fuse-based PLD where the AND array is programmable, but the OR array is fixed. To accommodate different logic functions with fixed number of OR gate, Boolean logic minimization techniques are adopted. PAL devices use K-maps to simplify the logic with reduced number of min-terms. Usually, PAL structure is associated with few macrocells that are generated by partitioning large array into small sections. A macrocell contains a register that can be configured to different flip-flops, namely D-FF, T-FF, etc. Consider, for example, realization of four functions (F_0, F_1, F_2, and F_3) with three variables A, B, and C as given in Equation (1.1) using PAL. Internal structure for this realization is shown in Figure 1.3.

$$F_0 = \sum m(0,1,2,6) = A'B' + BC'$$

$$F_1 = \sum m(2,3,5,6,7) = AC + B$$

$$F_2 = \sum m(0,1,3,7) = A'B' + BC$$

$$F_3 = \sum m(1,2,3,5,7) = C + A'B \tag{1.1}$$

Advantage of PAL structure over both PLA and PROM is that a register can be incorporated with this device. These registers can be used for feeding back the output signals toward input. Instead of using a number of min-terms multiple times to form other functions, intermediate results are fed back to the input using the register. This reduces the total number of min-terms required and simplifies the logic implementation.

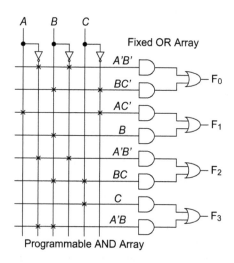

FIGURE 1.3 Internal structure for PAL. PAL, programmable array logic.

In case of PAL and PROM devices, one of the arrays is programmable and other is fixed, which limits its design flexibility and creates performance issue at high-volume applications.

1.9 PROGRAMMABLE LOGIC ARRAY

PLA is a subset of PLD that includes set of programmable AND array and OR array to achieve a variety of logic functions. These Boolean functions are usually presented in sum-of-product form. Min-terms are produced by the AND plane, whereas OR plane adds the required set of min-terms to create different functions. N-type metal oxide semiconductor transistors are generally used as switches to enable interconnections between input variables and AND plane or between min-terms and OR plane.

Each of the inputs is supplied to the AND plane in its normal and complementary form. Thus, for N inputs, 2^N combinations are possible that led to an AND plane with 2^N number of AND gates. Similarly, to realize P functions, with these N inputs, a programmable OR plane is involved that contains P number of OR gates. With increasing order of input variables, the number of min-terms and, hence, size of AND plane will increase sharply. This restricts the functionality of PLA. Figure 1.4 demonstrates the realization of four functions (F_0, F_1, F_2, and F_3) as given in Example 1 using PLA.

To compose different outputs, product term generated by an AND gate can be shared by any number of OR gates. For example, both functions F_0 and F_2 resulted from the OR plane share min-term A'B'.

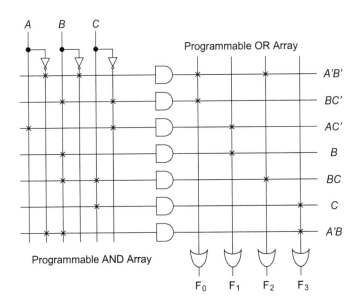

FIGURE 1.4   Internal architecture for programmable logic array.

PLA can be mask programmable or field programmable. It can be used to provide control over data path, as a counter, as a decoder, and as a BUS interface in programmed I/O. Main advantages of PLA approach are as follows:

- Accurate prediction of chip area, speed, and power dissipation is possible due to the regular structure of PLA.
- Two-level logic implementation enables higher programming flexibility. Such devices are suitable for implementing finite state machines for controller and sequencer circuits.

Main problems associated with PLA, PAL, or PROM are as follows:

- Array size grows as the implementation deals with complex logic functions. This causes degradation in performance and programming density.
- These devices perform only combinational logic implementations.

1.10   GENERIC ARRAY LOGIC DEVICES

The GAL device is a variety of electrically erasable programmable device that can be programmed or reprogrammed. This device contains a fixed OR plane and a programmable AND plane similar to a PAL structure. GAL devices use electrically erasable CMOS cells for making connection within the grid structure, i.e., between columns and rows instead of using fuses as in PAL devices. Further flexibility in GAL design is achieved by output logic macrocells (OLMCs). Programmable output cells incorporate combinational and sequential unit with output enable features. OLMC includes MUX, registers, and several feedback lines to enhance the programming capability. Additionally, GAL devices can directly be programmed on existing PAL JEDEC files. With the help of RAL device, a GAL device can be configured to emulate a PAL device.

A variety of GAL devices are available in market providing different performance based on variation of features such as speed, low power, supply voltage compatibility, operating frequency, OLMC count, and so on. GAL devices such as GAL16LV8 and GAL20LV8 from Lattice Semiconductor operate at 3.3 V and replace most of the 20- and 24-pin PAL devices. Similarly, GAL26CLV12 is featured as the world's fastest PLD operates at 4 ns. Use of CMOS technology lowers the power consumption of GAL devices. Examples of zero power–operated and low power–operated GAL from Lattice Semiconductor are 22LV10Z/ZD and 16LVC, respectively. Advantages obtained by GAL devices are usually at higher expense compared with PAL devices. Other manufactures such as Cypress Semiconductor, Atmel, and Excel Microelectronics are producing PLDs with added features to GAL.

1.11   COMPLEX PROGRAMMABLE ARRAY LOGIC

Implementing complex functions with SPLDs such as PLA or PAL demands a large number of min-terms. Increasing the number of min-terms makes the AND array size very large and creates problem during chip floor planning. Also, this lowers the speed of operation. This limits the uses of SPLD to SSI and MSI applications, dealing with the small number of inputs and outputs. To increase the speed of operation without increasing chip size exponentially, CPLDs are evolved.

CPLD structures consist of multiple SPLDs with the small number of programmable interconnection between them. Density of CPLDs is much larger compared with that of the PAL devices. While PALs include hundreds of logic gates, CPLDs contain thousand to ten thousand logic gates that make it suitable for medium complex logic implementation. CPLDs also contain registers enabling them to be used for sequential circuit design.

In general, a CPLD may be viewed to have a number of logic blocks, a dedicated I/O block and a switch matrix interconnecting them. I/O block consists of I/O elements that provide buffering for the input and output signals. The logic block is formed by several logic elements known as macrocells. A macrocell comprises AND and OR arrays, a dedicated flip-flop, and control signals for implementation of the desired combinatorial or sequential functions. Fast routing between macrocells leads to lower time delays within a logic block. Dedicated global clock lines leverage proper clock distribution among logic blocks and uniform timing properties. The modern CPLDs contain 32–1700 macrocells enabling designers to implement complex logic functions.

Major advantage of CPLD is that it starts operating as soon as the circuitry is powered up. In-system programmability, JTAG for boundary scan and debugging, interface to multiple logic levels (e.g., 1.8, 2.5, 3.3 V etc.), and nonvolatile nature of programs add to the advantages of CPLDs.

There are several manufacturers of CPLDs such as Intel (Altera), Xilinx, Lattice Semiconductor, Atmel, and Cypress Semiconductor. Architecture of Xilinx CoolRunner-II CPLD shown in Figure 1.5 details the interconnection between the main building blocks of the device. The Max II and Max V series of CPLDs from Intel (Altera) and CoolRunner series of Xilinx are very popular in the industry.

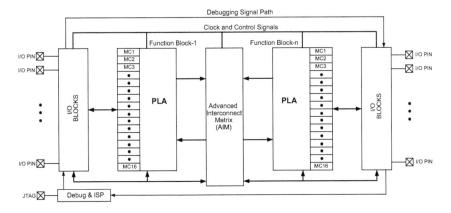

FIGURE 1.5 Xilinx CoolRunner-II CPLD architecture. CPLD, complex programmable logic device; PLA, programmable logic array.

1.12 FIELD-PROGRAMMABLE GATE ARRAY

As the name signifies, FPGA can be programmed in the field without intervention of the semiconductor foundry. A prefabricated silicon chip with array of logic cells along with interconnects is supplied by the manufacturer, over which the users specify their own designs. This design style is very useful for prototyping. Turnaround time is very small in this case compared with custom designs. Reconfigurability and reprogrammability features of FPGA make it unique from ASICs.

FPGAs are programmed using hardware description language such as Verilog and VHDL. A bit stream file is generated to convert the coded information to FPGA configuration. As the configuration is stored in RAM, when there is no power, the configuration is lost. Therefore, it must be configured every time when power is on. However, flash-based FPGAs overcome the volatility problem. The flexible nature of FPGA enables the embedded designer to program the FPGA-based product even after it is installed in the market.

Several manufacturers supply a wide variety of FPGAs to commercial market. Xilinx, Intel (Altera), Microsemi (Actel), and Lattice Semiconductor continue to be the major players in the market. FPGAs have been evolving at a fast rate with the advancement of VLSI technology. For example, Intel Agilex series of devices use 10-nm and Xilinx UltraSCALE+ series of devices use 16-nm process technology with multigate FinFET devices.

Contemporary FPGAs have migrated from being advanced PLDs to true system-on-chip since they have on-chip processor hardware (e.g., ARM Cortex, PowerPC), various accelerators, ADC, and DAC. Apart from the on-chip hardware resources, several software IP cores are available in the design suite. This has leveraged the embedded designers to explore new avenues in compact product design.

1.12.1 INTERNAL ARCHITECTURE OF FIELD-PROGRAMMABLE GATE ARRAY

Based on the technology used to configure the FPGA, the devices can be classified into three types.

- SRAM (static random access memory) based (volatile, needs reloading of the configuration bits upon power up from external memory)
- Flash based (nonvolatile, flash ROM contains the configuration bits within the device)
- Antifuse based (nonvolatile, programmed once)

Figure 1.6 demonstrates the internal architecture of FPGA. The FPGA architecture consists of three basic components, namely configurable logic blocks (CLBs), I/O blocks (IOBs), and programmable interconnects. Apart from this, distributed block memory, clock sources, and several intellectual property hardware elements (e.g. embedded processors and peripheral drivers) are also available on-chip.

1.12.1.1 Configurable Logic Blocks

CLBs are the basic building blocks of the FPGA arranged in array form. Each CLB is equipped with necessary hardware for implementing combinatorial and sequential

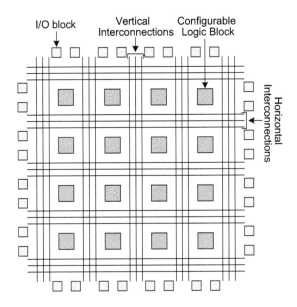

FIGURE 1.6   Architecture of Xilinx FPGA. FPGA, field-programmable gate array.

logic functions. Each CLB has look-up tables (LUTs), used for generation of logic functions. Flip-flops are provided for sequential circuit implementation. A combination of LUTs and memory elements provides ROM functions. Based on the complexity, CLBs also have carry structures, multiplexers, distributed RAM, and shift registers. Distributed RAM can be used in single port/dual port mode with one clock write and asynchronous read operations. The interconnection between LUT and registers in Xilinx Spartan 7 device is shown in Figure 1.7.

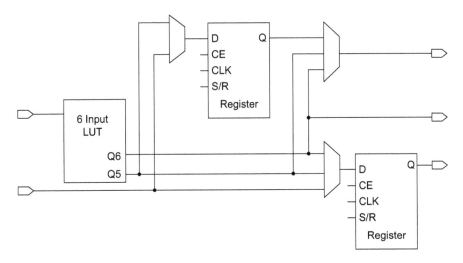

FIGURE 1.7   Xilinx Spartan-7 FPGA LUT and register connections. FPGA, field-programmable gate array; LUT, look-up table.

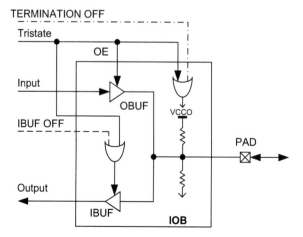

FIGURE 1.8   I/O buffers in IOB. IOB, I/O block.

For image and video processing applications, dedicated DSP slices are available in specific devices containing multipliers and accumulators capable of operating at frequencies in excess of 500 MHz. Preadder circuits greatly help in for symmetrical filter (e.g., FIR filter) design.

1.12.1.2   I/O Blocks

IOBs are special logic blocks placed around the device package providing interface to external I/O signals. The IOB provides the necessary interface for various signals in the range of 1.2–3.3 V. The IOB provides controlled buffering for both the input and output signals. In order to reduce power consumption, individual components are switched off during different operations. I/O pins are arranged to form a bank in the package. Figure 1.8 shows internal structure of I/O buffer in IOB.

Modern FPGAs support many I/O standards, e.g., high-speed transceiver logic, stub series terminated logic, low-voltage differential signaling, low-voltage CMOS, and reduced Swing differential signaling (RSDS). The design suite tool leverages in configuring these pins for specific type of interface.

1.12.1.3   Programmable Interconnects

The programmable interconnects establish connection between CLBs using an array of switch matrices and from CLBs to IOBs. Vertical and horizontal interconnects are programmed for the most efficient routing of signals. Direct connection between adjacent CLBs is provided. These interconnects comprise pass transistors, multiplexers, and tristate buffers. For implementation of complex logic functions, significant amount of interconnect resources are utilized.

1.12.2   DESIGN FLOW OF FIELD-PROGRAMMABLE GATE ARRAY

Design flow refers to the series of steps followed to transform the user specifications into a real hardware unit. Figure 1.9 shows the design flow of FPGA.

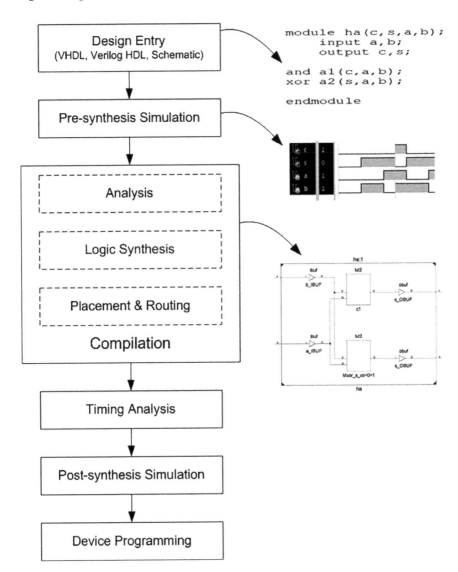

```
module ha(c,s,a,b);
     input a,b;
     output c,s;

and a1(c,a,b);
xor a2(s,a,b);

endmodule
```

FIGURE 1.9   FPGA design flow. FPGA, field-programmable gate array.

The user specifies the design in textual (HDL) or graphical (schematic) form using the design entry tool. Presynthesis simulation is carried out to verify the functionality of the logic described in HDL. This step helps in amending design errors, compatibility of parts, and specification issues.

The next step is compilation of the entered design comprising three substeps. In the analysis phase, the entered design is translated into an intermediate format. Different parts are linked together to generate the desired logic functions in the synthesis phase. The tool places the various parts and routes them on the target device. Also, the timing details are generated in the compilation phase. The timing analysis phase provides the clocking speed, intergate delay, critical path delay, setup, and hold times. These data are used by the designers to define the speed of the circuits. Postsynthesis simulation is run after synthesis of the design. It checks timing problems, race, and hazard conditions and helps in determining suitable clock frequency. A bit stream generated after place and route phase is used to program the target FPGA using a programmer and programming cable (USB or JTAG).

1.12.3   APPLICATIONS OF FIELD-PROGRAMMABLE GATE ARRAY IN MEDICAL IMAGING

Medical imaging equipment such as computerized tomography (CT) scanner, positron emission tomography, or magnetic resonance imaging (MRI) plays an important role to diagnose, analyze, and predict health condition of a patient. Accurate detection and diagnosis needs extraction of minute details of the image and demands high resolution. However, for analyzing and processing a high-resolution image at a fast rate, complex DSP algorithms are requires. Again, telemedicine systems require real-time data analysis and monitoring. This necessitates dedicated hardware for parallel computations, on-chip memory. Additionally, desirable features such as portability, low cost, low power, and long life medical instruments with ability to handle and update complex algorithms drive the use of FPGA. Nowadays, FPGAs are used in most of the medical instruments such as CT scan, MRI, and 3D ultrasound. FPGAs are used for measurement of body temperature, blood pressure, and respiratory rate; analysis of electrocardiogram, electroencephalography, and electromyography; monitoring blood oxygen; etc. Apart from these, FPGA finds its applications in wireless body sensor network, ultrasonography, diabetic monitoring, prediction of ventricular arrhythmia, detection of cardiac dysrythmia, and breast and brain cancer. Additionally, further advancement of FPGA technology allows MRI filtering and tumor characterization, medical surgery using FPGA robotics, and DNA sequence analysis operations.

1.13   SUMMARY

In this chapter, we briefly discussed different PLDs and their uses for varying complexity digital designs. These designs are categorized depending upon programming flexibility and performance. Merits and demerits of different PLDs along with their applications are outlined in each section. While factory-programmable logic devices lack programming flexibility, their permanent storage capability suits them for most of the firmware-based applications. On the contrary, field-programmable devices can be reconfigured a million number of times and well suited for prototyping with very less turnaround time. These devices save design time and production cost compared to custom designs or ASIC designs. Application of FPGA in biomedical engineering is enumerated for elaborated future study.

BIBLIOGRAPHY

Intel Corporation, 2019. Accessed on 25 November 2019, https://www.intel.com/content/dam/www/programmable/us/en/pdfs/literature/wp/wp-medical.pdf.

Intel Corporation, 2019. *INTEL FPGAs*. Accessed on 25 November 2019, https://www.intel.in/content/www/in/en/products/programmable/fpga.html.

Z. Navabi, 2005. *Digital Design and Implementation with Field Programmable Devices*. Boston, MA: Kluwer Academic Publishers.

V. A. Pedroni, 2004. *Circuit Design with VHDL*. Cambridge, MA: The MIT Press.

Rabaey, 2016. *Digital Integrated Circuits: A Design Perspective*. 2nd Ed. Pearson Education India.

Sung-Mo-Kang, 2003. *CMOS Digital Integrated Circuits, Analysis and Design*. Tata McGraw-Hill Education.

Xilinx, 2019. *FPGAs & 3D ICs*. Accessed on 25 November 2019, https://www.xilinx.com/products/silicon-devices/fpga.html.

2 Review of Digital Electronics Design

Reena Chandel
Dr. Y.S Parmar University of Horticulture and Forestry

Dushyant Kumar Singh and P. Raja
Lovely Professional University

CONTENTS

2.1 Introduction to Digital Design 18
    2.1.1 Analog versus Digital Design 18
        2.1.1.1 Analog Design 18
        2.1.1.2 Digital Design 18
2.2 Number Systems 19
    2.2.1 Binary 19
    2.2.2 Octal and Hexadecimal 19
2.3 Logic Families 19
    2.3.1 Digital Integrated Circuit Characteristics... 19
    2.3.2 Resistor–Transistor Logic 20
    2.3.3 Diode–Transistor Logic 20
    2.3.4 Emitter-Coupled Logic 20
    2.3.5 Transistor–Transistor Logic 22
    2.3.6 Complementary Metal Oxide Semiconductor Logic........................... 22
2.4 Combinational Logic 22
    2.4.1 Boolean Equation.. 22
    2.4.2 Introduction to Combinational Logic Circuits 24
    2.4.3 Analysis and Design Procedure... 24
    2.4.4 Adder and Subtractor.. 25
    2.4.5 Decoder.. 27
    2.4.6 Encoder.. 27
    2.4.7 Multiplexer and Demultiplexer 27
2.5 Sequential Circuits.. 29
    2.5.1 Introduction to Sequential Circuits... 29
    2.5.2 Steps Involved to Design a Sequential Circuit.................................... 29
    2.5.3 Types of Sequential Logic Circuits.. 30
        2.5.3.1 Comparison Table of Combinational and Sequential
                Logic Circuits 30

2.6   Storage Elements 30
         2.6.1   SR Flip-Flop 31
         2.6.2   D Flip-Flop 31
         2.6.3   JK Flip-Flop..31
         2.6.4   T Flip-Flop..32
2.7   Counters..33
         2.7.1   Asynchronous Counters...33
         2.7.2   Synchronous Counters...34
         2.7.3   Registers 34
2.8   Memory..34
         2.8.1   Read-Only Memory..35
         2.8.2   Random Access Memory..36
         2.8.3   Flash..36
         2.8.4   Optical and Magnetic 36
References..37

2.1   INTRODUCTION TO DIGITAL DESIGN

Circuit design is done with the analog or digital components or using both. This section discusses the difference between analog and digital designs.

2.1.1   ANALOG VERSUS DIGITAL DESIGN

2.1.1.1   Analog Design

Analog signal is continuous and real having infinite natures and forms. Any phenomena of real world in which we live such as temperature, humidity, fragrance, sound, and so on are all analog quantities with infinite number of possible values. Analog design requires every component to be positioned by hand, and thus, it is difficult. Figure 2.1 illustrates the analog signal.

2.1.1.2   Digital Design

In digital electronics, signals are represented by discrete values mainly in the combination of 0's and 1's. At any point of time, signal can be either 0 or 1. For digital

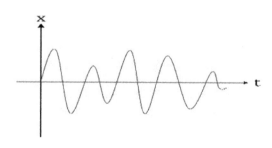

FIGURE 2.1   Analog signal.

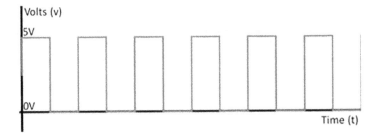

FIGURE 2.2 Digital signal.

design, automation techniques are available, and thus, they are very simple to design digital circuits [1]. Figure 2.2 illustrates the digital signal.

2.2 NUMBER SYSTEMS

2.2.1 BINARY

In digital system, binary number is a number system with base 2, which means that in binary system, there are dual symbols that are 0 and 1. In binary system, the value of the number is the sum of power of 2 for each 1 in number. The right most digit is with 2^0, the next representing 2^1, then 2^2, and so on.

For example,

$$100101_2 = [(1) \times 2^5] + [(0) \times 2^4] + [(0) \times 2^3] + [(1) \times 2^2] + [(0) \times 2^1] + [(1) \times 2^0]$$

$$100101_2 = [1 \times 32] + [0 \times 16] + [0 \times 8] + [1 \times 4] + [0 \times 2] + [1 \times 1]$$

$$100101_2 = 37_{10}$$

2.2.2 OCTAL AND HEXADECIMAL

In octal number system, base is 8, thus having 8 symbols from 0 to 7. In this, each digit is represented in the power 8 as elaborated in the following example:

$$764_8 = 7 \times 8^2 + 6 \times 8^1 + 4 \times 8^0 = 448 + 48 + 4 = 500_{10}$$

2.3 LOGIC FAMILIES

2.3.1 DIGITAL INTEGRATED CIRCUIT CHARACTERISTICS

Integrated circuits (ICs) are majorly classified into two types: one is analog and another one is digital IC. The digital ICs are separated on the basis of complication of the circuit or the number of transistors to construct the digital logic-circuit (IC).

The digital logic-circuit (IC) can be divided as per Table 2.1.

TABLE 2.1
Classification of Integrated Circuit

Level of Integration	Number of Active Devices Per Chip
Smallscale integration (SSI)	Less than 100
Medium-scale integration (MSI)	100–1,000
Large-scale integration (LSI)	1,000–100,000
Very large-scale integration (VLSI)	More than 100,000
Ultralarge-scale integration (ULSI)	More than 1 million

Various characteristics of digital ICs are as follows:

- **Speed of operation**: The speed is mentioned in terms of propagation delay and is measured at the 50% of the voltage levels of output.
- **Power dissipation**: The power dissipation can be calculated in terms of the amount of power dissipated by IC.
- **Figure of merit** (FoM): FoM is defined in numerical value to measure the effectiveness of the electronic components and circuits.
- **Fan-out**: The fan-out is defined as the output of a gate that can be a source to an extreme number of inputs.

2.3.2 RESISTOR–TRANSISTOR LOGIC

Resistor–transistor logic (RTL) or transistor–resistor logic consists of resistors and also transistors and was the most prevalent logic beforehand IC fabrication skills [2]. Basic RTL gate is NOR gate as shown in the following. Inputs are applied to NOR gate at 1 and 2 terminals. LOW logic level at both inputs energies the transistor into cutoff region and HIGH into saturation region. Both inputs as LOW will drive both the transistors into cutoff region and VCC, i.e., HIGH appears at the output. Similarly, HIGH logic level at both inputs drives the transistor into cut saturation region and LOW appears at the output (Figure 2.3).

2.3.3 DIODE–TRANSISTOR LOGIC

Diode–transistor logic (DTL) replaces the RTL family because of its greater fan-out. DTL mainly consists of diodes and transistors, and basic DTL device is NAND gate as shown in Figure 2.4.

DTL NAND gate has three inputs applied through three diodes, viz. The diode behavior when LOW logic is applied to its input. The transistor will go in cutoff region if any input is LOW that is one of the diodes in conducting and thereby output goes HIGH.

2.3.4 EMITTER-COUPLED LOGIC

In emitter-coupled logic (ECL), gates are implemented in disparity amplifier conformation because of which they are never determined into overload region, thus refining the speed of operation.

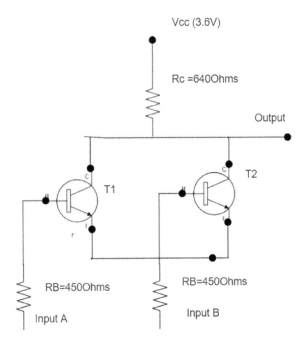

FIGURE 2.3   Resistor–transistor logic.

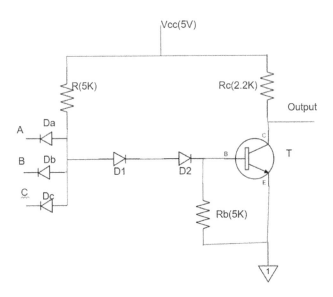

FIGURE 2.4   Diode–transistor logic.

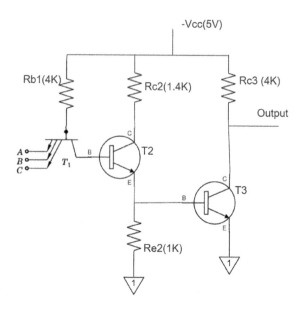

FIGURE 2.5   Transistor–transistor logic.

2.3.5   TRANSISTOR–TRANSISTOR LOGIC

Transistor–transistor logic (TTL) is the modification of DTL, and its basic gate is NAND gate as exposed in Figure 2.5.

2.3.6   COMPLEMENTARY METAL OXIDE SEMICONDUCTOR LOGIC

In complementary metal oxide semiconductor (CMOS), fabrication of N-MOSFET (n-type metal oxide semiconductor field-effect transistor) and P-MOSFET (p-type MOSFET) of enhancement mode is made up of same chip.

CMOS devices are known for their negligible power consumption and are most preferred in battery-operated devices. CMOS NAND gate is exposed in the following. If any of the inputs is asserted LOW, the transistor T3 or T4 is goes to ON state; therefore, concerning output of transistor is $+V_{cc}$, i.e., HIGH (Figure 2.6).

2.4   COMBINATIONAL LOGIC

2.4.1   BOOLEAN EQUATION

Boolean equation is used to minimize the digital circuits by dropping the amount of logic gates mandatory for the digital circuit. In Boolean algebra [2,3], each variable can have dual values as 0 or 1. Boolean algebra laws are tabulated in Table 2.2.

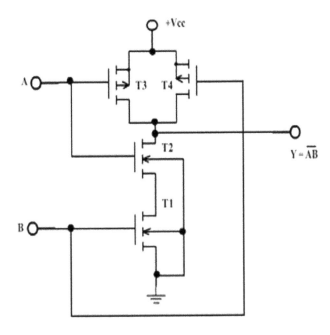

FIGURE 2.6 CMOS logic. CMOS, complementary metal oxide semiconductor.

TABLE 2.2
Boolean Algebra Expression

Boolean Expression	Description	Rule of Boolean Algebra
$P+1=1$	"P in concurrent with closed=CLOSED"	Annulment
$P+0=P$	"P in concurrent with open=P"	Identity
$P \cdot 1=P$	"X in series with closed=P"	Identity
$P \cdot 0=0$	"P in series with open=OPEN"	Annulment
$P+P=P$	"P in concurrent with P=P"	Idempotent
$P \cdot P=P$	"P in succession with P=P"	Idempotent
NOT $P=P$	"NOT not P=P"	Double negation
$P+\overline{P}=1$	"P in concurrent with NOT P=CLOSED"	Complement
$P \cdot \overline{P}=0$	"P in succession with NOT P=OPEN"	Complement
$P+Q=Q+P$	"P in concurrent with Q=Q in concurrent with P"	Commutative
$P \cdot Q=Q \cdot P$	"P in succession with Q=Q in succession with P"	Commutative
$\overline{P+Q}=\overline{P} \cdot \overline{Q}$	"Invert and replace OR with AND"	DeMorgan's theorem
$\overline{P \cdot Q}=\overline{P}+\overline{Q}$	"Invert and replace AND with OR"	DeMorgan's theorem

The basic laws of Boolean equations are as follows:

1. Commutative law:
The output of Boolean equation is same when the position of the input variable is changed.

Example: $P_s + Q_s = Q_s + P_s$ //(OR operator)
$P_s * Q_s = Q_s * P_s$ //(AND operator)

2. Associative law:
The output of three-variable Boolean equation remains same while interchanging the brackets on both sides for addition and multiplication.

Example: $P_s * (Q_s * R_s) = (P_s * Q_s) * R_s$ //(OR operator)
$P_s + (Q_s + R_s) = (P_s + Q_s) + R_s$ //(AND operator)

3. Distributive law:
The output of addition of two products remains same as that of sum of product of same variables.

Example: $P_s + Q_s\ R_s = (P_s + Q_s)\ (P_s + R_s)$ //(OR operator)
$P_s * (Q_s + R_s) = (P_s * Q_s) + (P_s * R_s)$ //(AND operator)

2.4.2 INTRODUCTION TO COMBINATIONAL LOGIC CIRCUITS

The combinational circuit [4] is defined as the digital circuit that has been designed or developed in the combination of basic digital components such as conventional gates and derived gates. The combinational circuit generates output based on present input value with respect to instantaneous time, and the circuit does not have any memory unit on it. The output of combinational circuit always depends on function of input signal as shown in Figure 2.7.

2.4.3 ANALYSIS AND DESIGN PROCEDURE

There are several steps involved to design an optimized combinational circuit as follows:

- The initial stage designer should clearly understand about the given problem statement.
- Determine the number of input line and output line required as per given specification.
- Create a truth table in accordance with logic of given problem statement.

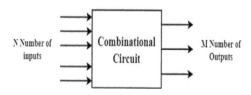

FIGURE 2.7 Combinational circuit.

- Simplify the Boolean expression for each output by using K-map reduction or any other reduction method.
- Sketch the digital logic circuit based on simplified Boolean expression, and verify the output logic through truth table.

2.4.4 ADDER AND SUBTRACTOR

Adder and subtractor is the circuit to addition and subtraction of input variables, and it generates the output of sum, carry and difference, and barrow respectively. Nowadays, adder and subtractor circuit is used in most of the processors and controllers (Figures 2.8 and 2.9 and Table 2.3).

$$S = (A \text{ XOR } B)$$

$$C = (A \text{ AND } B)$$

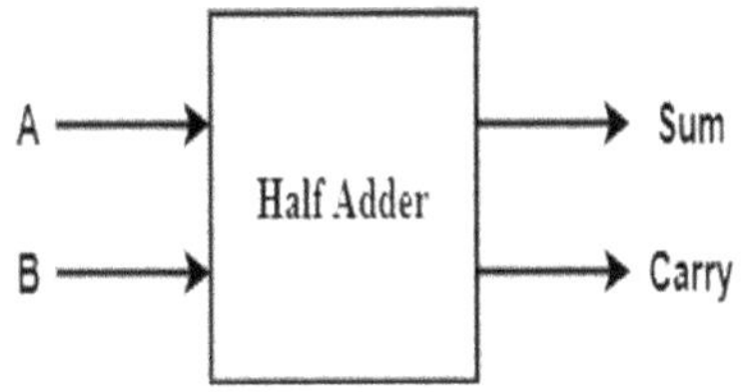

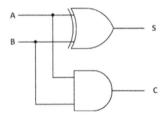

FIGURE 2.8 Block diagram of half adder.

FIGURE 2.9 Circuit diagram of half adder.

TABLE 2.3
Truth Table

Input		Output	
A_{in}	B_{in}	S_{out}	C_{out}
0	0	0	0
0	1	1	0
1	0	1	0
1	1	0	1

Similarly, for subtractor, AND gate is replaced with NAND gate, and truth table will be changed as per the Boolean equation. Full adder is the logic circuit to add three bits, namely two inputs and on carry of the previous output (Figures 2.10 and 2.11 and Table 2.4). Full adder is shown in Figures 2.10 and 2.11.

$$S = A'BX' + A'B'X + AB'X' + ABX$$

$$C = AB + BX + AX$$

Full adder circuit is used to achieve the adding of three inputs, including carry-in of the previous state; the output of the circuit will be verified through predefined possible combination of inputs. The grouping of dual half adder can achieve the whole

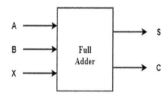

FIGURE 2.10 Block diagram of full adder.

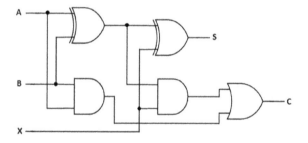

FIGURE 2.11 Circuit diagram of full adder.

TABLE 2.4
Truth Table of Full Adder

Input			Output	
A_{in}	B_{in}	X_{in}	S_{out}	C_{out}
0	0	0	0	0
0	0	1	1	0
0	1	0	1	0
0	1	1	0	1
1	0	0	1	0
1	0	1	0	1
1	1	0	0	1
1	1	1	1	1

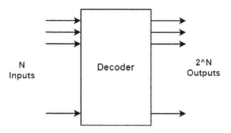

FIGURE 2.12   Block diagram of decoder.

full adder circuit; the first half adder gets the addition of first two inputs, and then the output of first half adder and remaining input get added and will generate the summation and carry output of the full adder.

2.4.5   DECODER

In digital electronics, information is represented in binary codes, and decoder is used to convert the coded information from one form to another. Decoder is a combinational circuit for converting n bit information to $2^{\wedge}n$ discrete outputs as shown in Figure 2.12.

2.4.6   ENCODER

Encoder is a combination circuit used to convert the binary information in $2^{\wedge}n$ lines to n output lines. Figure 2.13 shows the octal to binary encoder (Table 2.5).

2.4.7   MULTIPLEXER AND DEMULTIPLEXER

Multiplexer is used to select one output from many inputs with the help of select line. It has $2^{\wedge}N$ number of input lines, N number of select lines, and one output. The output will be fed with concern input line on the basis of select lines. Selected input line will be connected to the output line (Figure 2.14).

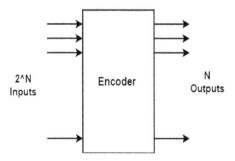

FIGURE 2.13   Block diagram of encoder.

TABLE 2.5
Truth Table of 8:3 Encoder

Inp7	Inp6	Inp5	Inp4	Inp3	Inp2	Inp1	Inp0	Xout1	Xout2	Xout3
0	0	0	0	0	0	0	1	0	0	0
0	0	0	0	0	0	1	0	0	0	1
0	0	0	0	0	1	0	0	0	1	0
0	0	0	0	1	0	0	0	0	1	1
0	0	0	1	0	0	0	0	1	0	0
0	0	1	0	0	0	0	0	1	0	1
0	1	0	0	0	0	0	0	1	1	0
1	0	0	0	0	0	0	0	1	1	1

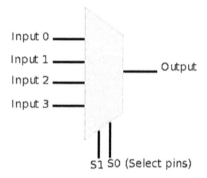

FIGURE 2.14   4:2 Multiplexer.

If multiplexer is of $2^n - 1$, then the number of select lines is n. For example,

- One selection line is needed to design a 2:1 multiplexer.
- Two selection lines are needed to design a 4:1 multiplexer.
- Three selection lines are needed to design a 8:1 multiplexer.
- Four selection lines are needed to design a 16:1 multiplexer.
- In general, N selection lines are required to design a $2^N:1$ multiplexer.

Demultiplexer is a digital circuit which function is just opposite to multiplexer. Demultiplexer has one input and many outputs and sends the signal from one input to one of the many outputs as selected by select lines (Figure 2.15). Representational image of multiplexer is given in Figure 2.15.

The number of select lines in demultiplexers depends on the number of outputs. For example,

- One selection line is needed to design a 1:2 demultiplexer.
- Two selection lines are needed to design a 1:4 demultiplexer.
- Three selection lines are needed to design a 1:8 demultiplexer.

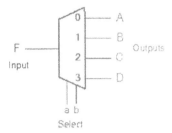

FIGURE 2.15   1:4 Demultiplexer.

- Four selection lines are needed to design a 1:16 demultiplexer.
- In general, N selection lines are required to design a 1:2^N multiplexer.

2.5   SEQUENTIAL CIRCUITS

2.5.1   INTRODUCTION TO SEQUENTIAL CIRCUITS

Sequential logic has memory element as against combination logic that does not have any memory element. Sequential circuit takes into account present as well as past input, and thus their output depends on current input as well as previous input (Figure 2.16).

2.5.2   STEPS INVOLVED TO DESIGN A SEQUENTIAL CIRCUIT

The following steps are involved to design a sequential circuit (Figure 2.17):

- First, analyze the given problem statement.
- Analyze and identify the states involved in problem statement.
- Draw the state diagram in accordance with identified states.
- Drive the excitation table for each state.
- Draw the state table as per the excitation table.
- Find out the Boolean expression/excitation equation as per the reduction methodology.

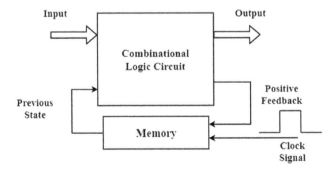

FIGURE 2.16   Block diagram of sequential circuit.

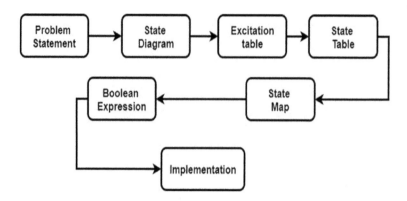

FIGURE 2.17   Design flow of sequential circuit.

- Implement/convert the Boolean expression/excitation equation into a digital sequential circuit.

2.5.3   TYPES OF SEQUENTIAL LOGIC CIRCUITS

See Figure 2.18.

2.5.3.1   Comparison Table of Combinational and Sequential Logic Circuits
See Table 2.6.

2.6   STORAGE ELEMENTS

In digital circuits, there are two memory elements:

- Latches
- Flip-flops

Latchet are very basic storage elements that operate with signal levels rather than signal transitions.

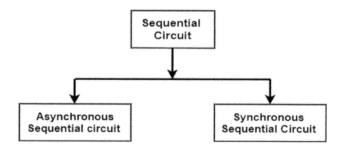

FIGURE 2.18   Types of sequential circuit.

TABLE 2.6

Difference Between Combinational and Sequential Circuits

Combinational Circuits	Sequential Circuits
Output is always depending on present input at the instantaneous of time.	Output is always depending on present state and previous input.
There is no need of memory unit.	Memory unit is required to store the previous state output.
This circuit is faster.	This circuit is slower.
It is easy to design this circuit.	It is difficult to design this circuit.
Examples: half adder, full adder, magnitude comparator, multiplexer, demultiplexer, etc.	Examples: flip-flop, register, counter, clocks, etc.

Latches controlled by a clock transition are called flip-flops; thus the flip-flops are clock edge sensitive. Different types of flip-flops are as follows:

- Set reset flip-flop
- D flip-flop
- JK flip-flop
- T flip-flop

2.6.1 SR Flip-Flop

SR flip-flop is set–reset the output of SR flip-flop and operates on positive- or negative-edge clock transition. The circuit diagram of SR flip-flop is shown in Figure 2.19 (Table 2.7).

2.6.2 D Flip-Flop

The D flip-flop is called delayed flip-flop that means one of the input will reach late to gate as compared with another input (Figure 2.20 and Table 2.8).

2.6.3 JK Flip-Flop

JK flip-flop functions on positive or negative clock change. JK flip-flop is the better-quality version of SR flip-flop taking care of unknown conditions in SR flip-flop (Figure 2.21 and Table 2.9).

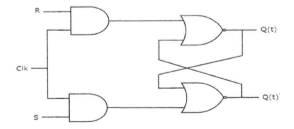

FIGURE 2.19 Circuit diagram of SR flip-flop.

TABLE 2.7
Truth Table of SR Flip-Flop

Input		Output	
S	R	Q	Q_{bar}
0	0	(NC)	(NC)
0	1	0	1
1	0	1	0
1	1	Unknown value	Unknown value

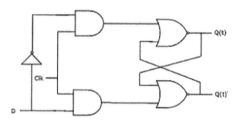

FIGURE 2.20 Circuit diagram of D flip-flop.

TABLE 2.8
Truth Table of D Flip-Flop

Input	Output
D	Q
0	0
1	1

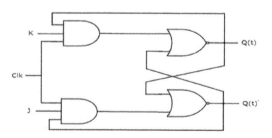

FIGURE 2.21 Circuit diagram of JK flip-flop.

2.6.4 T Flip-Flop

T flip-flop is toggle flip-flop and is basic version of JK flip-flop. T flip-flop functions on positive or negative clock change. The circuit diagram of T flip-flop is shown in Figure 2.22 (Table 2.10).

TABLE 2.9
Truth Table of JK Flip-Flop

Input		Output	
J	K	Q	Q$_{bar}$
0	0	Previous state (NC)	
0	1	0	1
1	0	1	0
1	1	Toggle state	

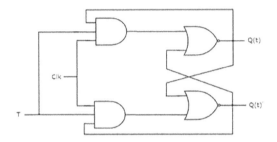

FIGURE 2.22 Circuit diagram of T flip-flop.

TABLE 2.10
Truth Table of T Flip-Flop

Input	Output
T	Q
0	1
1	0

2.7 COUNTERS

Digital circuit counters are used for the counting purpose. They count and store the number of events occurred or the amount of period a specific occurrence has happened using the clock signal.

2.7.1 ASYNCHRONOUS COUNTERS

Flip-flops are not connected to clock source in asynchronous sequential circuit; only the first flip-flop is applied with clock source. The clock inputs of further flip-flops are connected to output of the previous flip-flop. Therefore, the output of next flip-flop will change depending on the input of the previous flip-flop (Figure 2.23).

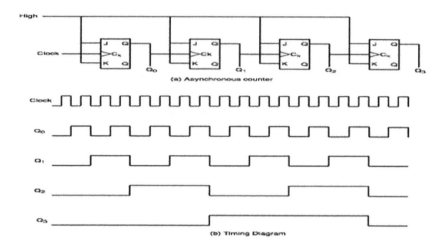

FIGURE 2.23 Asynchronous counter and its timing diagram.

In timing diagram, Q0 is varying at the growing edge of clock pulsation, Q1 is varying at the growing edge of Q0 as Q0 is clock pulse for next flip-flop, and so on. Thus, waves are produced from Q0 through Q3; henceforth, it is also known as *ripple* counter.

2.7.2 SYNCHRONOUS COUNTERS

In synchronous counters, there is universal clock, i.e., clock is connected to each flip-flop in the circuit, and the outputs of all flip-flop change in parallel. The synchronous counter can operate at a higher frequency because there is no cumulative delay as in the case of asynchronous counter in which previous flip-flop output is clock to the next flip-flop, leading to a delay in the circuit (Figure 2.24).

2.7.3 REGISTERS

Registers are the clocked sequential circuits designed using flip-flops and also consist of combinational circuits. Flip-flops are used to store the 1-bit information and combinational circuits to perform some simple tasks such as generating the signal for resetting the register (Figure 2.25).

2.8 MEMORY

A memory is referred as any device that is capable of storing information that can be retrieved as and when required [4]. Computer information is stored in digital format. A memory is neither a sequential circuit as memories are not clocked nor a combinational circuit as output of memory is dependent on previous input.

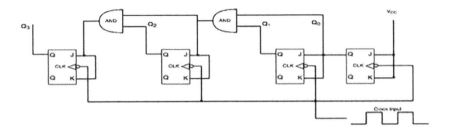

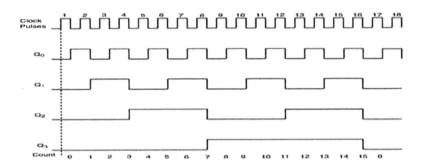

FIGURE 2.24 Synchronous counter and its timing diagram.

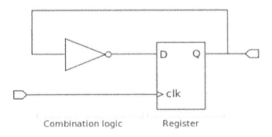

FIGURE 2.25 Register.

2.8.1 READ-ONLY MEMORY

Read-only memory (ROM) is the nonvolatile memory in which information is retained even after removing power. In computer, small amount of ROM is always there to store boot firmware, and for this reason, it is sometimes also called as firmware memory. Various types of ROM are as follows:

- ROM
- Programmable read-only memory (PROM)
- Erasable programmable read-only memory (EPROM)
- Electrically erasable programmable read-only memory (EEPROM)
- Flash memory

2.8.2 RANDOM ACCESS MEMORY

Random access memory (RAM) is the most popular memory on computer systems. Any location in the RAM can be accessed without touching the preceding memory location. RAM is a solid-state memory, and in most of the systems, it is used as the main memory.

The most widely two types of RAM are as follows:

- SRAM: Static RAM stores the bit information using six transistors, typically MOSFET, and is more expensive in comparison with the other RAM.
- DRAM: Dynamic RAM stores the bit information using transistors, typically MOSFET, and pair of capacitors. In this memory, transistor acts as a switch, and information is stored as charge in capacitor. These types of memories are less expensive.

Both SRAM and DRAM are volatile memories, which lose their information when the power is removed.

2.8.3 FLASH

Flash memory is used for storage and transfer of information between computers. It is a nonvolatile memory and can be electronically erased and reprogrammable. The types of flash memory are NOR and NAND. NAND is better for serial data access. Data read is faster in NOR, but data erase and write are faster in NAND.

2.8.4 OPTICAL AND MAGNETIC

Optical storage devices use LASER to read and write information. Different pattern of magnetization is used to store the data on magnetizable medium. The main differences in optical and magnetic memories are tabulated in Table 2.11.

TABLE 2.11
Optical and Magnetic Storage Devices

	Optical Storage Devices	Magnetic Storage Devices
1	The storage of this device has been done as a patterned image.	The storage of this device has been done as a magnetic form.
2	To read and write, LASER is required.	To read and write, heads are required.
3	Storage will not be disturbed due to magnetic field.	Storage will be disturbed due to magnetic field.
4	As technologies are improved, these can be readable, writable, and rewritable.	These are always readable and rewritable.
5	Drivers are required.	These contain built-in drivers.
6	These are easy to carry and safe.	These are difficult to carry.

REFERENCES

1. J. F. Warkerly, *Digital Design: Principles and Practices*, Pearson, USA, 2013.
2. T. L. Flyod, *Digital Fundamentals*, Pearson, USA, 2013.
3. S. Brown, Z. Vranesic, *Fundamental of Digital Logic Design with VHDL*, McGraw Hill, USA, 2007.
4. M. Morris Mano, M. D. Ciletti, *Digital Design: With An Introduction to Verilog HDL*, Pearson, USA, 2013.

3 Verilog HDL for Digital and Analog Design

Ananya Dastidar
College of Engineering and Technology

CONTENTS

Chapter Outline 40
3.1   Introduction to Hardware Description Languages 40
3.2   Verilog for Digital Design 41
     3.2.1   Verilog Language Basics 41
           3.2.1.1   Keywords 41
           3.2.1.2   Comment Line 41
           3.2.1.3   Whitespaces 42
           3.2.1.4   Identifier 42
           3.2.1.5   Variables 42
           3.2.1.6   Vector Data 44
           3.2.1.7   Numbers or Constant Values 45
           3.2.1.8   Parameter 45
           3.2.1.9   Sequential Statements 45
     3.2.2   Additional Constructs 49
           3.2.2.1   time value (#) 49
           3.2.2.2   @ (sensitivity_list) 49
           3.2.2.3   Generate 50
           3.2.2.4   Gate Primitives 50
           3.2.2.5   Tristate Gates 50
           3.2.2.6   Switch-Level Primitives 51
     3.2.3   Verilog Module Description 51
           3.2.3.1   Ports 52
     3.2.4   Operator Types 52
3.3   Modeling Types 54
     3.3.1   Behavioral or Algorithmic Model 54
           3.3.1.1   Blocking and Nonblocking Statements 55
     3.3.2   Dataflow or Register Transfer-Level Model 56
     3.3.3   Gate-Level or Structural Model 57
     3.3.4   Switch-Level Model 58
     3.3.5   Mixed Model 59
3.4   User-Defined Primitives 59
3.5   Test Bench 60

3.6   Verilog for Analog Design...62
3.7   Verilog-A Basics 63
3.8   Summary 66
References...66

CHAPTER OUTLINE

This chapter deals with the basics of Verilog hardware description language (Verilog HDL) and its special features toward implementation of digital and analog designs. It starts with a background on the emergence and importance of HDLs with a brief introduction to Verilog HDL. The features of Verilog for digital design include data types, model types, and so on. The next section discusses some programming concepts for analog electrical circuits using the features of Verilog-A. This chapter ends with summarizing the limitations and applicability of the Verilog Family of Languages.

The past few decades have seen the evolution of digital circuits from simple gates to complex system-on-chips (SoCs). From tubes to transistors to integrated circuits (ICs), design of digital circuits has come a long way. With the rise of very large-scale integrated circuit (VLSI) technology, complicated digital implementation of an IC not only became conceivable but also paved the way for automation in chip design. Depending on the functionality of the IC, they may be termed as digital IC, analog IC, or mixed-signal IC. Due to the complexity involved in IC design, verification of the correctness of the design under different conditions becomes a challenge, and the presence of millions of transistor on a single substrate makes detection of fabrica-tion faults an additional challenge. So a software solution for design specification, verification, and testing becomes essential to meet the requirements of the industry in terms of economics and technical viability of a product. Chip designers have thus turned to different tools such as electronic design automation (EDA) for both soft-ware and hardware designs in order to bring down the time-to-design and in turn the time-to-market for both circuit simulators and programming languages. These tools have proved to be indispensible in high complexity circuit design by allowing the transformation of the design specification to mask layout for transistor fabrication, verification, and testing.

3.1   INTRODUCTION TO HARDWARE DESCRIPTION LANGUAGES

Computer programing language used for simulation where the execution of instruc-tions is sequential in nature leads to the software implementation of any design (e.g., adder). However, in order to realize the hardware of an adder, these program-ming languages were not sufficient, and thus, the need for a tool for the synthesis and simulation of hardware to realize designs for VLSI circuits led to the era of HDLs. These HDLs made concurrent or parallel execution of instructions possible, which are inherent for any hardware design. Two popular HDLs in existence are the VHDL (VHSIC hardware description language), where VHSIC is the abbreviated form of very high-speed integrated circuit, and the Verilog HDL. While the Verilog HDL originated at Gateway Design Automation in 1983 as Verilog-XL, VHDL successively came into existence as a project under United States Department of Defense (Palnitkar 2012). The Verilog IEEE standard (IEEE 1364-1995) was adopted in 1995 and was revised to IEEE 1364-2001 in the year 2001 with some further minor modifications made in 2005. The IEEE 1364 standard defines the programming language interface that permits a bidirectional interface between C and Verilog (Navabi 1999).

VHDL models the behavior and structure of digital hardware designs. The first standardization of VHDL was approved in 1987 (IEEE 1076), and as per proposal, a revision was done every 5 years, which is widely used until today (Ashenden et al. 2008). The turnaround time for the design of digital circuits using both these HDL simulators is quite low, thereby leading to their acceptance among designers in both industry and academia alike. A HDL specifies the functionality and timing of the hardware. While it becomes more and more difficult to implement designs on the hardware directly, HDL provides an easier and economical design alternative. Apart from general-purpose Verilog HDL and VHDL, there exists another HDL, namely, System Verilog (an enhanced version of Verilog) that is capable of handling the complexities of SoC design. Standardized as IEEE 1800-2012, it has the capabilities of verification and development of test benches. System Verilog is a combination of capabilities of an HDL and hardware verification language. This unification allows register transfer-level (RTL), system-level, and structural-level implementations along with a verification of the functionality of the design (Sutherland et al. 2006).

3.2   VERILOG FOR DIGITAL DESIGN

Verilog HDL is a programmable HDL; thus, it is employed in the industry due to its ease of use and simplicity in understanding. It is an event-driven simulator that can emulate the hardware. It allows designing at multiple abstraction levels including gate-level, RTL as well as behavioral models. The Verilog HDL includes tools for simulations, synthesis, and fault as well as timing analysis, and its capability of automatic extraction of gate-level net list by the logic synthesis tool makes it a popular choice in the IC design industry. A host of simulators and synthesis tools are available as open source or as proprietary from leading EDA companies.

3.2.1   VERILOG LANGUAGE BASICS

3.2.1.1   Keywords
Keywords or constructs of Verilog HDL refer to the reserved words of the language that correspond to some special in-built function, for example, module, port, wire, loop, time, NAND, etc. including compiler directives, tasks, and functions. The use of these reserved keywords as an identifier is restricted.

3.2.1.2   Comment Line
Comment lines are necessary in any coding practices in order to provide additional information about the code that also plays an important role for documentation.

Shorter comment lines use double slash (//), whereas comments that extend to more than one line use a pair of slash and asterisks (/*....*/).

```
// Verilog Short Comment
/* Verilog Long Comment.........................that may span two lines */
```

3.2.1.3   Whitespaces
In order to increase the readability of the code, whitespaces are used. Whitespace operators include space, tab, or newline as separators while writing Verilog codes.

3.2.1.4   Identifier
Names of variables and other elements in a Verilog code are represented by identifiers where a valid identifier can be a combination of alphabet and numbers including underscore and the special symbol $. It cannot begin with a numbers and must not be a Verilog keyword.

```
Valid Identifiers: my_design1, mux_32X1
Invalid Identifiers: 1design, wire
```

Verilog is a case-sensitive language so the identifiers my_and and MY_AND are different.

3.2.1.5   Variables
Variables represent data types in Verilog as a *net* or *register* where *net* indicates the interconnection of two of more points of a circuit and the *registers* are path components that are capable of storing intermediate values.

3.2.1.5.1   Net
The net data type includes *wire* (most commonly used), *wand* (wired-AND), *wor* (wired-OR), *tri, supply0* (power supply connection), *supply1* (power supply connection), etc. We must declare a net as a signal when it models ports of a design module or when a net type variable is present on the left side of a continuous assignment statement (assign). Default value of net is high impedence (z).

wire
      The net data type *wire* represents the physical wire, and it models the interconnection between gates and other logic modules. A wire cannot store any value and so it may not be used inside a function or a block. An assign statement and the output of a design module drive a wire. Apart from the general wire, there are different wire types such as the *wand* (wired-AND) that inserts a AND gate in the interconnection point and *wor* (wired-OR) that inserts an OR gate in the interconnection points. The *tri* (three-state) is used for multiple drivers.

Example

```
wire [3:0] W; // an 4 bit vector (4 wires make up a 4 bit bus)
wire A, B; // simple wire
wor R; //insert a or gate at the interconnection
assign R=A;
assign R=B;// R= A or B
tri [1:0] T;// T can hold the value 0, 1 or Z
```

Power and Ground nets

The power and ground rails in a circuit-level design can be modeled with the keywords *supply1* and *supply0*, respectively. The *supply0* is used to represent wires that are connected to ground rails (logic 0), and *supply1* is used to represent wires that are connected to power rails (logic 1).

Example

```
supply1 vdd;
supply0 gnd;
```

3.2.1.5.2   Register

The register on the other hand holds the value of a procedural assignment statement until the next event triggers it. It is not a physical register but implies a variable-type register, and it finds use only in functions and procedural blocks. There are different register data types such as *reg* used for logic description, *integer* for loop calculations, *real* for system module design, and *time* for test benches value storage, of which the *reg* is the most frequently used register data type.

reg

The default value of reg variables and unconnected registers has "x" as initial value at the beginning of simulation, and it can be used to realize a combinational as well as a sequential logic. Specifying the size of the reg at the time of declaration is encouraged; otherwise, by default, it is 1-bit and an unsigned number in arithmetic expressions.

Example

```
reg r; // a single bit register variable
reg [5:0] r; // a 6-bit vector;
reg [7:0] s, t; // two 8 bit variables.
```

integer

They are used in general-purpose variables especially as index values in loops, constants, and parameters. Data are stored as signed numbers by default, but the default size of integer is 32 bits. In case the integer holds a constant, the minimum width needed is adjusted by the synthesis tool.

Example

```
integer i; // by default 32-bit integer
assign i=55; // a is adjusted as a 6-bit variable
```

real

It is used to store floating-point numbers where rounding off to nearest integer occurs when a real number is assigned to an integer variable.

Example

```
real eox, eps;
initial
begin
eox = 3.97;
eps = 8.854 e-14;
end
// now let us assign real value to an integer variable
integer int;
initial
int = eps; // int gets the value 8
```

time

It is used in test bench to hold the time of simulation. It is a nonsynthesizable construct and used only for simulation purposes. The system function "$time" gives the current time of simulation.

Example

```
time present_tme;
initial
present_tme = $time;
```

All Verilog data types under the value set can have the following possible values: logic 0 or false conditions (0), logic 1 or true conditions (1), unknown logic value (x), or the high impedance state (z). Time is a nonsynthesizable Verilog construct.

3.2.1.6 Vector Data

In order to implement an n-bit bus, we can make use of vectors to represent multibit busses. Both net- and reg-type variables are declared as vectors where the default size of 1 bit is considered if the bit width is not specified. The vectors declaration syntax is [MSB: LSB].

Example

```
input [3:0] a;
// a is a 4 bit vector with a[3] as MSB and a[0] as LSB
output reg [0:4] x;
/* x is an output port register having data type reg which is
a 5 bit vector with x [0] as MSB and x [4] as LSB */
```

Multidimensional arrays of any dimension can be declared in Verilog as

```
reg [63:0] bank_reg [32:0]; // 32 numbers of 64-bit registers
```

3.2.1.7 Numbers or Constant Values

A number in Verilog can operate on decimal, binary, octal, and hexadecimal numbers where 2's complement format for negative numbers representation is used. Both signed and unsigned integers and real numbers are a part of Verilog.

Syntax

```
<size> <radix> <value>
```

Example

```
2'b01 // 2 bit binary
8'hFF // 8 bit hexadecimal ...1111 1111
```

3.2.1.8 Parameter

For customization of a design module during the instantiation process, parameter construct can be used. The size of the parameter cannot be defined, but it is decided from the value of the constant assigned to the parameter.

Example

```
parameter X=5, Y=10;
parameter HOT=2'hFF, WARM=2'hAA, TREPID=2'h55, COLD=2'h11;
```

3.2.1.9 Sequential Statements

Sequential statements used inside procedural blocks, namely, *initial* and *always* blocks model both combinational and sequential designs. These procedural blocks include procedural assignment statements, some of which are as follows.

3.2.1.9.1 begin ... end

In order to group multiple statements inside a procedural block, the begin ... end constructs are used in Verilog HDL.

Syntax

```
begin
sequential_stmt_s1;
sequential_stmt_s2;
...
sequential_stmt_sn;
end
```

The *begin ... end* keywords are omitted when only a single procedural statement is present inside a block.

TABLE 3.1

Variation of if ... else Statement

Types of if ... else Statement	Examples
if (<expression1>) sequential_stmt;	if (a>b) z=1'b1;

if (<expression1>)
sequential_ stmt;
else
sequential_ stmt;
	if (a>b)
	z=1'b1;
	else
	z=1'b0;

if (<expression1>)
sequential_ stmt;
else if (<expression2>)
sequential_ stmt;
else if (<expression3>)
sequential_ stmt;
else default_ stmt;
	if (a>b)
	z=2'b10;
	else if (b>a)
	z=2'b01;
	else
	z=2'b11;

3.2.1.9.2   if ... else

It is a sequential statement that can occur alone or as a group of statements inside the "begin ... end" block depending to implement conditional flow of data in a design module. Depending on the type of conditional expression used, they may be expressed as any one of the following ways as seen in Table 3.1. If there are multiple sequential statements inside the if ... else statement, then they must be enclosed within a begin ... end construct.

When using if ... else statements to model combinational circuit, one must be careful to define the outputs for all possible input conditions; otherwise, the circuit will synthesize a latch.

3.2.1.9.3   case

We can have a single statement or a group of statements within the "begin ... end" construct inside a case statement. It can be used to replace a complex "if ... else".

Syntax

```
case (<case_expression>)
poss_expr1: sequential_ stmt;
poss_expr2: sequential_ stmt;
...
poss_exprn: sequential_ stmt;
default: default_ stmt;
endcase
```

In this statement, the case_expression is compared to given possibilities (poss_expr1, poss_expr2, ..., poss_exprn) in a sequential order. The default statement executes whenever none of the possibility matches.

Example

```
.......
parameter RED=1'd1, YELLOW=1'd2, GREEN=1'd3;
.......
case (colour)
RED: traffic=1'b0;
YELLOW: traffic=1'b0;
GREEN: traffic=1'b1;
default: traffic=1'b0; // default statement is used to cover
all unassigned conditions
endcase
.......
```

Case has two variants—"casez" and "casex." While all the "z" values are treated as don't cares in *casez*, both the "x" and "z" values are treated as don't cares in *casex* statement.

Example

```
reg [2:0] points;
integer score;
casex (points)
3'b1xx   : score = 3;
3'bx1x   : score = 2;
3'bxx1   : score = 1;
default : score = 0;
endcase
```

// If points is "3'bxz1", the third expression will give match, and the score will be 1.

3.2.1.9.4   Loop

There are four types of loops used for implementing repetitive evaluation of expressions in Verilog HDL—while, for, repeat, and forever. As long as the <expression> is true, the while loop continues.

Syntax

```
while (<expression>)
sequential_ stmt;
```

Example

```
while (a<b)
a=a+1;
```

The "for" loop consists of three parts—an initial value (initial_expr), an expression to see if the terminating condition is met (check_expr), and an update procedural assignment statement that will change the value of the control variable (update_expr). It executes as long as the expression check_expr is true.

Syntax

```
for (initial_expr; check_expr; update_expr)
sequential_statement;
```

Example

```
for (i=0; i<=50; i=i+1)
a[i] = 2'b11;
```

In order to execute a loop for a definite number of times, the construct "repeat" is preferred.

Syntax

```
repeat (<rep_value>)
sequential_statement;
```

The rep_value may be a variable, an expression, or a constant. Whenever the rep_value is a variable or an expression, its value at the start of the loop is considered as not updated during execution of the loop.

Example

```
repeat (20)
#10 clock = ~clock;
```

//Exactly 20 clock pulses generated.

The "forever" construct has no expression and executes endlessly until it encounters the $finish keyword in the test bench. It is equivalent to an always true "while" loop. The "forever" loop is typically used along with delay. In the absence of the same, the simulator executes the statement indefinitely without any advance in time; thereby the remaining design may never be executed.

Syntax

```
forever
sequential_statement;
```

3.1.2.1.5   Continuous Assignment Statement

The *assign* statement is used for continuous assignment where the output value is updated whenever any signal on the right of the assignment operator (=) changes, and these changes are continuously monitored. In other words, we may say that assign statement binds the "net"-type variable on the left of and the "net"- and "register"-type variables on the right of the assignment operator in the expression. There may be multiple assign statements in a module, and they precede the procedural descriptions.

Syntax

```
module (port description);
.................. . .
```

```
//assign statement
assign x_out = a_in and b_in;
assign y_out = c_in + d_in;
assign z_out = f [5];
//procedural statements
............ .
endmodule
```

In the above example, x_out, y_out, and z_out are of the type net, whereas a_in, b_in, c_in, d_in, and f may be of the type net or reg. The assign statement implements both combinational and sequential logic as can be seen in the following:

```
assign decoder [select] =data_in;// will implement a decoder
assign q = enable? d_in: q; // will implement a level
sensitive latch
```

3.2.2 ADDITIONAL CONSTRUCTS

3.2.2.1 time value (#)

This makes a block suspend for the given time_unit units of time. The time unit can be specified using the timescale command.

Example

```
a=#5 ~b;// not of b is assigned 5 time unit later
#10 c = a;/* this statement is executed after 10 time units
elapses after the execution of the first statement */
```

3.2.2.2 @ (sensitivity_list)

An @ in Verilog indicates that the block will be executed when a trigger occurs on any one of the sensitivity list parameters.

Example

```
input a_in, b_in, clock;
output c_out, d_out;
always @(posedge clock)
begin
c_out = a_in && b_in;
d_out=a_in || b_in;
end
```

Here, c _ out and d _ out get updated whenever a positive edge on the clock is encountered, i.e., clock is the sensitivity list parameter. The @(*) eliminates the problem of incomplete event expression list by adding all nets and variable that are read by a statement inside the block. In the previous example, if always @(posedge clock) is replaced with always @(*), the target variables c _ out and d _ out will be updated when any one of the clock a _ in or b _ in triggers.

3.2.2.3   Generate

In order to dynamically generate the Verilog code prior to simulation, the generate statement is used that allows us to create generic design, which can be updated as per our requirement. For example, we can design an 8-bit parallel adder from an n-bit parallel adder design code. The generate block is enclosed by the keywords *generate* and *endgenerate*. The *genvar* is a special variable associated with the generate loop.

Example

```
module xnor_bitwise (z, a, b);
parameter N = 8;
input [N-1:0] a, b;
output [N-1:0] z;
genvar g;
generate for (t=0; t<N; t=t+1)
begin xnorgen
xnor XNG (z[t], a[t], b[t]);
end
endgenerate
endmodule
```

This code generates an 8-bit xnor capable of performing bitwise xnor operations on a and b while providing the 8-bit result on z.

3.2.2.4   Gate Primitives

Verilog HDL provides logic gates as components known as gate primitives. The built-in gate primitives can be referenced using and, or, nand, nor, xor, and xnor keywords for use in Verilog for single or multiple inputs and a single output. Apart from these, we also have the not and buf primitives that implement single-input, single- or multiple-output not gate and buffer gate, respectively. They are used extensively in Verilog for digital circuit implementation.

Example

```
module my_gates( XZ ,Nin, Ain, Oin, Xin)
input Nin, Ain, Oin, Xin;
output XZ;
wire NZ,AZ, OZ;
not N1 (NZ, Nin);
and A1(AZ, Nin, Ain);
or O1(OZ, AZ, Oin);
xor X1(XZ,OZ, Xin);
endmodule
```

3.2.2.5   Tristate Gates

Verilog HDL also provides three state gates that have a single input, a single output, and a control signal which can either be active high or active low. These three state gates can either be a tristate buffer (tsb) or a tristate inverter (tsi). Verilog provides the following tristate gate primitives—bufif1 (active high control tsb), bufif0 (active low control tsb), notif1 (active high control tsi), and notif0 (active low control tsi).

Example

```
bufif1 buf1 (ZO, AIN, EN);
```

// ZO=A when EN is high and ZO=z when EN is low

```
notif0 n1 (ZO, AIN, EN);
```

// ZO=~Ain when EN bus high and goes tristate otherwise

3.2.2.6 Switch-Level Primitives

Just like built-in gate primitives, Verilog HDL provides some switch-level primitives for use in circuit-level or switch-level modeling. Verilog supports unidirectional or bidirectional ideal and resistive switches only. The ideal metal oxide semiconductor (MOS) switches include *nmos, pmos,* and *cmos,* whereas the ideal bidirectional switches include *tran, tranif0,* and *tranif1.* On the other hand, resistive MOS switches include *rnmos, rpmos,* and *rcmos,* and the resistive bidirectional switches include *rtran, rtranif0,* and *rtranif1.* Power and ground nets include *supply1* and *supply0* as discussed previously, and for implementation of pullup and pulldown, keywords *pullup* and *pulldown* are used, respectively.

Example

// Design of a 4×1 multiplier using bidirectional switches

```
module mux4to1 (out, s0, s1, i0, i1, i2, i3);
input s0, s1, i0, i1, i2, i3;
output out;
wire t0, t1, t2, t3;
tranif0 (i0, t0, s0); tranif0 (t0, out, s1);
tranif1 (i1, t1, s0); tranif0 (t1, out, s1);
tranif0 (i2, t2, s0); tranif1 (t2, out, s1);
tranif1 (i3, t3, s0); tranif1 (t3, out, s1);
endmodule
```

3.2.3 VERILOG MODULE DESCRIPTION

The design that needs to be implemented using any HDL is represented as a design module as seen in Figure 3.1. Each module has some inputs, outputs, and some functionality.

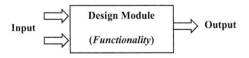

FIGURE 3.1   A Verilog design module.

The functionality is internal to the design, and the designer can alter as per requirement and for optimization. The fundamental building block of Verilog HDL is the module (analogous to functions in C). We can have one or multiple modules in a Verilog code (as we can have one or many functions in C), but a module description cannot contain other modules inside it, but one module may be instantiated by another module (similar to functions in C).

Module Description

```
module <module_id> (<module_port_list>);
...
<module_functionality>
...
...
endmodule
```

The module description begins with the reserved word *module* followed by a *module_id*, that is, any user-defined identifier. The *module_port_list* includes the input and output ports. The *module_functionality* includes the functionality of the design module that determines how the output changes its states or values in response to any change occurring at the input ports. The construct *endmodule* denotes the completion of the design specification process.

3.2.3.1   Ports

The ports are represented with a unique identifier, and the direction of the ports of a design module is represented using the Verilog constructs input, output, and inout to represent input signals, output signals, and bidirectional signals. While the input port is of a variable of type net or reg, the output port is configured as net of the data type wire, reg, wand, wor, or tri type, with the default being wire type. The inout port is of a variable of type net.

Example

```
module first_eg (a,b,c,d);
input c;
input [1:0] d;
output reg a;// the output port is configured as a register
output b;// the output port b is configure as a wire by
default
// continuous assignment statements
.............. .
// procedural assignment statement
.................... .
endmodule
```

3.2.4   Operator Types

Operators are special symbols or mathematical symbols that perform unary or binary operations. While relational operators operate on numbers and return either true or false (Boolean value), bitwise operators operate on bits and return a bit value. Reduction operators on the other hand work on single-word operands and return a single bit as output, and they operate on all the bits within the word. Table 3.2 shows the various operators and their use in Verilog HDL.

TABLE 3.2
Operator Types in Verilog HDL

Operators	Uses	Examples
Arithmetic Operators		
+	Unary (sign) positive	+a
−	Unary (sign) negative	−b
+	Binary plus (addition)	a+b
−	Binary minus (subtraction)	a−b
*	Multiplication	a * b
/	Division	a/b
%	Modulus	a%b
**	Exponentiation	a**3
Logical Operators		
!	Logical negation	!a
&&	Logical AND	a&&b
||	Logical OR	a||b
Bitwise Operators		
&	Bitwise AND	
|	Bitwise OR	
~&	Bitwise NAND	wire a, b, f1,f2;
~|	Bitwise NOR	assign f1 =~a|b;
^	Bitwise exclusive-OR	assign f2=a^b;
~^	Bitwise exclusive-NOR	
Relational Operators		
!=	Not equal	
==	Equal	a!=b
>=	greater or equal	a == b
<=	less or equal	a<b
>	Greater	a <= 0
<	Less	
Reduction Operators		
&	Bitwise AND	assign a =3'b011;
|	Bitwise OR	assign b =3'b110;
~&	Bitwise NAND	assign f1 =^a;
~|	Bitwise NOR	// f1 =0
^	Bitwise exclusive-OR	assign f2=&(a^b);
~^	Bitwise exclusive-NOR	// f2 =0

(*Continued*)

TABLE 3.2 *Continued*
Operator Types in Verilog HDL

Operators	Uses	Examples
Shift Operators		
>>	Shift right	wire [15:0] d, t;
<<	Shift left	assign t=d >> 3;
>>>	Arithmetic shift right	assign t=d >>> 2;
Conditional Operator		
cond_expr? true_expr : false_expr;	If cond_expr is true ("1") then true_expr else false_expr	z=sel?A : B
Concatenation Operator		
{…, …, …}	Joins together bits from two or more comma-separated expressions	wire [7:0] s, co; wire [8:0] so; assign so={co[7], s};
Replication Operator		
{n{m}}	Joins n copies of an expression m	assign f={3{2'b01}};

HDL, hardware description language.

3.3   MODELING TYPES

A Verilog description implements a hardware design at any one or a mix of the following abstraction levels—behavioral or algorithmic level, dataflow or RTL, gate level or structural level, and switch level. The algorithmic or behavioral architecture describes the response of a design to a set of inputs. The RTL or dataflow architecture models the flow of data between the registers the way in the same manner in which it flows in the design. The gate-level architecture models a design using gate primitives. Finally, the switch-level architecture models a design in terms of transistors and their interconnections.

3.3.1   BEHAVIORAL OR ALGORITHMIC MODEL

The behavioral or algorithmic modeling describes the behavior of a digital system that more or less represents the architecture of the design than the logic implementation of the design module as in case of gate-level modeling (later in the chapter). It is an algorithmic approach of the implementation of the design module, so the name algorithmic model in various literatures. It represents a higher abstraction level than the subsequent modeling methods. We usually make use of procedural assignment (procedural statements), which may be either blocking or nonblocking in nature. A procedural block represents a block of code that comprises sequential statements.

The two kinds of procedural blocks that are supported in Verilog are the "initial" block, which is executed only once at the start of simulation and, being nonsynthesizable, is used only in test benches, and the "always" block, which implements a continuous infinite loop. The "initial" block consists of multiple statements grouped inside a "begin … end" structure. Multiple initial blocks execute concurrently, and it specifies the stimulus provided to the design and the outputs displayed for a design under test (DUT).

The "always" block also has a group of statements bounded by "begin … end," and it models an indefinite activity of a digital logic design, e.g., clock signals are generated continuously. Like the initial block, multiple always blocks execute concurrently. The always block is associated with the @ *(event_ expression)* for both combinational and sequential designs. The event_expression may be associated with any type of variable such as *reg* or *wire*, but within an initial and always blocks, the type of variable to be assigned must be of the type *reg*. The reason is that in case of sequential designs, the block triggers for event conditions and remains idle for other conditions. In order to model this, the variable must be able to retain its past value that is possible only by the use of *reg*-type variables.

3.3.1.1 Blocking and Nonblocking Statements

As stated earlier, one or more sequential statements may be contained inside an "initial" or an "always" block. Multiple assignment statements inside a "begin … end" block may either execute sequentially or concurrently depending on the type of assignment.

Blocking (variable=expression;) or Nonblocking (variable <= expression;)

In the first type of statements, sequential assignment takes place (in the order in which they occur in a procedural block). In this case, the target variable of an assignment updates itself before the execution of the next sequential statement. They block the updating of the next statements in the same procedural block, hence the name blocking. Since they do not block the execution of statements in other procedural blocks, it is a recommended style for modeling combinational logic.

In nonblocking assignment statements, all the statements inside a procedural block irrespective of the order in which they occur update themselves once the simulation starts. Since they do not block the assignment of other procedural statements, they are referred as nonblocking statements. It allows concurrent procedural assignment and hence is recommended for sequential logic.

Consider the following code with the blocking statements:

```
integer p, q, r;
initial
begin
p = 20; q = 30; r = 5;//initial values
p = q + r;// p=30 + 5=35
q = p + 5;// q = 35+5= 40
r = p - q;// r=-35-40=-5
end
```

// Rewriting the code with nonblocking statements, we have:

```
integer p, q, r;
initial
begin
p = 20; q = 30; r = 5;//initial values
```

```
p <= q + r;// p=30+5=35
q <= p + 5;// q=20+5=25
r <= p - q;// r=20-30= -10
end
```

Mixing of blocking and nonblocking statements within an always block is not a rec-ommended coding practice. Moreover, one must be careful not to assign the same variable as the target of both blocking and nonblocking assignments.

```
a = a + 1;
a <= b;
```

Using the above two expressions in the same always block is not permissible.

Model a 2-to-1 mux using behavioral model that has two inputs represented by a 2-bit vector I, a select line represented by S, and a 1-bit output represented by F. The following code represents the 2-to-1 mux:

Example

```
module mux_2to1(F, S, I);
input [1:0] I;
input [1:0] S;
output reg F;
always @(S or I)// can be replaced with @(*)
begin
case (S)
1'b0: F=I[0];
1'b1: F=I[1];
default F=1'bx;
endcase
end
endmodule
```

//Using continuous assignment statement

```
assign F = I[S];
```

//Using conditional statement if-else

```
if (S) F = I[1];
else
F = I[0];
end
```

3.3.2   DATAFLOW OR REGISTER TRANSFER-LEVEL MODEL

The dataflow model represents the design module in a way that specifies the flow of data. In other words, it depicts how the data move between the hardware components of the design and how the processing takes place. This type of modeling uses logi-cal expressions for its implementation. Dataflow models presented using different techniques such as the continuous assignment statement, reduction operators, the conditional operator, relational operators, logical operators, bitwise operators, and shift operators are widely used.

This modeling approach is implemented using the continuous assignment statement *assign*, and we know that the *assign* statements usually precede the procedural descriptions.

// Use assign statement to implement the 2-to-1 mux

```
assign w = s? b: a;
```

//the assign statement is replaced as

```
assign w = (~s & a) | (s & b);
```

Here, w must be of the type net as it is continuously driven, whereas s, a, and b can be either net- or reg-type variables.

3.3.3 GATE-LEVEL OR STRUCTURAL MODEL

In this modeling approach, the implementation of the design module is in terms of logic gates and interconnections between them, and it resembles a schematic with interconnected components. A component activates whenever there is any change in the signal associated with the component. Multiple components may be activated simultaneously.

This modeling approach is a closer approximate of the physical implementation. Using these primitives, the structural-level implementation of a 2-to-1 multiplexer is given as in Figure 3.2.

Example

```
module mux_2to1(
input A, B;
input SEL;
output Z
);
wire t0;
wire t1, t2;
not N (t0, SEL);
and A0 (t1, SEL, A);
and A1 (t2, SEL, B);
or R (Z, t1, t2);
endmodule
```

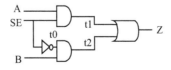

FIGURE 3.2   Logic gate representation of a 2-to-1 mux.

In the aforementioned method, we have used one not gate, two and gates, and one or gate. Designer has the knowledge of the internal functionality of the primitive prior to instantiation in any circuit design. These primitives connect in a manner that it realizes a multiplexer, and this type of modeling is referred to as structural modeling.

3.3.4 SWITCH-LEVEL MODEL

This modeling scheme is used to implement a design module in terms of transistors, switches, registers, and their interconnections. This model comes under the lowest abstraction level for Verilog. Switches in Verilog HDL are representation of transistors that can either conduct or open. These switches are governed by the four logic levels 0, 1, x, and z in order to determine the signal strength. It comprises a netlist of MOS transistors. This is not a common modeling tool for designers and may be used for leaf cell design only for hierarchical implementation. In order to implement a CMOS circuit-level implementation function $F = (\overline{AB+CD})$, we can make use of the following code:

Example

// CMOS implementation of a function F

```
module computeF (F, A, B, C, D);
input A, B, C, D;
output F;
wire W1, W2, W3, W4;
wire A_bar;
// power and ground rails
supply1 vdd;
supply0 gnd;
//generating A'
pmos P1 (A_bar, vdd, A);
nmos N1 (A_bar, gnd, A);
// p net
pmos p1(W1,vdd, A_bar);
pmos p2(W2,vdd, B);
pmos p3(F, W1,C);
pmos p4(F, W2, D);
//n net
nmos n1(F, W3, A_bar);
nmos n2(W3, gnd, B);
nmos n3(F, W4,C);
nmos n4(W4,gnd, D);
endmodule
```

Example

```
module mux_2to1 (F, S, I);
input S;
input [1:0]I;
output F;
```

```
wire SBAR;
not (SBAR, S);
cmos cmos_a (F, I[0], SBAR, S);
cmos cmos_b (F, I[1], S, SBAR);
endmodule
```

Switch-level modeling is not supported for synthesis in most simulators, as technology mapping is essential from a given library. Since most functional blocks are already present as primitives in the library, the transistor-level implementation of logic circuits is not popular.

3.3.5 MIXED MODEL

When the description of functionality of a Verilog Design Module is done using different modeling styles in the same code, such a modeling scheme is referred to as mixed modeling. Similarly, when simulating a mixed-model design, the simulation scheme is referred to as mixed-level simulation. We may also implement the 2-to-1 mux by a combination of dataflow and structural modeling.

Example

```
module mixed_andor (input a, b, c, d, output y);
assign y = (a & b) | (c & d);
endmodule
module mux_mixed (input A, B, SEL, output Z);
wire t0;
not U1 (t0, SEL);
mixed_andor U2 (A, SEL, t0, B, Z);
endmodule
```

The module mux_mixed incorporates structural as well as behavioral models to represent a 2×1 multiplexer. Synthesizable Verilog module must usually be written in structural model in order to implement the design as per specified requirements.

3.4 USER-DEFINED PRIMITIVES

User-defined primitives (UDPs) are customized primitives specified by the user with the help of lookup tables. They are used to specify the truth table for combinational logic and state table for sequential logic.

For combinational functions, truth table entries are specified as follows:

```
<input1> <input2> … <inputN> : <output>;
```

For sequential functions, state table entries specification is as follows:

```
<input1> <input2> … <inputN> : <present_state> : <next_state>
```

The UDP model does not model timing or process technology but models functionality only. A functional block with a single output only can be modeled using

UDP. For unspecified cases, the output is set to "x," whereas the don't cares in inputs are represented with a question mark (?).

Example

```
primitive udp_mux2to1 (F, A, B, S);
input S, A, B;
output F;
table
```

//	S	A	B		F
	0	0	0	:	0 ;
	0	0	1	:	0 ;
	0	1	0	:	1 ;
	0	1	1	:	1 ;
	1	0	0	:	0 ;
	1	0	1	:	1 ;
	1	1	0	:	0 ;
	1	1	1	:	1 ;

```
endtable
endprimitive
```

//The lookup table can be rewritten including don't care:

//	S	A	B		F
	0	0	?	:	0 ;
	0	1	?	:	1 ;
	1	?	0	:	0 ;
	1	?	1	:	1 ;

In sequential circuits, some output states are preserved for certain input conditions—in case of SR flip-flops, when S=0 and R=0, then the next state $Q_{n+1} = Q_n$. A "-" is used to represent this condition in the next state column in the lookup table of UDP.

3.5   TEST BENCH

After the design specification is complete, we must simulate our code for prelayout simulation or, in other words, formal verification of the design module. One of the important steps in IC design is verification, and prelayout and postlayout simulations using HDLs are methods for carrying out design verification. This is possible by writing a test bench, which is a high-level entity, used to provide signals such as clock, reset, and test vectors to a DUT and dump the outputs to enable the simulation of the design module. Test bench waveforms display the simulation results in graphical form. Test benches are procedural blocks that execute only once but are nonsynthesizable. The input and output of the DUT connect to the test bench. A test bench can include various simulator directives—$display, $monitor, $dumpfile, $finish, etc.

While $display ("<format>", expr1, expr2, ...) prints the immediate values of text or variables, the event-driven $monitor ("<format>", var1, var2, ...) prints whenever some variable in its expression list changes. The $finish is used to terminate a simulation process, and $dumpfile (<filename>) specifies the file that will be used for storing the values of the selected variables so that they can be graphically visualized later. The $dumpoff directive stops the dumping of variables, whereas $dumpon directive starts previously stopped dumping of variables.

Example

```
module tb_mux;
// Declaring Inputs
reg A;
reg B;
reg S;
// Declaring Outputs
wire F;
// Instantiate the Design Under Test (DUT)
mux2to1 dut (
    .A(A), //explicit association
    .B(B),
    .S(S),
    .F(F)
);
initial begin
// Apply Inputs
A = 0;
B = 0;
S = 0;
$monitor ($time, "A=%b, B=%b, S=%b, F=%b", A, B, S, F);
// Wait 50 ns
#50;
//Similarly apply Inputs and wait for 50 ns
A = 0; B = 0; S = 1; #50;
A = 0; B = 1; S = 0; #50;
A = 0; B = 1; S = 1; #50;
A = 1; B = 0; S = 0; #50;
A = 1; B= 0; S = 1; #50;
A = 1; B = 1; S = 0; #50;
A = 1; B = 1; S = 1; #50;
end
endmodule
```

The input ports of the module have been declared as reg, as they appear on left-hand side of the expressions appearing inside the initial block, whereas the output ports are declared as wire. The $monitor ($time, "A=%b, B=%b, S=%b, F=%b", A, B, S, F) will generate the current simulation time, i.e., value of A, B, S, and F in the simulation window in the following way (display format depends on the simulator):

```
0     A= 0,    B=0,    S=0,      F=0
50    A= 0,    B=0,    S=1,      F=0
100   A= 0,    B=1,    S=0,      F=0
150   A= 0,    B=1,    S=1,      F=1......and so on
```

While $monitor will display the value of its variable, the moment any one of the listed variables changes its value, $display will display the value of the listed variable irrespective of any change occurring in the value of the variables. The initial block in the test bench will execute only once, and the use of blocking statements will help for proper implementation.

Writing a Verilog with proper indention with suitable comments is essential in industry design. The following code uses behavioral style of modeling using the procedural block *always* that in turn includes the sequential statement case:

Example

//Design of an arithmetic and logic unit (ALU)

```
module ALU (res, a, b, oper); // module description
    input [7:0] a, b; // inputs
    input [2:0] oper; // operation to be performed
    output [7:0] res; // result
    reg [7:0] res; // res declared as reg as it is used
    inside always block
        always @ (*) // when either a, b or oper changes
        state always block is triggered
            begin
                case (oper) // operations to be performed on
                a and b
                    3'b000 : res = a;// data transfer
                    3'b001 : res = a + 1; // increment
                    3'b010 : res = a - 1; //decrement
                    3'b011 : res = a + b; // addition
                    3'b100 : res = a - b; // subtraction
                    3'b101 : res = a << 1; // shift left
                    3'b110 : res = a >> 1; // shift right
                    3'b111 : res = a^b; // xor
                endcase
            end
endmodule
```

Since the previous example has defined res for all input conditions inside the case statement, on synthesis, it will generate a combinational design of the ALU.

3.6   VERILOG FOR ANALOG DESIGN

Understanding the system-level performance, a system-level simulator capable of evaluating the performance of the design module prior to detailed circuit design and layout is necessary. Verilog-AMS is an example of such a top-down HDL used for its popularity in analog and mixed-signal implementations. The Verilog-AMS derived from the Verilog standard IEEE 1364-2005 is an analog equivalent of the Verilog HDL for describing analog and mixed systems. Verilog-A is a part of Verilog-AMS that contains the analog version of IEEE 1364-2005. This language implements algorithmic description of continuous-time systems, the syntax of which is similar to Verilog HDL specification. Mathematical description of the functionality of each module in terms of its input output ports and external parameters is an important feature of Verilog-A. Each component description is in terms of interconnected subcomponents to find its applications in various fields such as electrical, thermal, mechanical, and so on. Verilog-A utilizes mathematical expressions to define the electrical and nonelectrical behavior of a circuit.

3.7 VERILOG-A BASICS

The design module in Verilog-A is represented as a system that comprises one or more components wherein each component may or may not have subcomponents. A set of components in a system represents a hierarchical design, and the subcomponents known as primitive components are an important part of the design. A system reacts to external stimulus and produces response in response to these external stimuli. Like its digital counterpart, Verilog-A also uses built-in primitives for the circuit definition that include common circuit components such as resistors, capacitors, inductors, and semiconductor devices.

When a module refers to other modules in a design, we infer a structural model, whereas when a module uses equations for specifying a component in a design, we infer a behavioral model. A design that has a module containing both module instantiations and equations infers a mixed model. Sometimes, a design may have neither module instantiations nor equations; in such a case, the model is an empty module.

In Verilog-A, components are made of nodes and branches. While a node is a terminal, a branch is the connecting path between two nodes. In order to embody Kirchhoff's laws, the simulator follows the given rules with respect to the nodes and branches—the algebraic sum of flow in a node is zero (Kirchoff's flow law); the algebraic sum of all the branch potentials inside a loop at any point of time is zero (Kirchoff's potential law). Verilog-A HDL allows description of nets based on disciplines. A discipline may specify either potential or flow, or it may specify both potential and flow. In order to describe a component in Verilog-A, we must define its topology (nodes, branches) and the behavior of each branch.

// Potential signal-flow systems

```
analog
V(out) <+ 5.0 + V(in);
// Flow signal-flow systems
analog
V(out) <+ 5.0 + V(in);
```

In order to model a resistor with node terminals n1 and n2 having node voltages V(n1) and V(n2) and branch voltage V(b), the resistance is given as R Ω, whereas the current flowing in the branch is I(b).

Example

//Resistor 1

```
module resistor (n1, n2);
electrical n1, n2;
parameter real R=1;
branch (n1,nt2) b;
analog V(b) <+ R*I(b);
endmodule
```

The module description here also starts with a module id of the component (*resistor*) *followed by the* nodes. The keyword *electrical* represents voltage to be the potential and current to be the flow. The flow is positive if the flow is into the component. Parameter R of the components refers to the resistance value in order to instantiate the component. If the value of parameter is not specified, it is assigned the default value, and being a constant, this value remains fixed. The *branch* named *b* refers that a branch exists between n1 and n2. Here, the branch voltage V(b) is considered to be equal to V(n1) − V(n2) (as per the orders of the terminals mentioned). If node voltage of t1 is greater than that of t2, the branch voltage is positive. The current flowing in the branch is I(b). In this case, it is positive as the current flows from n1 to n2. This completes our topological description. In order to describe the branch flow relation, we must specify a relation between the voltage and current of the branch. The analog keyword denotes an analog process. Multiple statement would require a "begin … end." The behavior of a branch is specified by a contribution statement and a contribution operator (<+).

```
V(b) <+ R*I(b);
```

It states that "the voltage on branch res must equal the product of current through that branch and resistance R." The resistor can also be modeled without defining a branch explicitly.
    //Resistor 2

```
module resistor (n1, n2);
electrical n1, n2;
parameter real R=1;
analog V(n1,n2) <+ R*I(n1,n2);
endmodule
```

Let us look at some simple circuit models that are implemented using Verilog-A HDL.
    // Model a resistor with a dc source in series

Example

```
module port (n1, n2);
electrical n1, n2;
parameter real dc=0;
```

```
parameter real R=100;
branch (n1, n2) b;
analog begin
V(b) <+ R*I(b);
V(b) <+ dc;
/* the preceding two lines may be replaced by
analog V(b) <+ R*I(b) + dc;*/
end
endmodule
```

Example

// series RLC circuit may be modeled as

```
module rlc_in_ series (n1, n2);
electrical n1, n2;
parameter real R=10;
parameter real L=10;
parameter real C=10;
analog begin
    V(n1,n2) <+ R*I(n1,n2);
    V(n1,n2) <+ L*ddt(I(n1,n2));// ddt returns time derivative
    V(n1,n2) <+ idt(I(n1,n2))/c;// idt returns time integral
end
endmodule
```

//model of a parallel RLC circuit

```
analog begin
    I(n1,n2) <+ V(n1,n2)/r;
    I(n1,n2) <+ idt(V(n1,n2))/l;
    I(n1,n2) <+ c*ddt(V(n1,n2));
end
```

Example

// module description of a MOSFET

```
module nmost (g, s, d, sub);
inout electrical g, s, d, sub;
parameter integer nqsMod = 0 from [0:1];
end
endmodule
```

Verilog-A in collaboration with Spice and Verilog HDL has proved to be an excellent synthesis and simulation tool for IC design engineers. Specification, simulation, and verification of mixed-signal designs have become easier. They allow portability, compactness, robustness, and rapid prototyping in hardware implementation before actual tape-out.

3.8 SUMMARY

Verilog HDL provides many constructs, but many of them may not be synthesizable, such as *initial* block, *time, testbench, switch-level implementations, UDP,* and so on in specific synthesis tools. When modeling combinational designs, we must keep in mind to give explicit output for every input conditions; otherwise, we may get a sequential circuit. To produce synthesizable designs, we can use a netlist of Verilog built-in gate primitives, continuous assignments, and behavioral statements. We must avoid both feedback loops of any kind during combinational logic design and mixing of blocking and nonblocking assignments. So when designing a synthesizable logic, we must be careful to make use of only synthesizable constructs as available in the simulator under use. Once a synthesizable design is ready, testing of the design by giving signals to the synthesized design using a field-programmable gate array (FPGA) or application-specific integrated circuit (ASIC) or complex programmable Logic Devices (CPLD) can be performed.

A VLSI design is based on design verification test, and the Verilog Family of Languages provides a solution in handling each of these steps by minimizing the complexity of each step. The Verilog Family of Languages including Verilog HDL, Verilog-AMS, and Verilog-A has proved to be a popular EDA tool for both industry and academics. An excellent choice for both digital and analog designs and their features of simulation, synthesis, verification, and testability make them an excellent choice for IC design engineers. These languages allow the designer to implement designs in various abstraction levels. Compactness, closeness to C, and wide acceptance in industry are some major takeaways for Verilog HDL. Describing the behavior of electrical circuits, development of analog models, and facilitating model development using Verilog-A has proved to be quite useful. In order to handle the interface between real environment signals and the advantages of digital signal processing, the Verilog-AMS is a welcome addition for circuit designers.

REFERENCES

Ashenden, P. J. and Lewis, J. 2008. *VHDL-2008 Just the New Stuff.* Burlington, MA: Morgan Kauffmann.

Navabi, Z. 1999. *Verilog Digital System Design.* New York: Mc Graw Hill.

Palnitkar, S. 2003. *Verilog HDL: A Guide to Digital Design and Synthesis.* Upper Saddle River, NJ: Prentice Hall PTR.

Sutherland, S., Davidman, S. and Flake, P. 2004. *System Verilog for Design.* New York: Springer.

4 Introduction to Hardware Description Languages

Shasanka Sekhar Rout
GIET University

Salony Mahapatro
Asiczen Technologies

CONTENTS

4.1 Introduction 68
    4.1.1   Basic Principles of Hardware Description Languages 68
    4.1.2   Basic Concepts of Hardware Description Languages 69
            4.1.2.1   Timing and Concurrency 69
            4.1.2.2   Hardware Simulation Process 70
    4.1.3   Hardware Description Language Design Tool Suites 70
    4.1.4   Types of Hardware Description Languages 70
    4.1.5   Design Using Hardware Description Language 71
    4.1.6   Hardware Description Languages for Digital Design 71
    4.1.7   Very High-Scale Integrated Circuits Hardware Description
            Language 71
            4.1.7.1   Entity Declaration 72
            4.1.7.2   Architecture 72
    4.1.8   Verilog 74
            4.1.8.1   Lexical Tokens 74
            4.1.8.2   Data Types 75
            4.1.8.3   Timescale 75
            4.1.8.4   Continuous Assignments 76
             4.1.8.5   Procedural Assignment 76
            4.1.8.6   Procedures: Always and Initial Blocks 77
    4.1.9   Test Bench in Verilog 77
    4.1.10 System Verilog 78
    4.1.11 Test Bench Structure for System Verilog 78
            4.1.11.1 Design under Test 79
            4.1.11.2 Transaction 79
            4.1.11.3 Interface 80
            4.1.11.4 Generator 80
            4.1.11.5 Driver 80
            4.1.11.6 Monitor 81
            4.1.11.7 Scoreboard 81

       4.1.11.8   Test 82

       4.1.11.9   Environment 82

       4.1.11.10 Top 83

    4.1.12   Verilog-AMS 84

    4.1.13   Mini Project: Verilog and System Verilog 85

References 91

4.1 INTRODUCTION

Before getting started with learning about hardware description languages (HDLs), one must begin with *"Why?"*. Answering that in simple terms, it is to deal with the complexity in very large-scale integrated circuit (VLSI) design and its substantial reduction with the introduction of register level and transfer level of abstraction. Design methodologies using the traditional abstraction level, such as gate level and so on, are no longer sufficient in the era of complex VLSI design. Hence, a higher level of abstraction, i.e., register transfer logic (RTL) is introduced, which is also known as computer language or HDL. By tradition, digital system was a manual process of designing and capturing circuits with the use of schematic entry tools. This process had many demerits and is quickly being replaced by innovative methods.

In electronics, HDL is a particular computer language that helps to clarify the structure and performance of electronic circuits, mainly digital circuits. The HDL enables specific, formal explanation of electronic circuit that allows for automatic investigation in circuits simulation. A net list (electronic components specification with their connection) is allowed for synthesis of HDL description by HDL, which can be positioned and routed to generate a set of masks for generating an integrated circuit (IC). HDL looks like a C programming language. It is a textual explanation, which consists of expressions and control structures statements [1]. Nevertheless, HDLs are dissimilar from programming language as they openly consist of the perception of time. HDL forms an essential division of electronic design automation (EDA) systems particularly for composite circuits such as PLDs, ASICs, and so on. HDL has evolved in excess of time, and the standard has altered. However, this chapter focuses on essence of the core of the digital designs or design styles.

4.1.1 BASIC PRINCIPLES OF HARDWARE DESCRIPTION LANGUAGES

The spirit of using HDLs in the field of electronics occurred due to the simplification of the ever-growing complexity in the era of electronic designs or particularly in VLSI designs. Electronic designs can be declared as the method of transforming the behavioral explanation or structural explanation. After alteration into the structural depiction, there may be requirement of physical design, which attaches mainly with the back-end design flow elements such as choice of device size, interconnect lines routing, placements of block, and so on. While the software design has involved much before HDL came into sight, the available experiences of the software designers should be learnt for managing with the incrementing problems of difficulty. So, some points have to be adopted [2]:

- **Hierarchical design**: The complex circuits can design in parts of modules and submodules, etc.
- **Modular architecture**: The modification of fixed module should not impact in any other modules of the design.
- **Text-based explanation**: The standard parsing concepts can be used to handle depiction relatively than by the use of pictorial representation.
- **Reprocess of obtainable resources**: Reuse of the existing resources should be done for saving energy and time.

4.1.2   BASIC CONCEPTS OF HARDWARE DESCRIPTION LANGUAGES

The two major fundamental concepts of HDLs include the following:

    a. Timing and concurrency
    b. Hardware simulation process
- Analysis
- Elaboration
- Simulation

4.1.2.1   Timing and Concurrency

In actual circuits, logic gates have delays connected with them. Gate delays permit to specify delays throughout the logic circuits. There are mainly three types of delays in a logic gate starting from inputs to outputs.

- **Rise Delay**: It is connected with a gate output transition from any value to a 1.
- **Fall Delay**: It is related to a gate output transition from any value to a 0.
- **Turnoff Delay**: It is related to a gate output transition from any value to the high impedance value (z).

If there is only one particular delay, then that value is implemented for all the transitions. If there are two delays specified, then they belong to the value of rise and fall delay, where turn-off delay is the smallest among two delays. If there are all three delays given, they belong to rise, fall, and turnoff delay. If there are no delays given, then zero is the default value.

Electronic circuits are active at all times, where timing related with each event occurs. HDL is a language that explains the behaviors of electronic circuits, which means it exactly models the time and concurrency of electronic circuits. However, HDL must maintain the different delay times in the electronic circuits, such as inertial and transport delay. Inertial delays can be implemented to model delays throughout the capacitive networks. If the width of a pulse is smaller than the circuit's switching time, it shall not come into view at the circuit's output. The internal resistance and capacitance of the logic gates create the gate delays, so called inertial. Transport delays are implemented to model small length wires; hence, pulses are propagated with respect to the width of the pulse. A reproduction of the pulse will come into sight at the output after a particular delay time.

Event processing is the foundation of HDL simulation; hence, all HDL simulators must be event-driven simulators. There are three important concepts to event-driven simulation [3]:

   i. Simulation time
  ii. Delta time
 iii. Event processing

Simulation time: It is usually calculated as an integral multiple of a vital unit of time identified as resolution limit. The simulator focuses on the circuit time that has been modeled by the simulator.

Delta time: It is implemented for queuing up sequential actions. Delta delay is the time between two sequential actions. In a combinational logic circuit, all the elements have a 0 ns delay where all assignments will function at 0 ns with possibility of several delta delays. The simulator shall count up the quantity of delta times awaiting every signal is stable.

Event processing: The simulation cycle interchanges in between processing execution and event processing. Signals are updated as a group in event-processing element of the cycle, and then, processes are functioned in the process execution part.

4.1.2.2   Hardware Simulation Process

In HDL, hardware simulations occupy analysis, where the syntax of hardware description is checked and interpreted. Analysis does the analyzing of each design unit independently and placing analyzed units in a working library. Elaboration makes a complete circuit from a hierarchical depiction. Here, the design hierarchy is flattened as follows:

* Generate ports (for interfacing with others)
* Generate signals and processes

Which pieces of hardware are sensitive to which events is determined throughout the elaboration phase. This is simply known as a sensitivity list. For an event, one can rapidly obtain a record of all hardware which is sensitive to it.

4.1.3   HARDWARE DESCRIPTION LANGUAGE DESIGN TOOL SUITES

HDL simulator is a software package, which compiles and simulates the HDL expressions. Beyond the desktop level, enterprise-level simulator proposes fast simulation runtime, extra robust support for mixed language (VHDL and Verilog) simulation, and validating time-accurate gate-level simulation. The main signoff grade simulators are Cadence Incisive Enterprise Simulator, Synopsys, and Mentor ModelSim.

4.1.4   TYPES OF HARDWARE DESCRIPTION LANGUAGES

Examples of HDLs that enable the users to design analog circuits are analog HDL, Spectre HDL, Verilog-AMS (Verilog Analog and Mixed Signal), Verilog-A, and so on. Examples of HDLs that make to design digital circuits possible are Chisel, Hydra, KARL, System C, System Verilog, Verilog, VHDL (very high-speed integrated circuits HDL) [4], and so on. Examples of HDLs supporting for printed board circuits (PCBs) are PHDL (PCB HDL), EDA Solver, SkiDL, and so on.

4.1.5 Design Using Hardware Description Language

With the incremental growth of the efficiency gains that were observed as a result of using HDL, maximum of modern digital circuit design advances and revolves around it. The majority designs start with a set of requirements or a high-level architectural diagram. Control and decision structures are habitually prototyped in flowchart applications. The method of writing the HDL description is very much dependent on circuit's nature and the preference of designer for coding fashion. The HDL is purely the capture language starting with a high-level algorithmic explanation like a C++ mathematical model. The scripting language such as Perl is used to automatically produce repetitive circuit structures in the HDL language by the designers. Particular text editors propose features for automatic indentation, syntax-dependent coloration, and macro-based extension of the entity, architecture, and signal declaration. Then after, the HDL codes undergo a code review for further processing. The HDL description is focused to an array of automatic checkers in preparation for synthesis. The checker reports deviations from regular code guidelines, recognizes potential confusing code constructs before misconception occurs, and confirms for general logical coding errors such as shorted outputs and floating ports. This procedure aids in solving errors earlier than the code is synthesized [5].

In industry, HDL design normally ends at the synthesis phase. Once the synthesis tool has mapped the HDL description into a gate net list, the net list is accepted to the back-end stage. Depending on the physical technology (field-programmable gate array [FPGA], application-specific integrated circuit [ASIC] standard cell, ASIC gate array), HDLs may or may not play an important role in the back-end flow. Finally, an IC circuit is manufactured or programmed by using HDL design in different applications [6].

4.1.6 Hardware Description Languages for Digital Design

Most widely used HDLs around the globe are VHDL and Verilog for digital designing. But nowadays, System Verilog has also made its reputation in digital design field apart from verification. The basics of the above HDLs will be discussed briefly in the upcoming sections.

4.1.7 Very High-Scale Integrated Circuits Hardware Description Language

The programming language that is used for the digital systems modelling is represented in three styles, such as dataflow, behavioral, and structural. VHDL describes hardware modules using entities. Entity can be elaborated in the following parts: entity declaration, architecture, configuration, package declaration, and package body. The individual parts are presented in the following sections [7].

4.1.7.1 Entity Declaration

Entity declaration means to define the names, input–output signals, and operational modes of a hardware module. It should begin with "entity" and finish with "end" keywords, respectively. The direction is decided by the keywords "in," "out," or "inout," which are represented as follows [8]:

- Input port can read: in
- Output port can write: out
- Port can read and write: inout
- Port can read and write, but it has only one source: buffer

////// Syntax of Entity Declaration //////

```
entity entity_name is
    port declaration;
end entity_name;
```

4.1.7.2 Architecture

Architecture is described by using structural, behavioral, dataflow, or even also using mixed style. The entity name that is written for the architecture body should be specified. The architecture statements should be inside the "begin" and "end" keyword. Architecture declarative part contains variables, component declaration, or constants.

////// Syntax of Architecture Declaration //////

```
architecture architecture_name of entity_name
architecture_declaration_part;
begin
    statement;
end architecture_name;
```

4.1.7.2.1 Dataflow Modeling

In this modeling style, the flow of data throughout the entity is presented with concurrent signal. The parallel or concurrent statements in VHDL are WHEN and GENERATE. In addition, assignments by the use of only operators (AND, NOT, +, *, etc.) can also be employed to construct code. Also, a particular kind of keyword BLOCK can be used in this type of code. In concurrent code, the following keywords are used [9]:

- Operator
- The WHEN statements (WHEN/ELSE or WITH/SELECT/WHEN)
- The GENERATE statements
- The BLOCK statements

4.1.7.2.2 Behavioral Modeling

In this modeling style, the behavior of an entity as group of statements is executed sequentially in the particular order. Only statements placed within a PROCESS, PROCEDURE, or FUNCTION are sequential. PROCESSES, PROCEDURES, or FUNCTION are the only parts of the code that are executed in sequence. Therefore, any of these blocks is still concurrent with any additional statements located outside it. One significant aspect of behavior code is that it is not restricted to sequential logic. The behavior statements are IF, CASE, WAIT, and LOOP. VARIABLE is also limited, and it is only implemented in sequential code. VARIABLE can not at all be global, so its value never be approved directly.

4.1.7.2.3 Structural Modeling

In this modeling, an entity is defined as a group of interconnected components. A component instantiation declaration is a concurrent statement. Hence, the order of these statements is not so vital. The structural fashion of modeling explains only an interlinking of components (like black boxes), implying any behavior neither of the component itself nor of the entity that they together characterize. Here, architecture body is working in two parts:

- Declarative part (previous to the keyword begins)
- Statement part (subsequent to the keyword begins)

Example of VHDL code for logic gates and for 4-bit parity generator is presented as follows:

//////VHDL Code for Logic Gates

```
library ieee;
use ieee.std_logic_1164.all;
entity logic_gates is
   port (m,n : in std_logic;
              p0,p1,p2,p3,p4,p5,p6 : out std_logic);
end logic_gates;
architecture behavioural of logic_gates is
begin
   p0 <= m and n;
   p1 <= m or n;
   p2 <= m not n;
   p3 <= m xor n;
   p4 <= m xnor n;
  p5 <= m nand n;
  p6 <= m nor n;
end behavioural;
```

//////VHDL Code for 4-bit Parity Generator

```
library ieee;
use ieee.std_logic_1164.all;
```

```
entity 4bit_parity_gen is
    port (b0, b1, b2, b3: in std_logic; odd, even: out
std_logic);
end 4bit_parity_gen;
architecture structural_arch of 4bit_parity_gen is
begin
    process (b0, b1, b2, b3)
    if (b0 ='0' and b1 ='0' and b2 ='0' and b3 ='0')
        then odd <= "0";
        even <= "0";
    else
        odd <= (((b0 xor b1) xor b2) xor b3);
        even <= not (((b0 xor b1) xor b2) xor b3);
end structural_arch;
```

Here, the first two lines "library ieee; and use ieee.std_logic_1164.all;" are essential to include the standard packages and libraries, which contain all the declarations and predefined functions. The ports are to be declared inside the entity block followed by the architecture block containing the description of the hardware.

4.1.8   VERILOG

Verilog is the HDL used for describing the digital systems same as networking switches or microprocessors or a memory unit. By using HDLs, the clumsy task of explaining and representing digital hardware design at any level becomes very easy and productive. Verilog supports designing hardware at various levels of abstraction. The most widely used are behavioral level, RTL, and gate level [10].

4.1.8.1   Lexical Tokens

Verilog language resource text files are a flow of lexical tokens. It contains at least one or more characters, and every single character is accurately in one token. The fundamental lexical tokens that are used by the Verilog HDL are very much analogous with C language. It is very important to remember that Verilog is case sensitive and every keyword is defined in lower case. Lexical tokens that must be basically taken care of by writing Verilog HDL codes are white spaces, comments, identifiers, operators, and Verilog keywords.

a. **Whitespaces** include typeset for spaces or tabs or new lines. These are mostly ignored apart from when they are used to split between the tokens. These characters include blank space, tabs, new line, form feeds, and carriage returns.
b. **Comments** are represented in two forms: single-line comments (e.g., //Single line comment) and multiline comments (e.g., /* Multiline comments */).
c. **Identifiers** are the names that are used to identify the object, such as a function, module, or register. Identifiers start with alphabetical or underscore characters, e.g., V_erilog, v_hdl, _veriloghdl, etc. They are up to 1024 characters long.

d. **Operators** are exceptional characters used to function the variables. There are one, two, and sometimes three characters used to execute operation on variable, e.g., >, ~, &, =.

e. **Verilog keywords** are those that have a particular sense in Verilog, e.g., always, assign, for, case, if, while, wire, reg, and, nor, or, nand, module, and so on. The use of Verilog keywords should not be done as identifiers. They also consist of compiler directives, system tasks, and functions.

4.1.8.2 Data Types

Verilog includes four basic states or values. The entire data types used in Verilog have these states [11]:

0—logic zero or 0 V or false condition
1—logic one or 0.7 V or true condition
x—undefined or unknown logic state
z—high impedance state

The employ of x and z is negligible in cases of synthesis. The most focused data types are wire, register, input, output, inout, and integer.

4.1.8.3 Timescale

Verilog focuses on time units like a command to the software tool, by the use of a "timescale compiler directive." It has two parts: one is the time units and the other is the time precision that is to be used. The precision component conveys the information to the software tool how many decimal places of accuracy to be used by it. In the following code, 'timescale 1ns/10ps infers that the simulator tool is ordered to employ time units of 1 ns and precision of 10 ps, which is 2 decimal places relative to 1 ns. The 'timescale directive can be declared in one or many Verilog source files. Such directives having dissimilar values can be specified for unlike sections of a design. When such happens, the simulator tool must be able to determine the differences by the decision of the common denominator in all the specified time units and then scaling every delays in every section of the design to the common denominator. The 'timescale directive command is not bound to precise modules or to any exact file.

//////Verilog Snippet

```
'timescale 1ns/1ps
module sample_snippet (q_0, q_1, out);
input q_0, q_1;
output out;
wire q_0, q_1;
reg out;
integer c;
assign out = 4;
initial begin
    $display ("c = %d", c);    //display values
end
endmodule
```

4.1.8.4 Continuous Assignments

These assignments in a Verilog module are implemented for transmission values on wires, which is the job used exterior always or initial blocks. Such task is finished with an explicit assign statement or to allocate a value to a wire throughout its declaration. Continuous assignments are constantly executed at the instance of simulation. The sort of allocate statements does not influence it.

Example Snippet

```
module abc_dut0 ( a, b, c, out);
input a, b, c;
output out;
wire a, b, c;
reg out;
assign out = (a + c) ^ b; //continuous assignment of nets
endmodule
```

4.1.8.5 Procedural Assignment

This assignment is for updating reg, time, integer, and memory variables. Procedural assignments differ from continuous assignments as they drive the net variables and are evaluated and then updated whenever there is a change in input operand values, whereas the procedural assignments update the values of the register variables under power of the constructs surrounding the procedural. The right-hand side of a procedural assignment is any expression, which is to be evaluated to a value. The left-hand side is the variable that gets the assignment from the right-hand side. Procedural assignments are of two types: blocking assignments and nonblocking assignments.

a. The blocking procedural assignments (=) are those statements that are executed earlier than the execution of the statements that follow it in a sequential block. Such statements do not stop the execution of statements that are following it in the parallel blocks.

b. The nonblocking procedural assignment (<=) statements permit the scheduling of assignments devoid of obstructing the procedural flow. The employ of nonblocking procedural statements is done when it is needed to create multiple register assignments in the similar time frame devoid of requiring dependence upon each other.

//////Verilog Code to Continuous and Procedural Assignments

```
module assignment (out);
output out;
reg in_0, in_1, in_2;
wire in_3;
initial begin
in_0 = 1;
in_1 = 0;
in_2 = 1;
in_3 = 1;
end
```

```
    assign out = in_3;   ///continuous assignment
    always in_2 = #5  ~ in_2;   /// intra-assignment delay of
5 time units
    always @(posedge in_2)   begin
    in_0 <= in_1; // non-blocking assignment
    in_1 <= in_0; // won't show error since previous
assignment is not blocking current assignment
    in_3 = in_1;   //blocking assignment
    end
endmodule
```

4.1.8.6   Procedures: Always and Initial Blocks

Verilog HDL has specified some procedure blocks as the following: initial blocks, always blocks, task, and function. The initial and always blocks are workable at the start of simulation. The "initial" blocks execute only one time, and the statements declared within it end when the statement is ended, whereas the "always" blocks execute frequently and terminate only when the simulation is finished. There are no limits to the number of occurrences of initial and always blocks in a module.

4.1.9   TEST BENCH IN VERILOG

The purpose of test benches in HDL development is to test the design for various possible scenarios and to check whether all the functionalities are correct or not. The design and test bench code are presented in the below example. Test bench architecture in Verilog contains a top module that consists the instances of the design under test (DUT) and verification environment or stimuli to the DUT as shown in Figure 4.1.

Example Snippet

```
module sample_snippet_dut0 (
input clk,
```

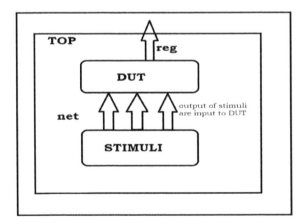

FIGURE 4.1   Test bench architecture in Verilog. DUT, design under test.

```
input [6:0] a,
input [6:0] b,
output reg [6:0] c,
output reg [6:0] d
);
always @ (posedgeclk) begin
d <= b + a;
if ( b> a)
c <= b - a;
else if (b < a)
c <= a - b;
end
endmodule
module sample_snippet_test0 ( );
wire [6:0] c, d;
reg [6:0] a, b;
integer i;
initial begin
clk = 0;
for (i = 0; i< 99; i ++) begin
#5 clk = ~ clk;
end
end
sample_snippet_dut0 d0 (.a(a), .b(b), .c(c), .d(d));    //dut
instantiation
initial begin
a = 7'b111000; b = 7'b1010110;
# 20 a = 7'b0000111; b = 7'b1100101;
end
endmodule
```

4.1.10   SYSTEM VERILOG

System Verilog is an extension of Verilog-HDL. It is a combination of concepts of multiple languages. System Verilog language components include many concepts from Verilog HDL, test bench constructs depending on Vera language, assertions from Open Vera, simulation interface for C and C++, and coverage application programming interface that supplies links to coverage metrics. It is a language that enabled designers to design and verify their IPs, circuits, etc. It was introduced majorly so as to reduce the complexity in coding and involve the concepts from C++ for programming the parts that need not be synthesized. Hence, it became the most widely used language for design verification throughout the world. But nowadays many new concepts are being introduced which makes it designer's favorite language to design as well [11].

4.1.11   TEST BENCH STRUCTURE FOR SYSTEM VERILOG

Test bench or, as we can say, verification environment is nothing but a group of classes, programs, modules, and components where each individual component has a

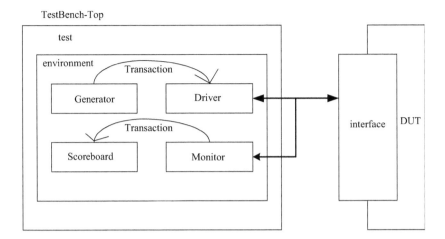

FIGURE 4.2 Test bench architecture in System Verilog.

specific operation. The basic components of a system Verilog test bench architecture are DUT, transaction, interface, generator, driver, monitor, scoreboard, test, environment, and top. The architecture is shown in Figure 4.2.

4.1.11.1 Design under Test

The DUT is the main design module that follows the specifications of the circuit being programmed.

```
//dff_dut.sv
`timescale 1ns/1ps
module dff_dut (din, clk, rst, dout);
input clk,rst,din;
output logic dout;
always @(posedge clk) begin
if(rst) begin
     dout <= 0;
end
   else begin
     dout <= din;
   end
  end
endmodule
```

4.1.11.2 Transaction

The transaction is a class that contains the elements or signals that are needed to be driven for testing purpose.

```
//dff_transaction.sv
class transaction; // class declaration
rand bit din; // rand keyword for randomizing the bits
```

```
rand logic rst; // randc for cyclic randomization
logic dout;
endclass
```

4.1.11.3 Interface

Interface is a component instantiated in the top module which contains the signals that connect the verification environment with the DUT.

//dff_interface.sv

```
interface dff_intf(input bit clk);
logic din, dout;
bit rst;
modport drv(input clk, output din, rst);
modport mon(input clk, dout);
endinterface
```

4.1.11.4 Generator

Generator generates the stimulus by creating and randomizing the transaction class and then sends it to the driver via mailbox.

//dff_generator.sv

```
`include "dff_transaction.sv"
class dff_gen;
  dff_trans t1; // class handle declaration
  mailbox gd_1;
  function new (mailbox gd_2);
    gd_1 = gd_2;
  endfunction
  task run();
    t1 = new; // allocating memory to class handle
    t1.randomize(); // calling in-built randomize function
    gd_1.put(t1);
  endtask
endclass
```

4.1.11.5 Driver

Driver class receives the stimulus or transactions from the generator class and then drives the packet-level data sequences within the transaction into pin level to the DUT.

//dff_driver.sv

```
`include "dff_transaction.sv"
class dff_drv;
  dff_trans t2;
  mailbox gd_3;
  mailbox ds_1;
  virtual dff_intf vif_1;
    function new (mailbox gd_4, mailbox ds_2, virtual
interface dff_intf vif_2);
```

```
        gd_3 = gd_4;
        ds_1 = ds_2;
        vif_1 = vif_2;
     endfunction
   task run();
        t2 = new;
        gd_3.get(t2);
        ds_1.put(t2);
        vif_1.din = t2.din;
        vif_1.rst = t2.rst;
  endtask
endclass
```

4.1.11.6 Monitor

Monitor looks the pin-level signaling on the interface signals and also alters them into packet level, which is to be sent to other components such as scoreboard, checker, and so on.

//dff_monitor.sv

```
`include " dff_transaction.sv"
class dff_mon;
  dff_trans t3;
  mailbox ms_1;
  virtual dff_intf vif_3;
    function new(mailbox ms_2, virtual dff_intf vif_4);
       ms_1 = ms_2;
       vif_3 = vif_4;
    endfunction
       task run();
          t3 = new;
          ms_1.put(t3);
          t3.dout = vif_3.dout;
       endtask
endclass
```

4.1.11.7 Scoreboard

Scoreboard is a class that receives the data items from monitor and compares them with the expected values.

//dff_scoreboard.sv

```
class dff_scb;
dff_trans t4, t5;
virtual dff_intf vif_5;
mailbox ds_3,ms_3;
function new (mailbox ds_4, mailbox ms_4, virtual dff_intf
vif_6);
      ds_3 = ds_4;
      ms_3 = ms_4;
```

```
      vif_5 = vif_6;
endfunction
task run();
   t4 = new;
   t5 = new;
   ms_3.get(t4);
   ds_3.get(t5);
   @(posedge vif_5.drv.clk) begin
         if (t4.rst) begin
                 t4.dout <= 0;
         end
         else begin
                 t4.dout <= t4.din;
         end
    end
    if( t4.dout == t5.dout) begin
        $display(" Scoreboard passed ");
    end
    else begin
    $display(" Scoreboard isn't passed ");
    end
endtask
endclass
```

4.1.11.8 Test

The test is liable for configuration of the test bench, initiation of the test bench components construction process, and initiation of the stimulus that is to be driven.

// dff test

```
`include "dff_env.sv"
program dff_test (dff_intf i0);
initial begin
dff_env en = new(i0);
en.build();
en.run();
end
endprogram
```

4.1.11.9 Environment

The environment is the top-level container class that groups the higher-level components.

//dff environment

```
`include "dff_trans.sv"
`include "dff_gen.sv"
`include "dff_drv.sv"
`include "dff_mon.sv"
```

```
`include "dff_scb.sv"
class dff_env;
dff_gen gen;
dff_drv drv;
dff_mon mon;
dff_scb scb;
virtual dff_intf vif;
mailbox ms, ds, gd;
function new (virtual dff_intf vif);
this.vif = vif;
endfunction
task build();
ms = new;
ds = new;
gd = new;
gen = new(gd);
drv = new(gd, ds, vif);
mon = new(ms, vif);
scb = new(ds, ms, vif);
endtask
task run();
gen.run();
drv.run();
mon.run();
scb.run();
#200 $finish;
endtask
endclass
```

4.1.11.10 Top

Top module is the topmost file that has the instances of the DUT and test bench and consists of their connections declared in interface component.

//dff_top

```
`timescale 1ns/1ps
`include "dff_dut.sv"
`include "dff_intf.sv"
`include "dff_test.sv"
module dff_top;
bit clk;
dff_intf i1(clk);
dff_dut d0 (.din(i1.din), .clk(i1.clk), .rst(i1.rst),
.dout(i1.dout));
dff_test t0 (i1);
initial begin
        i1.din=0;
#5      i1.din=1;
#25     i1.din=0;
#35     i1.din=1;
#50     i1.din=0;
```

```
#100      ;
end
initial begin
clk = 0;
for (int i = 0; i<256 ;i++)
begin
#5 clk = ~ clk;
end
end
endmodule
```

4.1.12 VERILOG-AMS

In analog and mixed-signal (A&MS) systems, Verilog-AMS plays as a behavioral language. It is a derivative from IEEE standard 1364-2005 Verilog HDL. It also captures the entire IEEE standard 1364-2005 Verilog HDL specification, analog equivalent for explaining analog systems (Verilog-A), and extensions together for mentioning the full Verilog-AMS. Verilog-AMS focuses on the designers of (A&MS) systems, IC creation, and applying of modules for high-level behavioral and structural descriptions of systems. It is applicable to both electrical and nonelectrical systems explanation as well as supportive for conservative and signal flow explanation.

It has the capacity to shorten design cycles and enlarge success of further mixed-signal ICs [12].

The main features involve the following:

- Both analog and digital signals can be stated in the identical module
- Rising verification of analog presentation at the design's top level
- Considerably diminishing the top-level simulation time
- Creating a suitable environment for architecture of any chip design

The basic necessities of Verilog-AMS code simulation include a Verilog-AMS code file and test bench through file extension of .vams. An example is shown here how Verilog-AMS is used for modeling the analog circuits [13].

////// Code for Ohm's Law

```
`include "disciplines.vams"
module res100ohm (
    resistor_in,
    resistor_out
);
inout resistor_in, resistor_out;
electrical resistor_in, resistor_out;
parameter real Resistor_Value = 100.0;
analog begin
V(resistor_in, resitor_out) <+ Resistor_Value * I(resistor_in,
resitor_out);
```

```
end
endmodule
```

4.1.13   MINI PROJECT: VERILOG AND SYSTEM VERILOG

After studying the complete chapter, one must have got an overall idea about HDLs. The main bullets focused are as follows:

- Simulation, synthesis, and elaboration
- Simulation tools
- Lexical conventions and semantics

Let us understand all the above with some mini projects.

---

1. Project Title: 16-Bit Heterogeneous Adder [14]
HDL Tool Suite: Xilinx Vivado Design Tool 2017.3
Language: Verilog

The main project window in VIVADO is shown in Figure 4.3. In *Sources* panel, the design file is clicked to get it in the coding area, and the programs are written for the design as per preferred HDL. Figures 4.4 and 4.5 represent editor window of Xilinx VIVADO and behavioral simulation result window, respectively.

Step 1: Hello VIVADO!!!
   Step 2: Writing the code

FIGURE 4.3   Main project window of Xilinx VIVADO.

FIGURE 4.4   Editor window of Xilinx VIVADO.

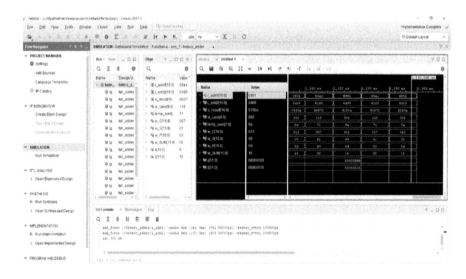

FIGURE 4.5   Behavioral simulation result.

Step 3: Simulation

In the *Flow Navigator* panel at extreme left of the window, *Run Simulation* is clicked to run the behavioral simulation of the design. There are different types of simulation present, which are categorized majorly as behavioral simulation, functional simulation, static timing analysis, gate-level simulation, switch-level simulation, and transistor-level or circuit-level simulation.

Step 4: Synthesis, elaboration, and implementation

The next steps after simulation are synthesis, elaboration, and implementation which provide a detailed information report regarding the area utilization, power consumed, timing report, DRC reports, etc. Such information is highly crucial for improvising a design. Placement and routing comes as next steps for dumping into FPGA boards where appropriate pins are allotted for each I/O signals or interfacing signals to interact with peer circuitry if available. The RTL-elaborated schematics at each step are shown in Figures 4.6–4.8.

The reports generated from these steps can be used in debugging. Also, a designer can write timing constraints and other constraint files scripted in ".*tcl*" to provide more information to the HDL tool suite about the design specifications (Figures 4.7 and 4.8).

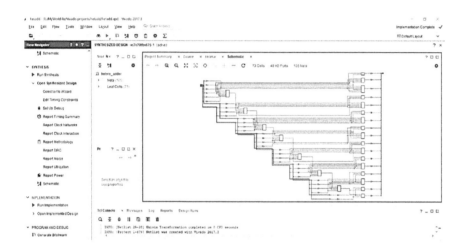

FIGURE 4.6   Synthesis schematic.

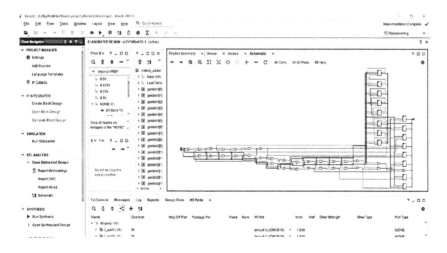

FIGURE 4.7   Elaborated schematic.

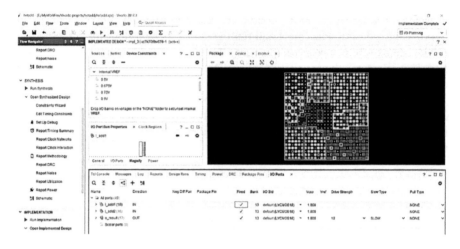

FIGURE 4.8   Implementation design view.

---

2. Project Title: Synchronized Dual-Port RAM
HDL Tool Suite: EDA Playground—Synopsys VCS 2019.06
Language: System Verilog

Step 1: Writing the code

EDA Playground is an open-source online simulator acting like miracle for the struggling designers who cannot avail the highly priced EDA tools licenses or HDL Tool Suites. It has various simulator platforms. Design codes and their testing environments can be written in EDA Playground (as shown in Figure 4.9) and run for error checks (errors can be in syntax or logic).

FIGURE 4.9   EDA Playground interface.

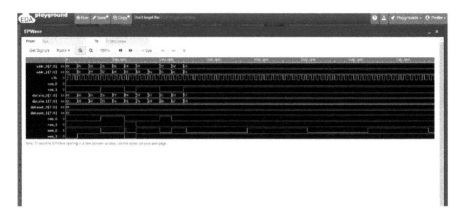

FIGURE 4.10   EPWave.

Step 2: Simulation

It is important to include *"$dumpfile("dump.vcd"); $dumpvars;"*in top module or *testbench.sv*in EDA Playground to run with EPWave enabled. Doing so pops up the simulation wave window as shown in Figure 4.10.

---

3. Project Title: D Flip-Flop
HDL Tool Suite: Questa Sim-64 10.3c
Language: System Verilog

Step 1: Writing the code

At first, design code and test bench component classes are written in any text editor. Figure 4.11 represents the generator, driver, interface, and transaction classes;

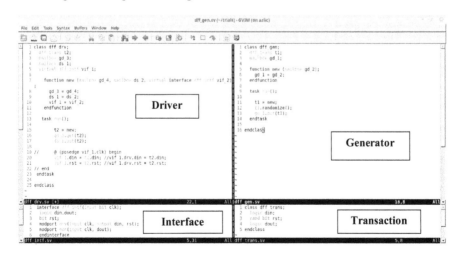

FIGURE 4.11   Generator, driver, interface, and transaction classes for D flip-flop verification environment.

Figure 4.12 shows scoreboard and environment classes; and Figure 4.13 presents DUT and top module and test, top and monitor classes for D flip-flop verification environment.

Step 2: Simulation

Next, the design file is compiled, and *vsim.wlf* is used to view the simulation waves as shown in Figure 4.14.

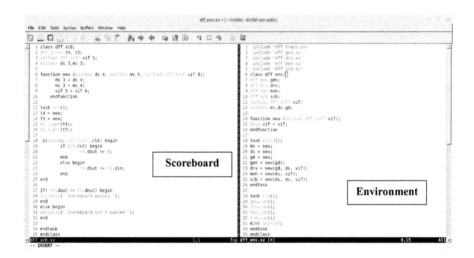

FIGURE 4.12   Scoreboard and environment classes for D flip-flop verification environment.

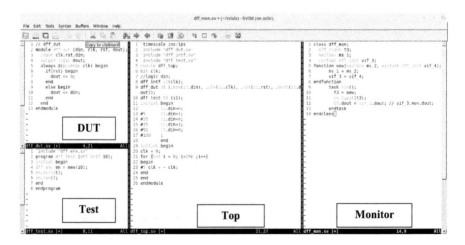

FIGURE 4.13   DUT and top module and test, top and monitor classes for D flip-flop verification environment. DUT, design under test.

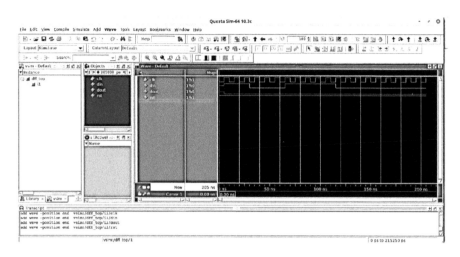

FIGURE 4.14   Simulation result for D flip-flop.

REFERENCES

1. IEEE Approved Draft. "Standard for VHDL Language Reference Manual, in IEEE P1076/D13", pp. 1–796, 2019.
2. https://en.wikipedia.org/wiki/Hardware_description_language.
3. R. W. Hartenstein, "Hardware Description Languages", Elsevier Science Publisher, Oxford, 1987.
4. J. P. Mermet, "Fundamentals and Standards in Hardware Description Languages", Series E, Springer Science, Barga, 1993.
5. C. D. Kloos, "Hardware Description Languages and their Applications: Specification, Modelling, Verification and Synthesis of Microelectronic Systems", Chapman & Hall, Toledo, OH, 1997.
6. C. Stumm, C. Brugger and N. When, "White Paper - Investigate the Hardware Description language Chisel", Kluedo, Kaiserslautern University of Technology, Germany, pp. 1–2, 2013.
7. S. Palnitkar, "Verilog HDL: A Guide to Digital Design and Synthesis", 2nd Edition, Prentice Hall PTR, Upper Saddle River, NJ, 2003.
8. D. E. Thomas and P. R. Moorby, "The Verilog® Hardware Description Language", 5th Edition, Kluwer Academic Publishers, New York, 2002.
9. IEEE Standard for Verilog Hardware Description Language, in IEEE Std 1364–2005 (Revision of IEEE Std 1364–2001), pp. 1–590, 2006. doi: 10.1109/IEEESTD.2006.99495.
10. S. Sutherland, S. Davidmann and P. Flake, "SystemVerilog for Design: A Guide to Using System Verilog for Hardware Design and Modeling", 2nd Ed., Springer, New York, 2006.
11. IEEE Standard for SystemVerilog-Unified Hardware Design, Specification, and Verification Language, in IEEE Std 1800–2017 (Revision of IEEE Std 1800–2012), pp. 1–1315, 2018. doi: 10.1109/IEEESTD.2018.8299595.
12. https://www.yumpu.com/en/document/view/11519359/verilog-ams-simulation-using mentor-and-cadence-tools-prateek.

13. F. Pecheux, C. Lallement and A. Vachoux, "VHDL-AMS and Verilog-AMS as alternative hardware description languages for efficient modeling of multidiscipline systems," in *IEEE Transactions on Computer-Aided Design of Integrated Circuits and Systems*, vol. 24, no. 2, pp. 204–225, 2005. doi: 10.1109/TCAD.2004.841071.

14. S. Mahapatro, K. C. Bhuyan, S. Acharya and A. Mishra, "Simulation and synthesis of Heterogeneous adder using VIVADO," in *International Journal for Research in Applied Science and Engineering Technology*, vol. 6, no. 3, pp. 964–976, 2018. doi: 10.22214/ijraset.2018.3154.

5 Introduction to Hardware Description Languages (HDLs)

P. Raja, Dushyant Kumar Singh, and Himani Jerath
Lovely Professional University

CONTENTS

5.1 Introduction 94
    5.1.1 Motivation...94
    5.1.2 Structure of Hardware Description Language....................................94
    5.1.3 History 94
    5.1.4 Hardware Description Language and Programming Languages 95
    5.1.5 Hardware Description Language Design Flow Expense...................96
5.2 Design Simulation, Debugging, and Verification with Hardware
    Description Languages 96
5.3 Introduction to VHDL 96
    5.3.1 Structure of Program 97
    5.3.2 VHDL Variables...98
    5.3.3 Functions, Libraries, and Packages 98
    5.3.4 Design Elements 99
        5.3.4.1 Structural 99
        5.3.4.2 Dataflow 100
        5.3.4.3 Behavioral 101
    5.3.5 Writing a Simple Code with VHDL 101
    5.3.6 Simulation and Synthesis... 103
5.4 Introduction to Verilog 104
    5.4.1 Structure of Program 104
    5.4.2 Module Declarations... 105
    5.4.3 Verilog Variables, Operators, and Directives 105
        5.4.3.1 Variable Data Types... 105
        5.4.3.2 Operators.. 105
    5.4.4 Design Elements 106
        5.4.4.1 Structural 106
        5.4.4.2 Dataflow 107
        5.4.4.3 Behavioral 107
        5.4.4.4 Writing a Simple Code with Verilog................................ 108
    5.4.5 Simulation and Synthesis... 109
References... 109

5.1 INTRODUCTION

Hardware description language (HDL) is a specialized programming language used to label the electronic and digital logic modules. The building process and also plan of the modules are made through HDL program. HDL enables the simulation output and building blocks of digital logic circuit by using textual representation of input, output, operation, and appearance. The compilation of HDL will convert textual format into gate level, and the same has been on the field-programmable gate array (FPGA) processor by using synthesis and implementation process. The language assists to define any digital circuit in the system of physical. There are three HDLs commonly used to define the behavior of the digital modules—Verilog, VHDL, and SystemC. The VHDL and Verilog are the oldest languages as compare with SystemC that is most and commonly used.

5.1.1 MOTIVATION

Because of the blast multilayered design of advanced electronic circuits since the 1970s, circuit creators required computerized rationale representations on the way to remain made at a raised-up level without involvement of a particular electronic component. HDLs are designed to realize the register transfer-level design.

5.1.2 STRUCTURE OF HARDWARE DESCRIPTION LANGUAGE

The construction and behavior of the digital electronic system over a period of time can be described in typical text-based languages such as HDLs. In simultaneous programming languages, the structure of statements incorporates clear credentials for cooperating simultaneously. The primary characteristic of HDL embraces an explicit belief of time. Dialects whose lone emblem is to direct circuit availability among a chain of command of squares are correctly named net-list dialects employed in electric PC-supported plan (CAD). HDL can be employed to direct plans in auxiliary, conduct, or enroll move-level models for an alike circuit helpfulness; in the previous dual cases, the synthesizer selects the engineering and rationale door design.

A program planned and executed the essential semantics of the dialect explanations, and mimicking the loan of period gives the equipment creator with the volume to validate a bit of equipment; sometime recently, it is completed actually. It is thus executable that stretches HDLs the fabrication of being program writing languages, when they are more accurately classified as determination dialects or demonstrating languages.

5.1.3 HISTORY

The conception of an HDL as a medium for style capture was first introduced in the 1950s; however, wide adoption by the look community did not begin till 1985. Traditionally, the event of software package programming languages stirred the evolution of HDLs. One example, among many, is that the artificial language APL that was used as a kind of style entry for a logic automation system was developed at IBM within the early 1960s. The notational conventions of APL were later utilized by researchers at the University of Arizona to style AHPL (A Hardware Programming Language).

Since its introduction within the early 1970s, AHPL has been hardly utilized in nonacademic applications, however, served as a good teaching tool in schoolroom environments. Within the three decades ranging from 1960, several HDLs such as DDL, ISPS, and Zeus were introduced. However, the utilization of those languages seldom exceeded analysis and tutorial applications. Part in reaction to the proliferation of HDLs and part because of its own desires, in 1980, the U.S. Department of Defense initiated the event of VHDL as a part of its VHSIC program (very high-speed integrated circuits).

The name VHDL comes from VHSIC hardware description language. The general objective of this effort was to style one language that may enable the look, documentation, and analysis of hardware at varied levels of abstraction. Moreover the intent was to create VHDL the typical HDL design projects and use of language as a way of communication between departments.

In December 1987, VHDL was accepted by the IEEE (IEE88, IEE94) as a customary description language used for modeling digital modules and systems. The emergence of such business normal marked the start of widespread adoption of HDLs by the planning community. In parallel with VHDL efforts, the Verilog alpha-lipoprotein originated in 1983 at the entry-style automation and was introduced into the market in 1985. Verilog was designed to deal with the necessities of circuit designers and at a similar time to be intuitive and as easy as doable. As a result, the language has fewer constructs than VHDL, its linguistics is not as advanced, and development of Verilog-based tools is soon achievable. Verilog became associate IEEE normal [IEE96] in 1995.

Today, Verilog and VHDL are used extensively and, maybe solely, by circuit designers all over. Several simulation and synthesis tools are developed and marketed that alter the analysis and realization of those lipoprotein models.

5.1.4 HARDWARE DESCRIPTION LANGUAGE AND PROGRAMMING LANGUAGES

HDL is exceptionally alike other programming languages but has still lot of differences. Most programming dialects are intrinsically procedural (single-strung), with restricted grammatical and semantic help to deal with simultaneous. HDLs, on the other hand, have the capability to typically manifold concurrent processes that mechanically perform individualism of one more. Any change to the procedure's input consequently triggers an update in the test system's procedure stack.

As HDLs and programming languages obtain ideas and things to see since one another, the edge among them is getting fewer precise. Nevertheless, unadulterated HDLs are inadmissible for universally beneficial application programming advancement, similarly as broadly useful programming vernaculars are inconvenient for signifying equipment.

The elevated level of reflection of SystemC models is appropriate to early design investigation, as compositional changes can be effectively assessed with little worry for signal-level execution issues.

5.1.5 Hardware Description Language Design Flow Expense

The alleged "front end" starts with making sense of the essential methodology and building hinders at the square graph level. Enormous rationale structures, similar to programming programs, are progressive, and VHDL and Verilog give you a decent system for characterizing modules and their interfaces. The subsequent stage is the genuine composition of HDL code for modules, their interfaces, and their inner subtleties. Despite the fact that you can utilize any word processor for this progression, the editorial manager incorporated into the HDL's instrument suite can make the activity somewhat simpler.

Figure 5.1 is representing the design flow of an HDL. There is a need to define design specification of a system and describe the behavior of the system by consuming any HDL (VHDL or Verilog). After synthesis, the code will be converted into register transfer-level logic that will be tested and checked for its logical synthesis. The next step is to generate the gate-level net list and perform logical verification and testing of the logical function of gate-level net list. If the specification and logical verification are correct, then it will proceed to implement the same logic on FPGA through floor planning and automatic placing and routing as per the specifications given by the programmer. Finally, the system will be implemented on a chip for further usage that is called fabricated modules.

5.2 DESIGN SIMULATION, DEBUGGING, AND VERIFICATION WITH HARDWARE DESCRIPTION LANGUAGES

Fundamental to HDL configuration is the capacity to recreate HDL programs. Reenactment permits a HDL depiction of a structure (called a model) to pass plan check, a significant achievement that approves the plan's expected capacity (particular) against the code usage in the HDL description. Furthermore, authorizations build examination. The designer could be trying different belongings with structure conclusions by constituting frequent variabilities of a dishonorable plan, the idea observing at their conduct of reenactment. Therefore, regeneration is elementary for rich HDL structure.

5.3 INTRODUCTION TO VHDL

This chapter speaks to the survey of VHDL that is an equipment depiction language that can be utilized to demonstrate a computerized framework at many degrees of reflection. It is contemporary as it consolidates the well-known and broadly utilized IEEE STD_LOGIC_1164 bundle. This language characterizes the punctuation as well as characterizes extremely clear reenactment semantic for every language building [2]. The look is depicted by dynamic understanding of a static investigation for the well-known equipment language VHDL. From a VHDL depiction, the investigation registers a superset of the state available all through any recreation run. These data are helpful inside the approval of well-being properties of equipment components. The improvement of the investigation depends on the conventional meaning of semantics for VHDL. Sufficiency with relevance this historical underpinnings is

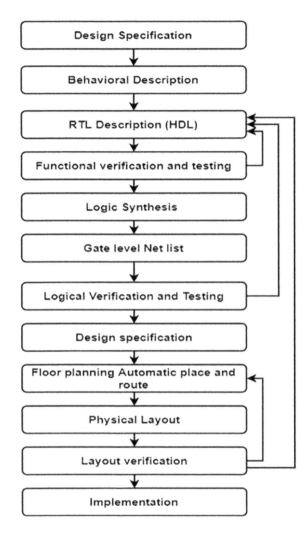

FIGURE 5.1 Design flow of HDL. HDL, hardware description language; RTL, register transfer-level.

showed up. Shifted systems grant a trade-off between the predetermined precision and, in this way, the estimation of a definitive algorithmic standard [1].

5.3.1 STRUCTURE OF PROGRAM

Each VHDL structure unit involves a substance assertion and at least one model. Every engineering characterizes an alternate usage or typical of a specified structure element. The element description characterizes the contributions toward, besides yields since the element, and a little conservative constraint utilized by the numerous traditions of the element.

Entity is the explanation of the interface among a design and its external environment. It can similarly stipulate the assertions and reports that are part of the design entity. A specified entity assertion might be collective via numerous design entities, respectively of which consumes a dissimilar architecture.

Syntax of Entity

```
Entity test is
    Generic declaration;
    Port declaration;
End test;
```

Example of port declaration: *portname: mode datatype;*

Generics are used to pass the specific value into entity, which has no direction to declare but includes timing of the part. The port declaration can be done with one of the modes, that is, in, out, buffer, and inout.

Architecture:

An architecture describes single specific employment of a design element on anticipated level of notion.

Syntax of Architecture

```
Architecture architect_test of testis
Declerations
begin
    Concurrent statements;
end architect_test
```

5.3.2   VHDL VARIABLES

Variables are substances that stock data resident to methods and subprograms. The values of variable could be modified through simulation, which can be declared as follows:

```
variable variablename : type;
variable variablename : type := initialvalue;
```

5.3.3   FUNCTIONS, LIBRARIES, AND PACKAGES

Functions are a piece of a gathering of structures in VHDL called subprograms. Functions are little areas of code that play out an activity that is reused all through your code. This serves to reduce the complexity of the code along with reusability.

Functions consistently utilize an arrival proclamation. They are commonly useful for doing a modest quantity of rationale or math and restoring an outcome. One extra note: holdup articulations cannot be utilized in a capacity.

Functions are subprograms in VHDL, which can be utilized for executing as often as possible utilized calculations. A function takes at least zero information esteems, and it generally restores a worth. Notwithstanding the arrival esteem, what separates a capacity from a method is that it cannot contain Wait explanations. This implies that capacities consistently devour zero reproduction time.

Syntax

```
function fun-name return type is
     declarations;
begin
     sequential statements;
end fun-name;
```

A VHDL package comprises subprograms, stable descriptions, or theoretically type descriptions to be applied all through at least one plan unit. Individual bundle contains a presentation segment, in which the accessible subprograms, constants, and types are noticeable, and a package body, in which the subprogram custom is considered, together with any privileged exploited constants and types. The assertion segment speaks to the part of the bundle that is understandable to the client of that package. Package declaration format is as follows:

```
package pack_test is
     constant declarations;
     type declarations;
     subprogram declarations;
end test_test;
```

5.3.4 DESIGN ELEMENTS

5.3.4.1 Structural

To make popular this representation, a component is portrayed as a portion of interrelated segments. A part launch articulation is a simultaneous declaration. Accordingly, the demand for these statements remains substantial. Auxiliary stylishness of exhibiting represents the interconnection of the elements, short of inferring any behavior of the sections themselves.

In structural displaying, engineering body is made out of two sections—the decisive part and the announcement part (after the catchphrase start).

VHDL Design for Half-Adder in Structural Modeling

```
library ieee;
use ieee.std_logic_1164.all;

entity andgate1 is
    port(x, y: in bit;
         rout: out bit);
end andgate1;

architecture comp1 of andgate1 is
begin
    rout <= x and y;
end comp1;
```

```
entity xorgate1 is
    port(x, y: in bit;
    rout: out bit);
end xorgate1;

architecture comp2 of xorgate1 is
begin
    rout <= x xor y;
end comp2;

entity h_adder1 is
    port(e, f: in bit;
    sum1, carry1: out bit);
end h_adder1;

architecture struct1 of h_adder1 is

component andgate1
    port(p, q: in bit;
    rout: out bit);
end component;

component xorgate1
    port(p, q: in bit;
    rout: out bit);
end component;

begin
u1 : andgate1 port map(p,q,carry1);
u2 : xorgate1 port map(p,rout,sum1);

end struct1;
```

5.3.4.2   Dataflow

The dataflow style and the behavior of the modules are expressed in the way data-flow will take place in module by using concurrent signal. The parallel statements are generated through WHEN and GENERATE. Apart from the assignment to construct, the code can also use operators.

VHDL Design of Half-Adder in DataFlow Modeling

```
Library ieee;
use ieee.std_logic_1164.all;
entity halfadder1 is
    port (e, f:in bit; sum1, carry1: out bit);
end halfadder1;
architecture data of halfadder1 is
begin
    sum1<= e xor f;
    carry1 <= e and f;
end data;
```

5.3.4.3 Behavioral

In this displaying style, the behavior of a substance as conventional of explanations performed successively in the prearranged demand. Unbiased deliveries put confidential a process function.

Procedures, FUNCTIONS, and PROCEDURES are the main areas of code that are executed successively. In any case, overall, any of these squares is as yet simultaneous with some other articulations set outside it. One substantial portion of behavior code is that it is not unnatural to consecutive rationale. Hypothesis consecutive circuits are impartial as combinational circuits.

VHDL Design for Half-Adder in Behavioral Modeling

```
library ieee;
use ieee.std_logic_1164.all;

entity halfadd1 is
port( d: in std_logic_vector(0 to 1);
  sum, carry: out std_logic);
end halfadd1;

architecture behav of halfadd1 is
begin
process(d)
begin

case d is
when "00" => sum <= '0';
when "11" => sum <= '0';
when others => sum <= '1';
end case;

case d is
when "11" => carry <= '1';
when others => carry <= '0';
end case;

end process;
end behav;
```

5.3.5 WRITING A SIMPLE CODE WITH VHDL

VHDL Design for 4:1 Multiplexor

```
library ieee;//Library
use ieee.std_logic_1164.all;//Library

entity Mux4_1 is
port( dir3:  in std_logic_vector(2 downto 0);
      dir2:  in std_logic_vector(2 downto 0);
```

```
            dir1:  in std_logic_vector(2 downto 0);
            dir0:  in std_logic_vector(2 downto 0);
            muxsel:in std_logic_vector(1 downto 0);
            muxout:out std_logic_vector(2 downto 0)
        );
        end Mux4_1;

        architecture behav2 of Mux4_1 is
        begin
        process (dir3, dir2, dir1, dir0, muxsel)
        begin

            case muxsel is
                when "00" => muxout <= I0;
                when "01" => muxout <= I1;
                when "10" => muxout <= I2;
                when "11" => muxout <= I3;
                when others => muxout <= "ZZZ";
            end case;
        end process;
        end behav2;
```

VHDL Design for 2:4 Decoder in Behavioral Modeling

```
library ieee;//Library
use ieee.std_logic_1164.all;//Library

entity DECODER2_4 is
    port( testinp: in std_logic_vector(1 downto 0);
          testout: out std_logic_vector(3 downto 0)
    );
end DECODER2_4;

architecture behav3 of DECODER2_4 is
begin


    process (testinp)
    begin

        case testinp is
        when "00" => testout <= "0001";
        when "01" => testout <= "0010";
        when "10" => testout <= "0100";
        when "11" => testout <= "1000";
        when others => testout <= "XXXX";
    end case;

    end process;

    end behav3;
```

VHDL Design for n-Bit Comparator

```
library ieee;
use ieee.std_logic_1164.all;

entity Comp is
generic(xtr: natural :=2);
port( p: in std_logic_vector(n-1 downto 0);
      q: in std_logic_vector(n-1 downto 0);
      lt: out std_logic;
      eq: out std_logic;
      gt: out std_logic
);
end Comp;

architecture behav4 of Comp is
begin
process(p,q)
begin
    if (p < q) then
            lt <= '1';
            eq <= '0';
            gt <= '0';
    elsif (p = q) then
            lt <= '0';
            eq <= '1';
            gt <= '0';
    else
            lt <= '0';
            eq <= '0';
            gt <= '1';
    end if;
end process;
end behav4;
```

5.3.6 SIMULATION AND SYNTHESIS

As of now referenced a few times, VHDL demonstration (or equipment displaying by and large) has in any event two uses: recreation with the end goal of confirmation and amalgamation for the programmed change of a moderately dynamic depiction into an assortment of doors from a library. The whole model of the equipment that one needs to fabricate is known as the structure under check (DUV).

A VHDL test system has different highlights to control the reenactment. A client can show the time stretch that the recreation should cover, the succession of test sign or improvements that ought to be given to the DUV, and so forth. Notwithstanding these offices, it is a superior plan to control the recreation, however, much as could reasonably be expected from VHDL itself. The upside of this is that it requires just insignificant information on the test system furthermore, which one gets free of the test system.

The sum of DUV and models that drive its information sources and procedure and its yields is known as a test bench. It is prescribed to manufacture a test bench that in any event comprises of the accompanying models:

The test system views a circuit as an assortment of sign and procedures. Sign can change in esteem after some time under the effect of procedures. A sign change is known as an exchange. In spite of the fact that equipment is parallel naturally, it is commonly reenacted on a successive machine. In one way or different, forms that are dynamic at the same time, just as sign that can change in esteem all the while, must be managed so that the contrasts among reproduction and the genuine world are as little as could reasonably be expected.

What the test system must do at a given minute is demonstrated through a rundown of activities that is arranged by time. This is the occasion list. "Occasion" is the assignment given to a sign change or a procedure actuation at a particular time. For instance, if the procedure that is dynamic at minute $t=t0$ experiences the announcement $a <=$ "1" after 10 ns, at that point exchange, $a <=$ "1" is put on the occasion list at minute $t=t0+10$ ns. An exchange never produces results promptly, not regardless of whether the code does not indicate any deferral (for instance, through a sign task without the catchphrase after). All things are considered; the exchange is put on the occasion list at minute $t=t0+\Delta$. Δ is equivalent to zero (or better: imperceptibly little), yet it permits forms that occur all the while to be requested in time. This is conceivable on the grounds that the accompanying applies: $0 < \Delta < 2\Delta$ ...

The thought of an imperceptibly little deferral in reproduction is known as a delta delay. The recreation begins with the development of the occasion list. All procedures in the VHDL portrayal are put in the correct situation in the rundown. (Most procedures start at time zero, applying the standard that a negligible time of Δ must happen between two initiations.) During recreation, the occasion list is prepared in the request for expanding time. New occasions that outcome from this are included in the occasion list at the correct position. The reenactment is finished when the occasion list gets vacant, when the reproduction is compelled to be ended by the activity of the client or by a blunder [3].

5.4   INTRODUCTION TO VERILOG

Acronym of Verilog is verifying logic. Phil Moorby of Gateway Design Automation developed it in 1985. Initially, the Verilog was invented as a simulation and synthesis language that was completed after thought process by Cadence Design Systems in 1990. Verilog HDL was a proprietary language, the property of Cadence Design Systems when it was finally opened to public in 1990. In the late 1980s, a new trend was observed when the designers started moving away from proprietary languages like n dot, HiLo, and Verilog and started using the U.S. Department of Defense standard VHDL. Verilog became an IEEE standard IEEE-1364 in 1995 (revised in 2000) [4]. The present rendition is IEEE standard 1800-2017.

5.4.1   STRUCTURE OF PROGRAM

There are two major HDLs, namely, VHDL and Verilog; the Verilog HDL is used in most environments and used to create a module through Verilog. Similar to VHDL,

Verilog describes the structure to be produced through design flow. Plans showed in HDL are innovation free, simple to structure and troubleshoot, and normally more meaningful than schematics, especially for huge circuits.

The different levels of abstraction in Verilog are as follows:

a. Based on algorithm.
b. Based on register transfer.
c. Based on gate.
d. Based on switch.

5.4.2 MODULE DECLARATIONS

In Verilog, module is the basic building block. The component or a gathering of lower-level design blocks can be a module. The handy of higher-level block provided by module concluded the aforementioned input and output connection. Communication of module established through module interface with external input and output devices. Ports can be classified as input, output, and bidirectional, and communication is possible between module and module interface through various ports.

```
Module dflipflop (q, qb, clk, d, rst);
input clk, d, rst;
output q, qb;
wire dl, dbl;
paramter delay1 = 3,
delay2 = delay1 + 1;
nand #delay1 n1(cf, dl, cbf),
n2(cbf, clk, cf, rst);
nand #delay2 n3(dl, d, dbl, rst),
n4(dbl, dl, clk, cbf),
n5(q, cbf, qb),
n6(qb, dbl, q, rst);
#500 force dff_lab.rst = 1;
#550 release dff_lab.rst;
Endmodule
```

5.4.3 VERILOG VARIABLES, OPERATORS, AND DIRECTIVES

5.4.3.1 Variable Data Types

There are two gatherings of types: "net information types" and "variable information types." An identifier of "variable information type" implies that it changes an incentive upon task and holds its incentive until another task. This is a customary programming language variable and is utilized in consecutive articulations [5].

5.4.3.2 Operators

See Table 5.1.

TABLE 5.1
List of Operators

Type of Operator	Operator	Description
Arithmetic operators	+, −, /	Arithmetic
	%	Modulus
Logical operators	!	NOT logic
	&&	AND logic
	||	OR logic
Bitwise operators	~	Bitwise NOT
	&	Bitwise AND
	|	Bitwise OR
	^	Bitwise XOR
	^~ ~^	Bitwise XNOR
Reduction operators	&	Reduction AND
	|	Reduction OR
	~&	Reduction NAND
	~|	Reduction NOR
	~^ ^~	Reduction XNOR
Equality operators	==	Logical equality
	!=	Logical inequality
Relation operators	>	Relational operators
	>=	
	<	
	<+	
Concatenation operator	{ }	Concatenation
Shift operators	<<	Left shift
	>>	Right shift
Conditional operator	{ }	

5.4.4   DESIGN ELEMENTS

5.4.4.1   Structural

The module is executed as far as rationale doors and interconnections between these entryways. It takes after a schematic drawing with parts associated with signals. An adjustment in the estimation of any information sign of a part initiates the segment. On the off chance that at least two segments are actuated simultaneously, they will play out their activities simultaneously too.

An auxiliary framework portrayal is nearer to the physical execution than social one yet it is progressively included in light of huge number of subtleties. Since rationale door is most prominent part, Verilog has a predefined set of rationale entryways known as natives. Any computerized circuit can be worked from these natives.

Verilog Design for 4:1 Multiplexer in Structure Modeling

```
module Mux_4_1(testinp, testsel, testout);
```

```
input [3:0] testinp;
input [1:0] testsel;
output testout;
wire NS0, NS1;
wire Y0, Y1, Y2, Y3;
not N1(NS0, testsel[0]);
not N2(NS1, testsel[1]);
and A1(Y0, testin[0], NS1, NS0);
and A2(Y1, testin[1], NS1, testsel[0]);
and A3(Y2, testin[2], testsel[1], NS0);
and A4(Y3, in[3], testsel[1], testsel[0]);
or O1(testout, Y0, Y1, Y2, Y3);
endmodule
```

5.4.4.2 Dataflow

Dataflow demonstration utilizes constant assignments and the watchword allocation. A persistent task is an explanation that doles out an incentive to remaining. The data type net list is working in V-HDL to express a bodily association among circuits and components. The worth relegated to net lists are determined by an articulation that utilizations operands.

For instance, expecting that factors remained declared, a 2:1 multiplexer through information test inputs A and B, test select input S and the output test Y. the output is represented through continues changes in output by executing this syntax testY=(testA and testS) | (testB and testS).

Dataflow representation of 2_4_decoder is appeared in HDLs underneath. The logic circuit is characterized with possibility of constant task explanations utilizing Boolean articulations, one for each yield.

Verilog Design for 2:4 Decoder in Data Modeling

```
module decoder_2_4 (testI0, testI1, En, testY);
input testI0, testI1, En;
output [3:0] testY;
assign testY[3] =~(~testI0 & ~testI1 & ~En);
assign testY[2] =~(~ testI0 & testI1 & ~En);
assign testY[1] =~(testI0 & ~ testI1 & ~En);
assign testY[0] =~(testI0 & testI1 & ~En);
Or assign D[3] =~(~testI0 & ~testI0 & ~En),
 D[2] =~(~testI0 & testI1 & ~En),
 D[1] =~(testI0 & ~testB & ~En),
 D[0] =~(testI0 & testI1 & ~E);
endmodule
```

5.4.4.3 Behavioral

Social modeling behavioral display speaks to advanced circuits at a practical and algorithm level. Behavioral model is utilized generally to represent consecutive modules. The conduct displaying idea is to be introduced for combinational circuits. Behavior description utilizes the catchphrase consistently pursued by a rundown of practical task explanations. The programmer need not know about gate level description of the module.

Verilog Design for 2:1 Multiplexer in Behavioral Modeling

```
module mux2_1_behav (I0, I1, sel, Y);
input I0, I1, sel;
output Y;
reg R;
always @(sel or I0 or I1)
if (sel == 1) Y=I0;
else Y=I1;
endmodule
```

5.4.4.4   Writing a Simple Code with Verilog

1. Verilog Design for 4:1 Multiplexer

```
module mux_4_1_beh(Out, tstI0, tstI1, tstI2, tstI3,
tstselect);
output Out;
input tstI0, tstI1, tstI2, tstI3;
input [1:0] testselect;
reg tstY;
always @(tstI0 or tstI1 or tstI2 or tstI3 or tstselect)
case (tstselect)
2'b00: Out = tstI0;
2'b01: Out = tstI1;
2'b10: Out = tstI2;
2'b11: Out = tstI3;
default: tstY = 0;
endcase
endmodule
```

2. Verilog Design for 2:4 Decoder

```
module decoder_2_4(tsto3, tsto2, tsto1, tsto0, tstI0, tstI1,
tste);
    output tsto3, tsto2, tsto1, tsto0;
    input tstI0, tstI1;
    input tste;
    reg tsto3, tsto2, tsto1, tsto0;

    always @(tstI0 or tstI1 or tste) begin
        if (tste == 1'b1)
        case ( {tstI0,i1} )
        2'b00: {tsto3, tsto2, tsto1, tsto0} = 4'b1110;
        2'b01: { tsto3, tsto2, tsto1, tsto0} = 4'b1101;
        2'b10: { tsto3, tsto2, tsto1, tsto0} = 4'b1011;
        2'b11: { tsto3, tsto2, tsto1, tsto0} = 4'b0111;
        default: { tsto3, tsto2, tsto1, tsto0} = 4'bxxxx;
        endcase
        If (tste == 0)
            {tsto3, tsto2, tsto1, tsto0} = 4'b1111;
        end
endmodule
```

3. Verilog Design for Full Adder

```
module Full-Adder(tstx, tsty, ci, cou, su);
input tstx, tsty, ci;
output cou, su;
wire tstw1, tstw2, tstw3, tstw4;
xor #(10) (tstw1, tstx, tsty);
xor #(10) (su, tstw1, ci);
and #(8) (tstw2, tstx, tsty);
and #(8) (tstw3, tstx, ci);
and #(8) (tstw4, tsty, ci);
or #(10, 8)(cou, tstw2, tstw3, tstw4);
endmodule
```

5.4.5 SIMULATION AND SYNTHESIS

It is not possible that the Verilog is almost right; its necessity is written so that it directs the incorporation instrument to produce great equipment, and in addition, the Verilog must be coordinated to the mannerisms of the exact union device actuality utilized. Synthesis is a procedure where a structure conduct that is displayed utilizing a HDL is converted into a usage comprising rationale doors. This is finished by a synthesis apparatus which is another product program.

Synthesis stands a widespread term frequently hand-me-down to portray overall dissimilar instruments. Synthesis is used to develop the internal architecture of a particular design, and design can be done by variant compiler.

Types of synthesis:

a. Behavioral
b. High-level
c. Register transfer-level

Simulation is the way toward utilizing a reproduction programming (test system) to check the useful accuracy of a computerized plan that is demonstrated utilizing an HDL (equipment portrayal language) such as Verilog.

REFERENCES

1. Grewal, A. (2018). Review Report on VHDL (VHSIC Hardware description language). *Language*, 27, 28.
2. LaMeres, B. J. (2019). *Introduction to Logic Circuits & Logic Design with VHDL.* Springer, New York.
3. Gazi, O. (2019). *A Tutorial Introduction to VHDL Programming.* Springer, New York.
4. Cavanagh, J. (2017). *Verilog HDL: Digital Design and Modeling.* CRC Press, Boca Raton, FL.
5. LaMeres, B. J. (2019). *Introduction to Logic Circuits & Logic Design with VHDL.* Springer, New York.

6 Emerging Trends in Nanoscale Semiconductor Devices

B. Vandana
KG Reddy College of Engineering and Technology

B. S. Patro, J. K. Das, and Sushanta Kumar Mohapatra
Kalinga Institute of Industrial Technology,
KIIT Deemed to be University

Suman Lata Tripathi
School of Electronics and Electrical Engineering,
Lovely Professional University

CONTENTS

6.1 Introduction 112
6.2 Background 112
6.3 Nanotechnology Emerging Improvements 115
    6.3.1 Technology Dependency (Bulk to Silicon-on-Insulator Technology)... 115
    6.3.2 Architectural Representation (Single- to MultiGate
        Field-Effect Transistor) 116
        6.3.2.1 Double-Gate Silicon-on-Insulator Metal Oxide
                Semiconductor Field-Effect Transistors............................. 117
        6.3.2.2 Triple-Gate Silicon-on-Insulator Metal Oxide
                Semiconductor Field-Effect Transistors............................. 118
        6.3.2.3 Surrounding-Gate Silicon-on-Insulator Metal Oxide
                Semiconductor Field-Effect Transistors............................. 118
        6.3.2.4 Other Multigate Metal Oxide Semiconductor
                Field-Effect Transistors.. 118
    6.3.3 Material Technology 118
        6.3.3.1 Strained Silicon... 119
        6.3.3.2 High-k Gate Dielectric and Metal Gate Electrodes........... 119
    6.3.4 Existing Metal Oxide Semiconductor Field-Effect Transistor
        Topologies at Nanoscale Regime... 119
        6.3.4.1 Junctionless Metal Oxide Semiconductor Field-Effect
                Transistor 119

      6.3.4.2 Tunnel Field-Effect Transistor 121

      6.3.4.3 Spintronics 121

      6.3.4.4 Memristor.. 122

      6.3.4.5 Graphene Transistors 123

      6.3.4.6 High-Electron Mobility Transistor 123

6.4   Conclusions... 124

References... 125

6.1 INTRODUCTION

Scaling the complementary metal oxide semiconductor (CMOS) device dimensions and the process technology has become more difficult for the semiconductor industry, if this approach reaches to 3 nm physical lengths by 2020 (Ryckaert 2019). Beyond the period of traditional CMOS scaling, various alternative electronic devices are required to integrate on a silicon platform with no change in its functionality (Patro and Vandana 2016; Deleonibus 2019). The first initiation of the technology was started in the micrometer range, as due to Moore's law prediction, the devices were forcefully pushed to nanoscale. Before the era of nanoscale technology, microelectronics innovates various technological benefits that are still compatible with the nanoscale regime, and this plays a crucial role in microsystem designs.

The microelectronics industries have faced two conventional difficulties to enhance the integrated circuit (IC) technology beyond the CMOS scaling. The CMOS needs to be extending further with the consideration of scaling factors. The challenges associated are high speed, dense, nonvolatile memory, and information processing substantially. The references of alternate electronic devices include one-dimensional structure (carbon nanotubes and compound semiconductor devices), resonant tunneling diodes, single-electron transistors, and molecular and spin devices. However, these concepts represent charge-based logic; restrictions of scaling by thermodynamic will be minimum switching energy per binary operation. Beyond this limit, new challenges need to be implemented that will compensate electronic charge and extend the scaling of information processing technology. This chapter extensively provides an outlook toward emerging new devices and associated technologies. These technologies may help microelectronics to serve as a link between conventional CMOS and beyond CMOS scaling. This chapter discusses new technology trends that exist for present-generation metal oxide semiconductor field-effect transistor (MOSFET) at the nanoscale regime.

6.2 BACKGROUND

In the present era, the most leading industries in the commercial world are "semiconductor industries." This has interpreted a vast revolution in the growth of technology for various applications. Due to its inherent features, this made human life easier with the invention of portable devices for communication systems, medical sensors instrument applications, military applications for coding/decoding exchange of information, and many more that are able to access using Internet of things. Besides these applications, a complex circuitry designed with nanoscale devices is incorporated on a printed circuit board (PCB) to make the system size small. The nanoscale devices play a vital role in emerging technology and trends in device innovations; scaling semiconductor devices to nanoscale was started with Gordon Moore. Later came up with Moore's law known as "cramping number of transistors for every 18 months."

Nanoelectronics is demarcated as nanotechnology and allows integration with purely electronic devices, electronic chips, and circuits. As the years passed, Moore's law further extended to "More than Moore" (MtM) domain of development (Arden et al. 2010). The diversification factor of Moore's law inculcates scaling feature, and the road to gigascale systems can be described as the "More Moore" domain of development ("Nanoelectronics Applications Moving Simply at Nanoscale," n.d.). Scaling the feature size and packing density will still continue in following Moore's lore and enjoy the present innovations in social life. In accordance with this, a new solution is important to face the challenges for nanodevices.

Basically, the effort made by the R&D for advance chip technology has been a driving force within three to four decades, and this results in better performance and high packing density. The manufacturing of ICs is processed through the top-down approach, and the fabrication tools are improved with each technology generation. The major intention is that if the size of the components is scaled to nanoscale, physical effects would stop the tools from working properly. The fabrication challenges of the top-down silicon (Si) technology are considered with three fundamental design limits: (i) transistor scalability, (ii) performance, and (iii) power dissipation (Kuo and Lou, n.d.).

Out of three limitations, the most important design is transistor scalability. As per the VLSI domain, CMOS is the major technology for integrated systems, the technology especially suitable for ICs which operates with low power at low supply voltage. The literature study makes us understand why scaling is necessary for the semiconductor industry and the downscaling channel length (L_{CH}) as the years pass; it is reported that the L_{CH} is about 180 nm in 2000 and the L_{CH} is of 5 nm in 2020.

Due to the vast scaling of L_{CH}, the considerations of the electric field (E-field) are limited, and the power supply is also scaled down from 1.8 to 0.7 V. Therefore, the low supply voltage is necessary for next-generation low-power CMOS circuits (Kuo and Lou, n.d.).

The most straightforward path to meet the low-power systems is the MOS device that lowers the power supply voltage. Due to continuous scaling, device geometry at nanoscale leads to several leakage mechanisms. The challenges associated to reduce short-channel effects (SCEs) are shown in Figure 6.1b. Nanoscale transistors usually require low supply voltage to reduce the internal E-field and consume low power. This intends to reduce threshold voltage (V_{TH}) that substantially increases leakage currents (I_{OFF}). As the drain voltage increases, the depletion region of channel drain widens and increases drain current (Adan et al. 1998). This causes to increase OFF current, due to which the channel surface induces barrier lowering effect known as drain-induced barrier lowering, or due to channel punch-through effects. All the adverse effects that cause to reduce V_{TH} in scaled devices are known as SCEs. The SCEs are immune to channel length scaling and reduced oxide thickness. This further increases E-field and the scaled oxide thickness, and high E-field destroys the infinity input impedance, which affects the MOS transistor performance.

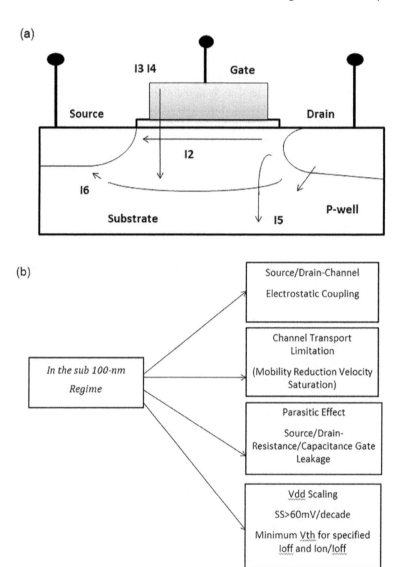

FIGURE 6.1 (a) Various leakage currents. (b) Challenges to reduce SCEs. SCEs, short-channel effects.

The projections such as device physical dimensions, supply voltage, and power consumption are taken according to the International Technology Roadmap for Semiconductors (ITRS) (Moore and others 1998; "No Title," n.d.). The various SCEs show an overview of the intrinsic leakage currents of a MOSFET device as shown in Figure 6.1a. The SCEs are source/drain current (I_{SD}), source/drain extension current (I_{SDE}), or ($I1$). The subthreshold current (I_{SubVTH}) or ($I2$) arises between source and drain at zero gate voltage. The gate leakage (I_{GATE}) or ($I3, I4$) flows through the oxide dielectric to the silicon substrate in vertical direction. Gate-induced drain leakage (I_{GIDL}) or (*I*5) flow from the corresponding space charge regions in Si and the channel current ($I_{CHANNEL}$) or (*I*6) across the channel depletion region, which is responsible for the space charges, are briefly described in Roy et al. (2003). Accordingly, our research moved forward to find out the possible ways to reduce the SCEs. However, the emerging trends activated by the ITRS are utilized by semiconductor industries.

6.3   NANOTECHNOLOGY EMERGING IMPROVEMENTS

The previous section highlighted many issues when the transistors are downscaled. In this section, the improvements activated for the present-generation nanotechnology included in new material and products are discussed. Technology impacts commercially worldwide in the future. The ultraperception classifications are precision engineering, microengineering (MST/MEMS/MOEMS), nanoscience, and nanotechnology. The general possibilities that a nanotechnology encourages Si industries are technology dependency (bulk to silicon-on-insulator [SOI] technology), architectural representation (single- to multigate), and material technology (composite materials). The engineering approaches to reduce electrostatic control at scaled L_{CH} as the generations of technology increase, and the discussion is given in the following. The section discusses the various adherence factors for present-generation nanoscale technology.

- New materials with higher mobility for enhancing transport (SiGe, Ge, GaAs, InGaAs, etc.)
- New gate stack to reduce tunneling and gate leakage (high-k dielectrics)
- New contact materials (metal and semiconductor)
- New S/D structures (e.g., raised S/D)
- Improved device architecture (SOI, UTB, DG, 3D FinFET, trigate to improve SCEs)

6.3.1   TECHNOLOGY DEPENDENCY
(BULK TO SILICON-ON-INSULATOR TECHNOLOGY)

The long-channel bulk MOSFET is the most effective technology introduced in the 1970s, due to its adequate features (such as a switch, low parasitic effects, integration density, and simple process manufacturing); it still follows Moore's law and makes it possible to produce complex circuits in the economic world. The conduction mechanism of a simple N-type MOSFET operates with applied gate and drain voltages; the region between source and drain is the channel, and the surface tends to invert with the flow of majority charge carriers across the source to drain. The formation of the channel is actually due to the depletion of charge carriers. However, if the channel length is shortened (below100 nm), the device dimensions are scaled, but several SCEs are associated with the device technology. Accordingly, the operation mechanism is the same as that of long-channel devices, but the depletion depth moves downward to the surface substrate causing leakage current issues in bulk MOSFET in Figure 6.2a. To avoid this, the oxide isolation structure, i.e., SOI in Figure 6.2b has superior capability in good

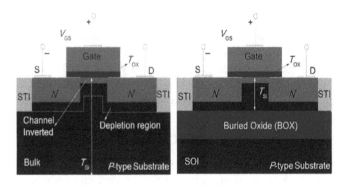

FIGURE 6.2 Cross-section view of (a) bulk and (b) SOI MOSFET. SOI, silicon-on-insulator; MOSFET, metal oxide semiconductor field-effect transistor.

radiation hardness, no latchup, and high device density. The oxide layer isolates the body from the substrate and induces no reverse-biased junctions. The junction capacitances are small where it reduces leakage currents and results in high-speed performance. In biased condition, if the body terminal is not added, the device is referred as floating-body SOI MOSFET. The floating body generates serious issues in the device, which are suppressed through body control techniques.

In bulk MOSFET, the dependency of body effect occurs due to the change in the V_{TH} on the back gate bias. The back gate bias effect occurs due to the depletion width at the oxide interface and the substrate. When the back gate bias becomes more negative, the depletion region widens under the gate, which thereby requires high gate voltage (V_{GS}) to invert the channel, and hence this effect increases the V_{TH}. SOI technology separates the active region from the substrate through the isolation technique. Therefore, the V_{TH} is less dependent on the back gate bias as compared with bulk technology.

The formation of SOI MOSFET technology is achieved from partially depleted (PD) SOI MOSFET with a single gate and fully depleted (FD) SOI MOSFET with multigate. PDSOI MOSFETs are successful at early silicon-on-sapphire devices, which are useful for the niche applications for radiation-hardened or high-temperature electronics. This device has become the mainstream for the semiconductor manufacturer and has been started to use to fabricate high-performance microprocessors. To operate a low-voltage PDSOI device, the contacts are created across the gate and floating body; such a contact improves subthreshold slope (SS) and drives current of the devices. The FDSOI MOSFET has a better electrostatic coupling between gate and channel. This device is successful in various applications such as low-voltage, low-power to RF ICs (Vandana, Das, et al. 2017).

6.3.2 ARCHITECTURAL REPRESENTATION (SINGLE- TO MULTIGATE FIELD-EFFECT TRANSISTOR)

In this context, the device architecture moves forward from single- to the multigate structure to have better channel controllability through the gate. To have a reduced SCEs, the SOI MOSFET evolved from a classical single-gate device into a 3D device

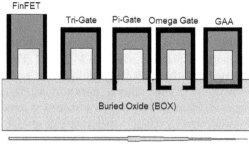

FIGURE 6.3 Schematic view of multigate SOI MOSFET. SOI, silicon-on-insulator; MOSFET, metal oxide semiconductor field-effect transistor.

with a multigate structure. The various device representation is (double-triple or quadruple gate device), usually double gate refers to a single gate electrode is present on two opposite side of the device. The triple gate is used as a single-gate electrode that is taken from three sides of the gate. Other representation is multiple independent gate FET (MIGFET) in which the gate electrodes bias separately with different potentials. The possible approaches of multigate SOI MOSFET are shown in Figure 6.3 such as FinFET, trigate, pi-gate, omega-gate, and gate-all-around (GAA) gate.

6.3.2.1 Double-Gate Silicon-on-Insulator Metal Oxide Semiconductor Field-Effect Transistors

The article on double-gate MOSFET was first published by T. Sekigawa and Y. Hayashi in 1984 (Sekigawa and Hayashi 1984). The paper discusses an idea of SCEs reduction, which influences due to drain E-field on the channel. The device architecture is designed by sandwiching an FDSOI between two gate electrodes connected together. The first fabricated DGSOI MOSFET was the fully depleted lean-channel transistor (DELTA) in 1989 (Hisamoto et al. 1989) in which the device is designed using narrow Si island known as a finger, leg, or fin. The FinFET structure is similar to the DELTA with the presence of a hard mask dielectric layer on top of the Si fin. This prevents parasitic inversion channel at corners of the device.

The other representation of DGSOI MOSFET includes GAA and is a planar MOSFET that is wrapped around the channel. The other MOSFETs such as silicon-on-nothing (SON), multifin XMOS (MFXMOS), the triangular-wire SOI MOSFET, and DELTA channel SOI MOSFET are implemented. The GAA device was a DG structure, even though the gate is wrapped all around the Si body where the Si island is much wider and the device has a width-to-height ratio that is near to unity.

In the MIGFET, due to independent gate potential, the V_{TH} is modulated with the applied voltage to another gate; this effect is the same as that of FDSOI MOSFET. Since the channel is fully depleted and the gates are perfectly aligned and symmetric in nature, the device modulation is feasible. This signal modulation circuit helps to design low-power mixers. Low power is achieved due to reduced transistor count and rail-to-rail transistor stack.

6.3.2.2   Triple-Gate Silicon-on-Insulator Metal Oxide
Semiconductor Field-Effect Transistors

The MOSFET with thin Si film with a gate on three sides is known as trigate; this implementation also includes quantum wire SOI MOSFET. The electrostatic integrity is enhanced by extending the sidewall portions of the gate electrodes to some depth in the buried oxide (BOX); the other trigates such as Pi-gate and omega-gate (Colinge 2007; Yang et al. 2002, 2004) are very effective in improving electrostatic integrity. The use of strain effect and high-k dielectric gate insulator further improves the drive current device (Krivokapic et al. 2003; Andrieu et al. 2006).

6.3.2.3   Surrounding-Gate Silicon-on-Insulator Metal Oxide
Semiconductor Field-Effect Transistors

The structure implies wrapping gates all around on the vertical Si pillar (Miyano et al. 1992), which theoretically proves the possible control of the channel region by the gates. This increases the drive current per unit area. Multiple surrounding gates are stacked on top of one another, while sharing common gate source–drain electrodes, and hence are known as multibridge channel MOSFET (MBCFET) (Lee et al. 2004; Yoon et al. 2004), nanobeam stacked channels, and twin-Si nanowire (TSNWFET) (Park 2006).

When positive bias is applied to the underlying Si and the negative bias is applied to the gate electrode, an inversion layer is created at the surface of the substrate and the accumulation layer is generated at the other interface. The metastable dip (MSD) is observed. With the time-dependent and hysteresis effect, the transconductance increases and then decreases and rises randomly.

6.3.2.4   Other Multigate Metal Oxide Semiconductor
Field-Effect Transistors

A thin planar film SOI device with the trigate transistor is an inverted T channel FET (ITFET) (Zhang et al. 2006; Mathew et al. 2005). It is a planar horizontal and vertical channel in a single device. It provides the transistor action at the space provided between the fins (the unused space in multigate FET). These additional channels increase the current drive. The numerical simulations reveal different turn-on mechanism in various parts of the device. At first, the device turns on at the corners. Afterward, the vertical channel exists at the surface of the planar regions. The device consists of seven corner elements. These elements allow substantial current to each ITFET, which is more than a planar device of almost the same area.

6.3.3   Material Technology

Advance materials often provide performance improvement and remove barriers for scaling. The most successful material that is continuing with the generations of technology includes SOI devices, and for back-end interconnects, copper and aluminum materials are in use. Various new materials such as strained Si, compound materials, high-k gate dielectrics, and metal gate electrodes are coming into existence for new device topologies.

6.3.3.1 Strained Silicon

The well-known representation of mobility enhancement is achieved through a strain-induced effect (Tiwari et al. 1997). The method to obtain a strain effect is by growing strained silicon epitaxial on top of the relaxed SiGe substrate (Rim et al. 1998). Due to strain effect, mobility enhances up to 70% for nFET structures even at high fields and 30% in moderate fields for pFET (Wong 2002). The fraction of mobility enhancement is converted to drive current for short-channel devices. The strained silicon is an excellent topology. Because without changing the basic device structure, it improvises the device performance. This helps in eliminating the worries of integrating several technologies. Combining the benefits of strained silicon with SOI technology has bright future (Mizuno et al. 2002; Huang et al. 2001). The major challenges for strained silicon FET are enhancing the performance at shorter channel lengths under high-field transport, preventing strain relaxation during fabrication, and reducing material defects (Vandana, Patro, et al. 2017).

6.3.3.2 High-k Gate Dielectric and Metal Gate Electrodes

At the nanoscale, the gate insulator plays a major requirement for device performance. However, reducing the gate insulator thickness leads to direct tunneling. This increases the standby power exponentially due to gate current. The solution to overcome the effects is that replacing gate insulator with high dielectric materials (high-k) gate insulator with a thick physical thickness and thin electrical thickness has been the focus of intense research for several years. Process development has mellowed to a point where uniform film deposition can be achieved either on H-terminated (HF last) Si substrate or ultrathin thermally grown oxide-based interlayer. For maintaining conventional CMOS process flow, thermal stability with respect to silicon is the key criterion. Thermally stable Al_2O_3 nFET with conventional polysilicon gates has been discussed (Wong 2002).

At the time of process flow, the poor thermal stability of the high-k gate dielectric tends to achieve a low thermal cycle. This can replace the metal gate electrode process. Also, due to the polysilicon gate depletion layer, the metal gates avoid the gate capacitance loss. The present MOS devices consider metal gates because the work function is close to the conduction band for n-type and valence band for p-type (Wong 2002).

6.3.4 Existing Metal Oxide Semiconductor Field-Effect Transistor Topologies at Nanoscale Regime

6.3.4.1 Junctionless Metal Oxide Semiconductor Field-Effect Transistor

According to Denna's scaling theory, the control over SCEs is achieved through novel device architectures. The changes in S/D channel regions forming an unconventional device structure are discussed in the literature. The new scaling methodology of MOSFET having the channel controllability with double-gate 2D-FinFET is reported in Fiegna et al. (1993), and its scaling theory is discussed in Suzuki et al. (1993) using SOI platform. The facts of semiconductor geometrical dimensions

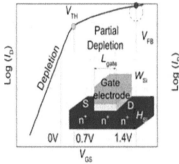 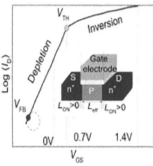

FIGURE 6.4   I–V characteristics in log scale, and the inset figures represent n-type MOSFET without and with junctions: (a) JLT and (b) IMT. MOSFET, metal oxide semiconductor field-effect transistor; JLT, junctionless transistor.

(Majkusiak et al.1998) and the challenges facing (FD) SOI transistor are due to the scaling dimensions of the active Si channel region.

In the case of single-gate FDSOI devices, in order to maintain full substrate depletion under gate control, the T_{Si} needs to be about a one-third to a half of the electrical L_G (Yan et al. 1992). The important concerns of multigate devices depend on dynamic performance; the I_D is multiplied by a factor of 2. For the same reason, the C_{GG} is twice to that of a single gate, and thus, the same intrinsic propagation delay ($C_{GG}.V_{DD}/I_{ON}$) is obtained. The intrinsic power delay product (PDP) ($C_{GG}.V_{DD}/2$) of the device is doubled if the devices are designed with the same width. Double-gate devices have to operate at a $V_{DD}/2$ time lower than that of a single gate, in order to have the same PDP at a given transistor width. To reach the same $C_{GG}.V_{DD}/I_{ON}$, the drive current should be, at $V_{DD}/2$, two times the I_D obtained with a single-gate device at V_{DD}.

At the nanoscale regime, the emerging semiconductor technology possesses various advance interpretations for CMOS applications. One among them is the junctionless transistor (JLT) structure, which imposes a good candidate for scaling down the SCEs and improvises the drive current at scaled L_G. Figure 6.4 represents the conventional MOSFET with inversion mode of operation and the MOS transistor with no inversion mode of operation and no formation of junctions across the S/C/D and hence is known as JLT from Figure 6.4a.

From inset figure, the S/D/C is highly doped for n-type MOSFET usually known as a gated resistor, and the I–V characteristics are nearly the same as IMT, but the principle of operation is different to that of IMT. At the subthreshold condition, the channel seems to be fully depleted. As the applied gate bias, the work function difference between gate and channel makes the bands flat at V_{FB}, and the channel is partially depleted where the conduction takes place at channel (Colinge et al. 2010). The formation of NPN at S/C/D from the inset Figure 6.4 b is an IMT structure. The *I–V* characteristic in log scale represents full depletion at subthreshold condition, but due to the V_{FB}, small amount of current flows giving rise to leakage currents. With the applied gate bias, the electrons are accumulated at the top surface of the gate, and the channel slowly turns to be inverted from p-type to n-type forming a channel region from source to drain (Vandana et al. 2018; Vandana, Mohapatra, et al. 2017; Vandana et al. 2019).

6.3.4.2 Tunnel Field-Effect Transistor

The device that signifies power saving is obtained using low-voltage tunnel FET (TFET). The structure is similar to that of MOSFET, but the switching mechanism is different, and TFETs are the promising candidate of low-power electronics. The concept was proposed by Chang et al., while working at IBM (Chang and Esaki 1977); later in 2004, Joerg Appenzeller and group reported that TFET was implemented using carbon nanotube channel and a subthreshold swing of just 40 mV/decade (Appenzeller et al. 2004).

The basic TFET structure is similar to a MOSFET except that the source and drain terminals of a TFET are doped of opposite type as shown in Figure 6.5. A common TFET device structure consists of a P-I-N (p-type, intrinsic, n-type) junction. Here, the gate terminal controls the electrostatic potential of the intrinsic region. According to the constant field scaling, reducing SS is essential. The transistor speed is proportional to the subthreshold swing. The speed and the maximum subthreshold leakage thereby define SS with minimal V_{TH}. Scaling V_{TH} has become a major task for the technology developers and got stuck where the supply voltage cannot be scaled. In general, for high-performance devices, the supply voltage should be at least three times the threshold voltage. Due to some technical reasons, the processor speed has not been improved as fast as before. The instigation of a mass-producible TFET device will allow the industry to continue the scaling trends from the 1990s, where processor frequency-doubled every 3 years. This can be achieved by fabricating TFET with a slope below 63 mV/decade.

6.3.4.3 Spintronics

Spintronics is coined from the words "spin" and "electronics." This is a branch of physics with the study of the spin of the electrons and its magnetic moment along with its fundamental charge. Electrons spin is represented by two types, i.e., up and down. This is a good alternative against the present transistors, which requires a controlled flow of electrons for turning on and off the transistors. Unlike transistors, changing the spin requires less energy and hence less power consumption by the devices. The spin of the electrons is measured as it generates tiny electric fields.

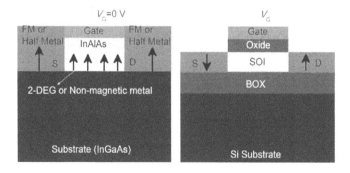

FIGURE 6.5 Schematic representation of (a) Spin-FET and (b) Spin-MOSFET; the arrows indicate the direction of magnetization of the S and D contacts. MOSFET, metal oxide semiconductor field-effect transistor.

Also, these spintronics-based devices can be used with the help of commonly available metals such as iron, aluminum, and so on. The independent behavior of the spin of the electrons from the energy helps the spintronics device to be nonvolatile in nature, and hence, data storage can be done even after the loss of power.

There are many devices developed to perform similar functions as that of transistors. The spin valve is one of them. It was first invented by Dieny et al. in 1991. This revolutionizes the industries that deal with storage devices and magnetic sensors. In a similar way, giant magnetoresistance (GMR)–based and tunneling magnetoresistance (TMR)–based devices are working. IBM, Honeywell, Freescale, SanDisk, and so on are some of the companies that are consistently marketing and improvising the technology associated with these types of devices. Magnetic tunnel junction and ferroelectric tunnel junction are also some of the most important spintronic devices that are used for logic and memory-based applications.

Figure 6.5 shows the proposed Spin-FET and Spin-MOSFET (Joshi 2016). The best example of today's equipment in the market is organic light-emitting diode that shows good visual effects and consumes very little power as compared with other televisions. So, it can be inferred that spintronics-based devices have a very good future.

Here, V_G is the gate voltage. When V_G is zero, the injected spins that are transmitted through the 2-DEG layer reach at the collector with the same polarization. When $V_G \gg 0$, the precession of the electrons is controlled with electric field; here S and D regions have opposite direction of magnetization and do not require spin precession of spin-polarized electrons in the channel (Datta and Das 1990; Zhang et al. 2014; Joshi 2016).

6.3.4.4 Memristor

The word "memristor" is coined from two words, i.e., memory and resistor. This is categorized as the fourth fundamental electrical component after resistor, capacitor, and inductor. The theoretical concept was first proposed by L. Chua in 1971.

In Figure 6.6, a two-terminal memristor-based device is shown. Depending upon the voltage applied, the flow of electrons is controlled by the thickness of the material and hence can work as a transistor in switching-based applications. Owing to its nonvolatile nature and no leakage power, it is considered to be a better replacement against a transistor. Applications such as memory storage and designing logic gates are some of the areas where memristors can work in a robust manner because of its high-speed switching.

Memristor can be classified into two sections on the basis of the way it works: firstly, ionic thin film and molecular memristor, which works on the basis of its

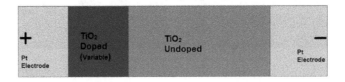

FIGURE 6.6   A memristor device.

material composition and properties of the lattices, and secondly, magnetic and spin-based memristor, which depends on the orientation or spin of the electrons of the material used for functionalities in various memory storage-based applications. This device is in a developing stage, and many industries are focusing on developing cost-effective and long-lasting memristor-based systems.

6.3.4.5 Graphene Transistors

Graphene transistor was first built by Prof. A. Geim in 2007. It is a two-dimensional graphite, which is an allotrope of carbon. This device is very much capable to replace the transistors made up of semiconductor material like silicon. Its processing time is very fast and can able to withstand high temperatures. Since it is a two-dimensional structure, the transistors made from it are of very small size, which can reduce the lines on the forehead of the many scientists regarding reducing the channel length of the semiconductor-based transistors. It works at very low voltage, and hence, power consumption can be reduced further. The other name of graphene is single-electron transistor as it allows one electron at a time.

Figure 6.7 shows the graphene lattice arrangement in a hexagonal manner. Graphene-based transistors can help to build transistors of channel size that of an electron, and hence, terahertz speed of conduction of electrons can be achieved. This can able to counter the adverse effects a transistor is facing, which are termed as SCEs. The dimensional reduction like graphene has made the scientists focus their minds on the other two-dimensional materials such as germanene, silicene, stanene, and so on. Although the stable material generation is not that easy, still these two-dimensional materials have a huge scope to play in the future in further increasing the number of transistor counts in the IC.

6.3.4.6 High-Electron Mobility Transistor

The present technology is demanding higher speed devices for high-speed communications. The device speed can be enhanced by decreasing the channel size. But while miniaturization, many SCEs are becoming the hindrances to build such transistors. High-electron mobility transistor (HEMT) has the capability to overcome

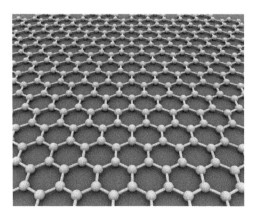

FIGURE 6.7 A hexagonal planar graphene structure.

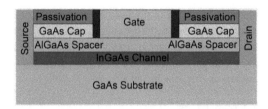

FIGURE 6.8 Cross section of a GaAs/AlGaAs/InGaAs pHEMT. HEMT, high-electron mobility transistor.

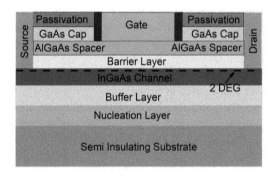

FIGURE 6.9 Schematic cross section of an HEMT. HEMT, high-electron mobility transistor.

such type of limitations faced by these transistors. It incorporates heterojunctions that are formed with the help of different materials of different bandgaps. These heterojunctions confine the electrons in the required quantum well and avoid any chance of scattering effects. That is why HEMT is also termed as a heterostructure field-effect transistor. Some commonly used heterojunctions are GaAs, AlGaN, AlGaAs, and so on.

Figure 6.8 shows a structure of pseudomorphic HEMT device, which is an advanced form of HEMT. These devices have a variety of applications. Since HEMT has better noise performance, so it is used in low noise amplifiers such as satellite receivers, defense industries producing low-noise oscillators, radar communications, and so on. It can operate on very low voltage and can handle a large amount of current. Hence, these are used in power amplifiers. A two-dimensional structure of a HEMT device is shown in Figure 6.9.

6.4 CONCLUSIONS

This chapter focuses on the various technologies available, which will be of major attention in the coming years. This is due to the fact that the transistor size miniaturization has come to its saturation point. Also, these scaled channel transistors are facing lots of problems. The devices that are discussed have lots of advantages such as they are operating at very low voltages and hence consume very low power.

Also, these can work in systems where high-speed computation is required. Some of the devices do not depend upon the flow of electrons for switching on and off the device but the spin of the electrons. These behaviors make these devices nonvolatile in nature and hence can store the data for a longer period of time.

REFERENCES

Adan, A O, T Naka, A Kagisawa, and H Shimizu. 1998. "SOI as a Mainstream IC Technology." In *1998 IEEE International SOI Conference Proceedings*, 9–12.

Andrieu, F, C Dupré, F Rochette, O Faynot, L Tosti, C Buj, E Rouchouze, et al. 2006. "25nm Short and Narrow Strained FDSOI with TiN/HfO2 Gate Stack." In *Proceedings of Symposium on VLSI Technology, 2006. Digest of Technical Papers*, 134–135.

Appenzeller, J, Y-M Lin, J Knoch, and Ph Avouris. 2004. "Band-to-Band Tunneling in Carbon Nanotube Field-Effect Transistors." *Physical Review Letters* 93 (19). APS: 196805.

Arden, W, M Brillouët, P Cogez, M Graef, B Huizing, and R Mahnkopf. 2010. "More-than-Moore White Paper." *Version* 2: 14.

Chang, L L, and L Esaki. 1977. "Tunnel Triode-a Tunneling Base Transistor." *Applied Physics Letters* 31 (10). AIP: 687–89.

Colinge, J-P. 2007. "Multi-Gate SOI MOSFETs." *Microelectronic Engineering* 84 (9–10). Elsevier: 2071–2076.

Colinge, J-P, C-W Lee, A Afzalian, N D Akhavan, R Yan, I Ferain, P Razavi, et al. 2010. "Nanowire Transistors Without Junctions." *Nature Nanotechnology* 5 (3). Nature Publishing Group: 225–229. doi:10.1038/nnano.2010.15.

Cooper, D R, B D'Anjou, N Ghattamaneni, B Harack, M Hilke, A Horth, N Majlis, et al. 2012. "Experimental Review of Graphene." *ISRN Condensed Matter Physics* 2012. Hindawi Publishing Corporation.

Datta, S, and B Das. 1990. "Electronic Analog of the Electro-Optic Modulator." *Applied Physics Letters* 56 (7). AIP: 665–667.

Deleonibus, S. 2019. *Electronic Devices Architectures for the NANO-CMOS Era*. CRC Press.

Fiegna, C, H Iwai, T Wada, T Saito, E Sangiorgi, and B Ricco. 1993. "A New Scaling Methodology for the 0.1-0.025 m MOSFET." In *Proceedings of Symposium on VLSI Technology*, 33.

Hisamoto, D, T Kaga, Y Kawamoto, and E Takeda. 1989. "A Fully Depleted Lean-Channel Transistor (DELTA)-a Novel Vertical Ultra Thin SOI MOSFET." In *Proceedings of Technical Digest - International Electron Devices Meeting, 1989. IEDM'89*, 833–836.

http://www.Seas.Ucla.Edu/Cmoslab/People.Html. n.d.

Huang, L-J, J O Chu, S Goma, C P D'Emic, S J Koester, D F Canaperi, P M Mooney, et al. 2001. "Carrier Mobility Enhancement in Strained Si-on-Insulator Fabricated by Wafer Bonding." In *Proceedings of Symposium on VLSI Technology, 2001. Digest of Technical Papers*, 57–58.

Joshi, V K. 2016. "Spintronics: A Contemporary Review of Emerging Electronics Devices." *Engineering Science and Technology, an International Journal* 19 (3). Elsevier: 1503–1513.

Krivokapic, Z, C Tabery, W Maszara, Q Xiang, and M R Lin. 2003. "High Performance 45nm CMOS Technology with 20nm Multi-Gate Devices." *Solid State Devices and Materials* 1998 Business Center for Academic Societies: 760–761.

Kuo, J B, and J-H Lou. n.d. "Low-Voltage CMOS VLSI Circuits."

Lee, S-Y, S-M Kim, E-J Yoon, C W Oh, I Chung, D Park, and K Kim. 2004. "Three-Dimensional MBCFET as an Ultimate Transistor." *IEEE Electron Device Letters* 25 (4). IEEE: 217–19.

Majkusiak, B, T Janik, and J Walczak. 1998. "Semiconductor Thickness Effects in the Double-Gate SOI MOSFET." *IEEE Transactions on Electron Devices* 45 (5). IEEE: 1127–1134.

Mathew, L, M Sadd, S Kalpat, M Zavala, T Stephens, R Mora, S Bagchi, C Parker, J Vasek, and D Sing. 2005. "Inverted T Channel FET (ITFET)-Fabrication and Characteristics of Vertical-Horizontal, Thin Body, Multi-Gate, Multi-Orientation Devices, ITFET SRAM Bit-Cell Operation. A Novel Technology for 45nm and beyond CMOS." In *Proceedings on Technical Digest IEEE International Electron Devices Meeting, IEDM,* 713–716.

Miyano, S, M Hirose, and F Masuoka. 1992. "Numerical Analysis of a Cylindrical Thin-Pillar Transistor (CYNTHIA)." *IEEE Transactions on Electron Devices* 39 (8). IEEE: 1876–1881.

Mizuno, T, N Sugiyama, T Tezuka, T Numata, and S Takagi. 2002. "High Performance CMOS Operation of Strained-SOI MOSFETs Using Thin Film SiGe-on-Insulator Substrate." In *Proceedings of Symposium on VLSI Technology, 2002. Digest of Technical Papers,* 106–107.

Moore, G E, et al. 1998. "Cramming More Components onto Integrated Circuits." *Proceedings of the IEEE* 86(1): 82–85.

Park, D. 2006. "3 Dimensional GAA Transitors: Twin Silicon Nanowire MOSFET and Multi-Bridge-Channel MOSFET." In *Proceedings of International SOI Conference, 2006 IEEE,* 131–134.

Patro, B S, and B Vandana. 2016. "Low Power Strategies for beyond Moore's Law Era: Low Power Device Technologies." *Design and Modeling of Low Power VLSI Systems.* IGI Global, 27.

Rim, K, J L Hoyt, and J F Gibbons. 1998. "Transconductance Enhancement in Deep Submicron Strained Si N-MOSFETs." In *Proceedings of Technical Digest - International Electron Devices Meeting, 1998. IEDM'98,* 707–710.

Roy, K., S Mukhopadhyay, and Hamid Mahmoodi-Meimand. 2003. "Leakage Current Mechanisms and Leakage Reduction Techniques in Deep-Submicrometer CMOS Circuits." *Proceedings of the IEEE* 91 (2). IEEE: 305–327.

Ryckaert, J. 2019. "Scaling below 3nm Node: The 3D CMOS Integration Paradigm (Conference Presentation)." In *Advanced Etch Technology for Nanopatterning VIII,* 10963:109630O.

Sekigawa, T, and Y Hayashi. 1984. "Calculated Threshold-Voltage Characteristics of an XMOS Transistor Having an Additional Bottom Gate." *Solid-State Electronics* 27 (8–9). Pergamon: 827–828.

Semiconductor Industry Association. 2015. "The International Technology Roadmap for Semiconductors."

Suzuki, K, T Tanaka, Y Tosaka, H Horie, and Y Arimoto. 1993. "Scaling Theory for Double-Gate SOI MOSFET's." *IEEE Transactions on Electron Devices* 40 (12). IEEE: 2326–2329.

Tiwari, S, M V Fischetti, P M Mooney, and J J Welser. 1997. "Hole Mobility Improvement in Silicon-on-Insulator and Bulk Silicon Transistors Using Local Strain." In *Proceedings of Technical Digest - International Electron Devices Meeting, 1997. IEDM'97,* 939–941.

Vandana, B, J K Das, B S Patro, and S K Mohapatra. 2017. "Exploration towards Electrostatic Integrity for SiGe on Insulator (SG-OI) on Junctionless Channel Transistor (JLCT)." *Facta Universitatis, Series: Electronics and Energetics* 30(3): 383–390.

Vandana, B, S K Mohapatra, J K Das, and B S Patro. 2017. "Prospects of 2D Junctionless Channel Transistor (JLCT) Towards Analog and RF Metrics Using Si and SiGe in Device Layer." *Journal of Low Power Electronics* 13 (3). American Scientific Publishers: 536–544.

Vandana, B, Prashant Parashar, B S Patro, K P Pradhan, S K Mohapatra, and J K Das. 2019. "Mole Fraction Dependency Electrical Performances of Extremely Thin SiGe on Insulator Junctionless Channel Transistor (SG-OI JLCT)." *Advances in Signal Processing and Communication.* Springer: 573–581.

Vandana, B, B S Patro, J K Das, and S K Mohapatra. 2017. "Physical Insight of Junctionless Transistor with Simulation Study of Strained Channel." *ECTI Transactions on Electrical Engineering, Electronics, and Communications* 15 (1): 1–7.

Vandana, B, B S Patro, J K Das, B K Kaushik, and S K Mohapatra. 2018. "Inverted 'T'Junctionless FinFET (ITJL FinFET): Performance Estimation through Device Geometry Variation." *ECS Journal of Solid State Science and Technology* 7 (4). The Electrochemical Society: Q52–Q59.

Wong, H-SP. 2002. "Beyond the Conventional Transistor." *IBM Journal of Research and Development* 46 (2.3) IBM: 133–68.

Yan, R-H, A Ourmazd, and K F Lee. 1992. "Scaling the Si MOSFET: From Bulk to SOI to Bulk." *IEEE Transactions on Electron Devices* 39 (7). IEEE: 1704–1710.

Yang, F-L, H-Y Chen, F-C Chen, C-C Huang, C-Y Chang, H-K Chiu, C-C Lee, et al. 2002. "25 Nm CMOS Omega FETs." In *Proceedings of Digest. International Electron Devices Meeting,* 255–258. doi:10.1109/IEDM.2002.1175826.

Yang, F-L, D-H Lee, H-Y Chen, C-Y Chang, S-D Liu, C-C Huang, T-X Chung, et al. 2004. "5nm-Gate Nanowire FinFET." In *Proceedings of Symposium on VLSI Technology,* 2004. Digest of Technical Papers, 196–197.

Yoon, E-J, S-Y Lee, S-M Kim, M-S Kim, S H Kim, L Ming, S Suk, et al. 2004. "Sub 30 Nm Multi-Bridge-Channel MOSFET (MBCFET) with Metal Gate Electrode for Ultra High Performance Application." In *Proceedings of Technical Digest - International Electron Devices Meeting, 2004.* IEDM. IEEE, 627–630.

Zhang, W, J G Fossum, and L Mathew. 2006. "The ITFET: A Novel FinFET-Based Hybrid Device." *IEEE Transactions on Electron Devices* 53 (9). IEEE: 2335–2343.

Zhang, Y, W Zhao, J-O Klein, W Kang, D Querlioz, Y Zhang, D Ravelosona, and C Chappert. 2014. "Spintronics for Low-Power Computing." In *Proceedings of Design, Automation and Test in Europe Conference and Exhibition (DATE), 2014,* 1–6.

7 Design Challenges and Solutions in CMOS-Based FET

Madhusmita Mishra
NIT Rourkela

Abhishek Kumar
IIT Jodhpur

CONTENTS

7.1 Introduction 129
7.2 Moore's Law and the International Technology Roadmap for
     Semiconductors.. 131
7.3 CMOS Scaling Challenges and Solutions with New FET Geometries........ 132
7.4 NanoDevices Beyond Complementary Metal Oxide Semiconductor 136
7.5 Technical Challenges and Solutions 142
7.6 Conclusion 144
Acknowledgements.. 144
References.. 144

7.1 INTRODUCTION

Recently, the microelectronic industry has got massive benefits from the metal oxide semiconductor field-effect transistor (MOSFET) scaling down. The shrinkage of transistors to measurements beneath 100 nm empowers a vast number of transistors to be located on a solitary chip. Transistor scaling gave the advantages of low-cost manufacturing, high-speed data transfer, and multitasking ability.

Moore's "law" and the International Technology Roadmap for Semiconductors (ITRS) are complimenting each other since the early 1990s. Moore's "law" is primarily founded on the integration of the number of transistors into a microchip. The ITRS, on the other hand, empowers the semiconductor industry to translate this reflection into the real world. So carefulness is needed while explaining "Moore's law," as a physical or numerical law. This initiates to discuss the role of this on the growth of the semiconductor industry. So, in this chapter, Moore's law is hashed out concisely in conjunction with the ITRS primarily from the MOSFET scaling viewpoint [1–3].

The majority of the integrated circuits (IC) employ complementary metal oxide semiconductor (CMOS) technology. Scaling down of MOSFET from µm to nm

129

has the advantages of cheaper circuits, smaller capacitances, higher IC speeds, less power consumption, etc. But scaling down results in short gate length, which in turn leads to the problems as follows:

- The whole design is dominated by static and dynamic power consumption.
- There is a need for constant monitoring of overheating and possible evaporation.

Another point could be crucial from a fabrication perspective. An oxide thickness of around 1 nm is required for adequate channel control via the gate. 1 nm matches to only a couple of atomic layers, so scaling down results in increased leakage gate current. This maximum tolerable measure of the gate leakage current density has a significant impact, while high-k material is required to substitute oxy-nitride for the gate dielectric to meet the maximum gate leakage prerequisite. Several techniques have been utilized to extenuate the retracts of CMOS scaling hashed out above. They are effective to hold up the CMOS scaling limits leastwise to the 45-nm node. We have discussed them in this chapter. But further scaling down after 45 nm was facing problems due to large static and dynamic power losses. This was due to the higher threshold voltage. Using subthreshold conduction, it would lead to the less current passing through the channel. Hence, few researchers proposed to have a different geometry for gate in MOSFET. The gate-all-around structures and multiple gates in a FET were proposed to overcome these shortfalls. These techniques are coming under "new FET geometries." We have given a glance at it in this chapter [4–8].

Since dimensional scaling of CMOS transistors is attaining their fundamental physical limits, various researches have been effectively executed to determine a substitute means to keep on following Moore's law. Amid these attempts, several forms of substitute logic devices and memory-alleged "beyond CMOS devices" have been advised [9–11,15]. These nanodevices exploit the quantum mechanical phenomena and ballistic transport characteristics underneath lower supply voltage and thence low power usage. Those devices are anticipated to be employed for ultrahigh-density integrated electronic computers owing to their enormously little size. In this chapter, a section contributes the general overview for the assuring emerging logic nanodevices, for instance, nanowire (NW) transistors, carbon nanotube field-effect transistors (CNTFETs), graphene nanoribbon (GNR) transistors [12,20], organic field-effect transistors (OFETs) [13,14], spin transistors, and single-electron transistors (SETs) based on their fundamental functioning maxims, the current condition of improvement, and challenges for commercialization [15–17].

Although nanoelectronics allows integrating billions of devices into an individual chip, there is a chance of getting both defects and variations throughout manufacture and chip operations. This chapter provides a glance on technical challenges for nanoelectronics technology from the device as well as system architecture level point of view [18,19].

The following sections are organized as follows. In Section 7.2, Moore's law is hashed out concisely in conjunction with the ITRS primarily from the MOSFET scaling viewpoint. Section 7.3 discusses CMOS scaling challenges and solutions with new FET geometries. Section 7.4 gives details on nanodevices beyond CMOS.

Section 7.5 highlights the technical challenges faced by nanodevices and possible solutions. Finally, concluding remarks are given.

7.2 MOORE'S LAW AND THE INTERNATIONAL TECHNOLOGY ROADMAP FOR SEMICONDUCTORS

Recently, following Moore's "law," around a large portion of a billion transistors could be integrated on a solitary microprocessor, as shown in Figure 7.1.

The scaling of MOSFETs has overturned the semiconductor industry to realize the immensely complex devices and systems. Figure 7.1 shows that during the 1970s, the number of components per chip was doubled per 12 months, and for sure, it proceeded to the mid-1980s. The next year and a half time period of doubling of the chip components is an alteration in accordance with the ITRS editions and the real state of the industry at that time.

The significances of Moore's law resulted in enhancing functionality, reduction in cost per function, and improved performance for all newly generated ICs. As per the ITRS, the definition of functionality is "the number of bits contained in a DRAM chip or the counted logic transistors in a microprocessor unit." Integration of a large number of individual components in a single chip results in the increasing functionality per chip unitedly with the increasing function densities (functions/area). Delay of data flow occurs as a result of the closing off of individual functions on separated integrated systems. The increased functionality minimizes this delay. The cost per function reduction is based on the principle that the goal of all manufacturing units is overstating the profit with understated production cost. The electronics industry is not the only one regarding this. The significant entailment of Moore's law is reducing the cost of manufacturing per function and concurrently increasing the functionality per chip [1–3,5].

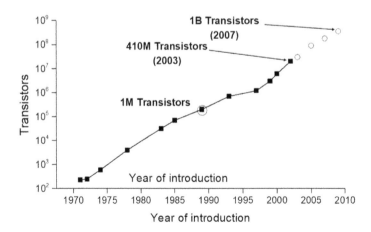

FIGURE 7.1 Visual perception of Moore's law.

Speed of operation is the key to measure the performance factor. Maximum input is the off-chip frequency, and the signal frequency is the output to high-performance devices. The on-chip local frequency is not raising quicker than off-chip frequency according to the recent edition of the ITRS.

The frequency of operation is highly dependent on switching speed of CMOS logic components that are constituted by transistors. The frequency hence is inversely proportional to the delay time that is taken to propagate the signal through the CMOS logic circuit. In a technology node, the delay time of an inverter can be used as an approximation of delay time for that technology, which empirically is calculated as

CMOS delay time = gate capacitance (drain voltage/drain saturation current)

While scaling the device, since CMOS delay time is inversely proportional to the factor by which it is scaled, it allows faster circuit operations. The packing density of transistors is inversely proportional to the total chip area, and hence, the density will increase by a factor of κ^2, where $\kappa \approx \sqrt{2}$ is the scaling constant [1–5].

7.3 CMOS SCALING CHALLENGES AND SOLUTIONS WITH NEW FET GEOMETRIES

Several techniques to mitigate the drawbacks of CMOS scaling are hashed out here. They are good to hold up the CMOS scaling limits leastwise to the 45-nm node [6–11].

A. Scaling around 90 nm:

It became possible to do scaling around 90 nm by utilizing the strained silicon technology. In strained silicon, within a layer of silicon, there is stretching of silicon atoms beyond their reasonable interatomic distance. When these silicon atoms are moved further apart, then the atomic forces are brought down, which intervene with the electron movement by the transistors, and hence the improved mobility leads to superior chip performance along with less energy consumption. More recent techniques for inducing strain lie in the fact of giving doping of lattice atoms such as germanium and carbon to the source and drain.

B. Scaling around 65 nm:

Since from the commencement of silicon MOSFET, the favored gate insulator is SiO_2. The thicker the gate oxide, the lower the gate tunneling current. Due to more thin oxide, the current raises, and this, in turn, raises the speed. It is not simple to manufacture thin oxide. Too thin oxide results in a very high electric field in the oxide, and this, in turn, can lead to a destructive breakdown. In the case of thinner silicon dioxide (SiO_2) film than 1.5 nm, tunneling leakage gate current gets serious. This leakage current can be diminished by a factor of 10 by adding nitrogen into SiO_2.

C. Scaling around 45 nm:

Achieving the maximum gate leakage limit is the ultimate target in design. In case when the gate dielectric oxy-nitride is unable to meet this limit, the favored way for achieving this limit is to use a "high-k" dielectric instead of the oxy-nitride. For scaling around 45 nm, the rule is to enclose metal gates and substitute SiO_2 by dielectric materials of high-k. Zirconium dioxide (ZrO_2), hafnium oxide (HfO_2), and aluminum oxide (Al_2O_3) have higher relative dielectric constants than SiO_2. In case of high-k gate dielectric, there exists always smaller gate leakage current for huge energy barrier among silicon and the dielectric.

D. Problems beyond 45 nm:

Transit frequency f_t is generally used as a speed performance parameter of transistors. This is the frequency where the generalized small-signal current gain is unity. At high electrical fields, due to short channel and saturation effects, the gate current (i_g)–based scaling of f_t has put down to $\sim i_g - 1$ all over the 45-nm node. Consequently, the scaling factor is getting doubled during the scaling from 90 to 45 nm. So a decrease in i_g by a factor of 2 builds the f_t by a factor of 4. Albeit, due to very high drain–source small-signal output conductance (g_{ds}) in case of aggressively scaled CMOS circuits and many "beyond CMOS" accesses, irrespective of the load g_L, no small-signal voltage gain can be attained any longer. The higher intrinsic voltage gain g_m/g_{ds} restricts the maximum attainable voltage gain in a circuit, where g_m is the transconductance. As a result, there is an incompatibility with high-speed architectures of circuits having voltage gain–dependent functionality. So it is nonmeaningful of utilizing the benchmark parameter as only f_t. Hence, in addition to f_t, at least the intrinsic gain g_m/g_{ds} must be considered as a parameter. Another efficient parameter for performance can be the maximum frequency of oscillation f_{max}, because without utilizing any impractical terminations, it admits entire transistor parasitics. Next, performance parameters can be the noise figure and linearity. At 10-nm node, the strong effect of scaling on f_t can be less sounded out with $\sim i_g^{-x}$ and $0 < x < 1$, to a greater degree or extent. To void unreasonable electrical fields in the channel, drain–source voltage (V_{ds}) has to be reduced with i_g, leading to a V_{ds} around 0.5 V for the 10-nm node. Furthermore, due to disproportionate scaling of the threshold voltage V_{th}, the value of $V_{ds} - V_{th}$ is importantly reduced, and this, in turn, challenges the driving of the succeeding stage. At 45-nm node, the value of g_m/g_{ds} is being decreased from 15 to 5 because of an increase in g_{ds}. Ongoing of this phenomenon, calculation at 10 nm will lead to nearly unity intrinsic gain. The possible solution is to use gate-all-around structures and multiple gates in a FET. The main aim of both the structures is to maximize gate-to-channel capacitance and minimize drain-to-channel capacitance. Figure 7.2 illustrates the scaling disputes in planar devices.

E. Scaling around 22 nm:

At 22 nm, for the structures, gate length scaling can be achieved beyond the boundary of mainstream MOSFET. With the lessen oxide thickness, the gate acquires exact command on the channel, just merely right at the Si surface. Reduction in channel length leads to shortened drain-to-channel distance. More control of the drain than the gate is experienced along some distant paths below the Si surface where the control of the gate is weak since it is far away. As $V_{gs} < V_t$, an N-channel MOSFET acquires the off state.

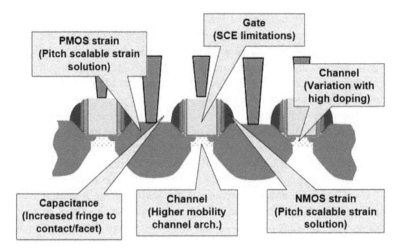

FIGURE 7.2 Scaling disputes in planar devices. SCE, short-channel effect; NMOS, n-type metal oxide semiconductor.

Albeit, there will be a leakage current streaming among the drain and source. This also makes an issue of serious power consumption in standby operation. Hence, there is a requirement of maximizing the gate-to-channel capacitance while minimizing the drain-to-channel capacitance. As a result, there will be a reduction in drain leakage current. Drain-to-source and the drain-to channel distance gets reduced when the channel length is reduced. This, in turn, increases the influence of the drain. The leakage paths are still under the surface, although oxide thicknesses are thin. This is because the gate has a perfect control on the Si surface, and deep in the channel, the drain has more control. For eradicating deeply submerged leakage, the gate should have more control over the channel. This is attained by giving gate control from several sides of the channel. Due to the very thin Si film, there is less probability of the existence of leakage path far from one of the gates. More than one gate structure is called multigate MOSFET or sometimes FinFET.

- FinFET

    Considering any technology node, the FinFET provides more advantages over its planar counterpart as stated in the following:

    - It provides better electrostatic control over the channel along with easy "choking off" of channel. FinFETs show nearly ideal subthreshold behavior, which is not easily achieved in planar technology without significant design endeavor.

    - It provides reduced short-channel effects, which is too complicated in planar technology and gives rise to an enormous impact on gate length variations and therefore on electrical performance.

    - It has high integration density or 3D structure due to its vertical channel orientation, which delivers more performance per linear

"W" than planar even after the isolation dead area among the fins is taken into consideration.

- It has a minimal variability, mainly variability arising from random dopant fluctuation mainly due to doping-free channels. Also, variability associated with line-edge roughness (LER), the random difference of gate line edges from the intended ideal shape, that results in nonuniform channel lengths is lower in FinFETs.
- FinFET requires much lower dopant concentrations in the channel region.
- Since the gate is defined from the top of the fin, the predominant part of the gate is defined by etching processes, which have very low LER.
- Furthermore, it provides the ability to operate at much lower supply voltages and extends voltage scaling, which was leveling off in CMOS devices, and allows further static and dynamic power savings.
- FinFET device technology is the most assuring device technology for broadening Moore's law all the way down to 10 nm or even toward 7 nm. It is fully compatible with CMOS in both bulk and silicon-on-insulator varieties. Also, FinFET technology is fully compatible with the CMOS back-end design processes, reducing the need for new FinFET-specific developments in that area.

Albeit, no new technology is completely free of risk or challenges. The challenges faced by FinFET technology are as follows:

- The modeling of FinFET is a significantly more complicated process. Both precise FinFET parasitic extraction and generation of good yet compact SPICE models are challenging to a greater extent with FinFETs than with planar devices. For most design activities, the mentioned complexities are transparent to the designer. Albeit, there remain optimization challenges for the circuit designer who wishes to utilize FinFET technology.
- Analog, as well as digital, design optimization becomes more complicated due to limited selection in channel length range and finite granularity of the fin width. Although the desired fin width can be generated by ganging together many fins, the length and width cannot be selected freely. This is because of the 3D structure of FinFETs, which does not allow controlling variability for the high-aspect-ratio processes with nonuniform pitches or locally varying pitches. Thus, there are a significant number of restricted design rules for FinFETs.
- FinFET structure has better gate control and lower threshold voltage with less leakage. But, at lower technology node, say, below 10-nm node, the issue of leakage begins again. This contributes to many other issues such as threshold flattening, an increase in power density, and thermal dissipation.

On balance, FinFET technology offers a bright future of device scaling. It is an indispensable technology for high-performance (HP) and power-sensitive applications ranging from mobile smartphones to enterprise computing and networking. The technology introduces new design challenges that can be properly addressed with the growing knowledge and experience of designing with FinFETs to ensure design success. FinFET structure is less efficient in terms of heat dissipations, as heat can quickly be accumulated on the fins. These concerns can lead to a new class of design rule—design for thermal, unlike other design rules such as design for manufacturability. As these devices are approaching their limitations, there is a need for modification in device structure, replacing existing silicon material with new materials. Among them, CNTFET, gate-all-around NW FET, or FinFETs with compound semiconductors may prove as promising solutions.

Below 10 nm, scaling down is likely but becomes difficult and costly. According to the ITRS, there is a unanimity between researchers that CMOS devices will stop scaling by 2020. Some of the potential technologies that can replace the current generation of CMOS, for ultimate scaling down, are CNTs, SETs, and OFETs. These alternatives form a completely new branch of science, namely nanoelectronics.

7.4   NANODEVICES BEYOND COMPLEMENTARY METAL OXIDE SEMICONDUCTOR

Over the years in continuation of challenges discussed earlier, a huge number of nanoelectronic devices have been added up in the field of device technology to substitute the prevailing silicon-based FETs. Out of them, only a handful of devices seem to be standing out in the test of device geometry, operation, physics, and, most importantly, feasibility and economics. These include NW or CNT transistors, graphene FETs, SETs, and spin transistors. These devices offer the size of transistors in atomic levels and can hence offer very reliable and low power circuits. [6,7,10,12].

A. NW Transistors:

It is a type of transistor where the channel is replaced or created with very thin (few atomic length) wire made up of material different than a silicon substrate. It has been extensively researched, and many material alternatives and different geometries have been proposed. NW FETs, also known as gate-all-around or surrounding-gate FETs, offer better scaling opportunities as their nonplanar geometry offers better electrostatic control of channel than in conventional FETs. Researchers and industrialists are garnering attention in NW research and development while banking on several key factors, including their high-yield reproducible electronic properties, higher carrier mobility, smooth surfaces, and producing radial and axial NW heterostructures. Apart from these, the better scaling factor (below 10 nm) and cost-effective bottom-up fabrication have been the primary focus in industrial acceptability. However, smaller NW diameters create quantum confinement, resulting in the change of inversion charge from the surface to bulk inversion. These fabrication imperfections cause variations in actual NW dimensions, leading to perturbations in charge carrier potential and scattering, which causes degradation in charge transport characteristics. Also, any variation in the NW diameter will change the FET threshold voltage. Reduction in these variability is one key challenges which will make NW FETs a feasible technology. It is difficult to mathematically model NW transistors because of quantum confinement effects. Modeling NW FETs based on nonequilibrium Green Function or Monte-Carlo approaches were possessing significant research potential. However, physics associated with the operation of NW FETs requires to be in the form of well-articulated simple, compact models and comprising of ballistic transport and realistic subband parameters, which can be used for the development of SPICE-like circuit design simulators.

B. CNTFETs:

The most prominent and researched alternative of a silicon-channel FET are graphene-channel transistors, which are a particular type of NW FETs. CNTs are made up of graphene, which is a two-dimensional honeycomb lattice structure of carbon atom sheets rolled up into cylinders. Their chirality (direction of rolling) determines their electrical conductivity, which can vary between conductor and semiconductor. Threshold voltage can also be easily controlled, as the bandgap potential is inversely proportional to their rolling diameter. These FETs show excellent material properties, such as large current carrying ability, excellent thermal and mechanical stability, and very good thermal conductivity. These make CNTFETs also a prominent candidate for future interconnects. Semiconducting nanotubes are also advantageous as channel materials in HP FETs. In addition to channel and interconnect materials, high-k dielectrics can easily be incorporated in CNTFETs as it possesses absence of dangling bonds. Exhibiting nearly identical V–I characteristics by NMOS and PMOS made them substantially advantageous for CMOS circuits designing. Their advantages over Si-based semiconductor in the industry are as follows:

1. CNTFETs showed significant improvements in device performance metrics, especially for low-power and high-speed applications.
2. As the device operation physics is mostly the same as that of Si-based counterparts, most of the CMOS design infrastructure will be reused.

But to successfully incorporate this technology on industrial standards, the following challenges would need to be solved.

- Still, the process for synthesis and growth of nanotubes with identical geometry and chirality are in development phase.
- Purification of seed material to a high degree is still not feasible on industrial standards. Even when a lot of research has been done in finalizing the feasible process, they are still far from industry standards.
- Moreover, better control on gate voltage through ultrathin high-k dielectrics with eminently abrupt doping profiles is needed for the fabrication process.

C. GNR Transistor:

GNRs are necessarily one atomic layer of carbon tightly arranged into a 2D honeycomb lattice. For the sake of simplicity, it can be thought of as an unrolled CNT sheet. It is an essential building block for all the grapheme-based FETs of other dimensionalities. For ballistic transport of electrons, it possesses high carrier mobility, and for fast switching, it provides high carrier velocity. Production of wafer-scale graphene films along with fully planar processing technology for devices has ensured high integration potential and high packing density than conventional CMOS fabrication technologies. 2D graphene has exhibited zero-bandgap, semimetallic characteristics. The graphene materials have also shown some flexibility in tuning antenna response. The tuning can be done either by applying changes in doping profile at the time of manufacturing and also by changing external DC voltage. This property of graphene material is utilized in tuning the response of nanoscale devices where manufacturing modifications in the structures are difficult. Several THz and optical devices have been designed using the graphene (Figure 7.3).

Albeit, by patterning graphene into a couple of nanometer-wide GNRs, a bandgap can be brought on. Similar to the CNT cases, they have energy bandgap, which is inversely proportional to their width. There is a requirement of width confinement down to the sub-10 nm scale to open a bandgap that is sufficient for room temperature transistor operation. Unlike CNTs, which shows metallic and semiconducting materials behavior, recent samples of chemically acquired sub-10-nm GNRs have depicted all-semiconducting behavior. Under the presumption that hydrogen atoms passivate the edges of ribbons, GNRs are of two main types: armchair-edge and zigzag-edge GNRs (AGNRs and ZGNRs). ZGNRs are predicted to be metallic by a simple tight-binding model, but a bandgap exists in more advanced and spin-unrestricted simulations [21]. For digital circuit applications, the preferred structures are AGNRs and ZGNRs as the channel

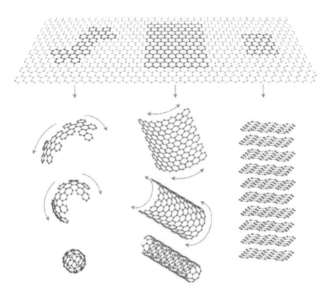

FIGURE 7.3 Features of graphene material.

material. AGNRs have an electronic structure that is nearly associated with that of zigzag CNTs.

While the GNR material assures ultrasmall, fast, and low-energy FETs, there are two fundamental effects of variability and defects: leakage and low noise margins are substantial. Owing to the atomically thin and nanometer-wide geometries of GNRs, variability and defects have a more significant impact on GNRFET circuit performance and reliability as compared with the conventional silicon devices. Variability results from the difficulty of control of the GNR width, edge roughness, or oxide thickness on fabrication. Defects may occur on fabrication by a charged impurity in the gate insulator or a lattice vacancy and thus create a significant performance variation or a nonfunctional device. Hence, it is required to identify and study each of the variability and defect mechanisms. Also, the models built up for GNRs and GNRFETs should be capable of anticipating their effects in isolation as well as together systematically. Hence variability, defects, and parasitics effects must be cautiously watched in the performance assessment and design optimization for graphene-based electronics technology. A comparison of the characteristics of high-speed graphene, CNT, and NW FETs is given in Table 7.1.

D. OFET:

It is a FET using an organic semiconductor in its channel. OFETs can be developed by vacuum evaporation of small molecules or by solution casting of polymers or small molecules or by the mechanical transfer of a peeled single-crystalline organic layer onto a substrate. These devices have been formulated to actualize low-cost, large-area electronic products. OFETs have been fabricated with several device geometries. The most usually used device geometry is a bottom gate with top drain- and source electrodes, since this geometry is similar to the thin-film silicon transistor using thermally grown Si/SiO_2 oxide as a gate dielectric. Organic polymers, say, poly(methyl-methacrylate) (PMMA), can be applied as a dielectric, too.

TABLE 7.1
Comparison of Graphene, CNT, and Nanowire FETs [10]

Properties	Graphene	CNT	Nanowire
Transport	Two-dimensional	One-dimensional, Quantum effects	Quasi one-dimensional for extremely small diameters, otherwise three-dimensional
Potential for high linearity	Device width of few nanometers is required	Yes, provides	Provides if device diameter is very small
Ambipolar properties	Possible	Possible	Possible
On/off ratio	Poor since no or small bandgap	Currently poor due to metallic tubes	Good

CNT, carbon nanotube; FET, field-effect transistor.

The flexibility and versatility of OFETs have enabled them for a wide range of new applications, for instance, in the field of wearable systems, bioelectronics, or as driving elements in a flat panel display. Impressive progress has been achieved in the design, synthesis, and processing of organic semiconductors in the past few years. Organic semiconductors can be more attractive owing to comparable functioning as traditional amorphous inorganic semiconductor materials and their near-infinite tunability.

Recently, more attention is given toward the solution-processable, air-stable HP organic n-type semiconductor. Meanwhile, some critical issues still need further investigation such as operational stability, low-cost and large-area fabrication process, device integration, as well as functionalization in sensor fields. The study of the electronic defect structure of organic semiconductors is the main area of research investigation. Along with device optimization, producing novel organic semiconductor materials and employing thin-film alignment techniques are other directions to achieve HP devices. It is anticipated that by combining correct organic semiconductor materials and earmark fabrication techniques, HP devices for various applications could be obtained. Recent research has highlighted that vertical organic transistors give assurance of increasing the performance of flexible organic transistors while keeping their low-cost advantage. Albeit vertical organic transistors have not been adopted for inorganic semiconductors, they may show some key advantages for organic semiconductors [13,14,20].

E. Spin Transistors:

A spin transistor or spintronic transistor is a magnetically sensitive transistor. This is currently still being developed. It is also named for spintronics. Spin electronics caters to the analysis of the intrinsic spin of the electron and its related magnetic moment, in addition to its primal electronic charge, in solid-state devices. In spintronics, along with charge state, electron spins are put upon as a further degree of freedom with implications in the efficiency of data storage and transfer. Spintronic systems are regularly acknowledged in dilute magnetic semiconductors and Heusler alloys. They are quite compelling in the field of quantum computing and neuromorphic computing.

One advantage over regular transistors is that these spin states can be modified without the application of an electric current. This permits for the detection of hardware that is much smaller and sensitive. The raised sensitivity of spin transistors is also being studied in producing more sensitive automotive sensors. A second advantage is that due to the semipermanent spin of an electron, it can be used as a means of creating cost-effective nonvolatile solid-state storage that does not need the constant application of current to sustain. It is the technologies that are being explored for magnetic random access memory. Because of its high potential for practical use in the computer world, spin transistors are currently being researched in various firms throughout the world, such as in England and in Sweden [15].

F. SETs:

Due to their tiny size and reasonable low-power dissipation rate, SETs are desirable devices for future large-scale integration. The SET consists of three terminals, for example, drain, gate, source, and the second gate is optional. SET has a tiny conductive island coupled to a gate electrode with gate capacitance C_G. The operation of SET is based on islands that are small conducting particles surrounded by insulating material. An island can be charged only by a small number of electrons. Due to the insulating layer that circumvents the island, the transport of electrons to and from the island is possible only by tunneling effect. Thus, the voltage of each island is only in the form of quantized values. Source and drain electrodes are joined to the island through a tunnel barrier (junction). The tunnel barrier that controls the motion of every single electron has two conductors separated by thin layer, and it is modelled as tunneling resistances R_{DS} and junction capacitances C_{DS}. The increased gate bias attracts electrons to the island only through either drain or source tunnel barrier, and the number of electrons in the island only has a fixed integer. Therefore, the increased gate bias makes electrons flow one by one when a small voltage is applied between the source and drain electrodes by means of the "Coulomb blockade" phenomenon. Figure 7.4 shows the schematic of a SET.

Its operation is alike to that of MOSFET except that electron conduction takes place in one electron at a time, whereas in MOSFET, many electrons simultaneously take part in the conduction. With proper gate voltage application, the potential energy of the conduction island is made low enough to allow one electron from the source to tunnel to the conduction island. With the potential energy of the drain lower than that of the conduction island, the electron then tunnels to the other side to reach the drain. With the conduction island empty and the potential lower again, the process repeats. The SET fabrication process in silicon is CMOS compatible. The use of SETs

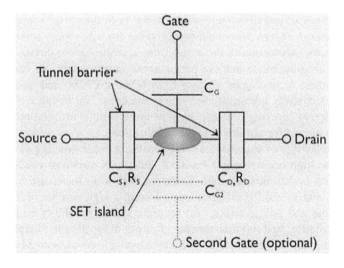

FIGURE 7.4 Schematic of basic SET. SET, single-electron transistor.

combined with MOS transistors allows a compact realization of basic logic functions that exhibit periodic transfer characteristics.

However, significant threshold voltage variation continues to impede the realization of large-scale SET circuits, making it difficult for SETs to compete directly with CMOS devices used to implement Boolean logic operations. Engineering breakthroughs are needed to eliminate the size and background charge fluctuations to suppress the threshold voltage variations. The majority of the SET circuits demonstrated to date employ so-called "voltage state logic," where a bit is represented by the voltage of the capacitor charged by many electrons. The problem of the low fan-out for this scheme can be overcome by reducing the capacitance and/or by combining with conventional FET circuits. The use of a single electron as a source of random number generation has been limited to laboratory demonstrations. The problem of the limited fan-out, which is caused by using only a single electron in the truly single-electron devices, may be solved by innovative circuit designs such as the binary decision diagram [6,7].

7.5 TECHNICAL CHALLENGES AND SOLUTIONS

The applications and limitations of nanoelectronic devices are summarized in Table 7.2. Owing to extremely small size, quantum-mechanical effects such as ballistic transport, tunneling effects, and quantum interference start to predominate in nanodevice operation. Hence, a good understanding of quantum-mechanical phenomena is required to exploit the properties of nanodevices. 1D structure devices, for instance, NW transistors and CNTFETs are hopeful since they show higher I_{ON}/I_{OFF} ratio rate at lower supply voltage. Because of their lacks of reliability, it is not potential for revolutionary nanoelectronic devices to substitute CMOS technology entirely without compulsory needs for those devices. For connecting tiny devices, interconnects are required having less than 10 nm diameter. But due to scaling down of metallic interconnect wires, both the surface scattering from the boundaries of ultranarrow conductors and the grain boundary scattering might block electronic conduction in the wires. Consequently, slower circuit results, and hence, it is not possible to achieve performance improvement. So several potential interconnect technologies ranging from NWs to CNTs, and quantum wires have been extensively investigated to satisfy the needs, for instance, low resistivity, large current carrying capacity, easy of fabrication, and isolation with low-k dielectric materials for ultrahigh-density applications.

High unreliability of nanodevices is the constraint in constructing nanoelectronic systems. The high occurrence of transient faults and variations results in higher dynamic fault occurrences as well. Therefore, following issues are needed to be investigated and solved: (i) effective way of using a high number of nanodevices, (ii) the offline and online errors, (iii) accurate system reliability without performance degradation, and (iv) maintenance of power dissipation in tolerable level. To minimize the drawbacks, the efforts include mapping more algorithms into parallel forms, determining alternative computation and information representation models, and designing fault-tolerant, reconfigurable, and power-efficient systems.

TABLE 7.2

Applications and Limitations of Nanoelectronics Devices

Device	Applications	Limitations
Nanowire transistors	In faster and smaller more dense processors	• Scattering of electrons through wire lateral surface when propagation length is smaller than average mean free path for electrons in given material, resulting in smaller than expected current. • Nonuniformity in nanowire geometry is the available industrial process technology, making it difficult for mass production
Graphene nanoribbon transistors	• Appeared as an alternate to silicon-channels FETs with high sensitivity • Finds usefulness in certain applications including bio- and chemical sensing applications	• Control of geometry • Process control is still not achieved • So it is ineligible for mass production
Carbon nanotube field-effect transistors	• Attractive as future interconnects • Very attractive to Si-based semiconductor industry	• Tighter gate control through ultrathin high-k gate dielectrics is demanded • The starting material must be purified
Single electron transistors	• Useful for implementing binary logic circuits • Useful for implementing multivalued (MV) logic circuits and binary MV mixed logic circuits • Useful for scientific instrumentation and metrology	Implementation is still in theoretical and experimental phases
Organic field-effect transistors	• Very useful in biodegradable devices • Achieving good applications in sensors and LEDs and flexible transistors	• Large in size • More power required for operation
Spin transistors	Very good for memory devices	Implementation is still in theoretical and experimental phases. Few working chips have been developed

In addition to manufacturing defects, nanodevices may be hypersensitive to persistent changes over their operational lifetime due to ionizing-particle strikes, ageing, movement of individual atoms, and temperature-induced variations. Hence, novel fault and defect tolerance circuit design techniques and methodologies have to be built up for nanoelectronic low-power, high-speed, and high-density systems.

7.6 CONCLUSION

In this chapter, we have reviewed various scaling down aspects of MOSFET, possible new FET geometries, need for nanoelectronic devices, and challenges to the design of nanoelectronic devices with some solutions. Although substantial transistor challenges arise for technologies past 45 nm, various possible solutions are being explored to drive Moore's law forward. Although nanoelectronics is the solution for technologies past 22 nm, it is essential to build up a novel knowledge and reliability prototype for nanoelectronics to enable industries for predicting, optimizing, and designing direct reliability and performance of nanoelectronics.

ACKNOWLEDGEMENTS

The authors acknowledge Mohan Krishna (fourth-year ECE student) and Goutham (second-year ECE student) of MITS (Andhra Pradesh) for their support in chunking of materials.

REFERENCES

1. Hiroshi, I, May, 2016. End of the scaling theory and Moore's law. IEEE (IWJT) pp: 1–4.
2. Zeitzoff, P.M., and Chung, J.E., 2005. A perspective from the 2003 ITRS: MOSFET scaling trends, challenges, and potential solutions. *IEEE Circuits and Devices Magazine*, 21(1), pp: 4–15.
3. Meindl, J.D., 2003. Beyond Moore's law: The interconnect era. *Computing in Science & Engineering*, 5(1), pp: 20–24.
4. Kuhn, K.J., May, 2009. Moore's Law past 32nm: Future challenges in device scaling. *13th International Workshop on Computational Electronics (IEEE)*, pp: 1–6.
5. ITRS Reports online: http://www.itrs2.net/itrs-reports.html.
6. M.T. Abuelma'atti, 2012. MOSFET scaling crisis and the evolution of nanoelectronic devices: The need for paradigm shift in electronics engineering education. *6th International Forum on Engineering Education (ELSEVIER)*, pp: 432–437.
7. Y-B Kim, 2010. Review paper: Challenges for nanoscale MOSFETs and emerging nanoelectronics. *Transactions on Electrical and Electronic Materials*, 11(3), pp: 93–105.
8. Razavieh, A., Zeitzoff, P., and Nowak, E.J., 2019. Challenges and limitations of CMOS scaling for FinFET and beyond architectures. *IEEE Transactions on Nanotechnology*, 18, pp: 999–1004.
9. Horowitz, M., Alon, E., Patil, D., Naffziger, S., Kumar, R., and Bernstein, K., 2005. Scaling, power, and the future of CMOS. IEEE International Electron Devices Meeting (IEDM Technical Digest), pp: 7–14.
10. Ellinger, F., Claus, M., Schröter, M., and Carta, C., 2011. Review of advanced and beyond CMOS FET technologies for radio frequency circuit design. *IMOC (IEEE)*, pp: 347–351.

11. Vora, P.H., and Lad, R., 2017. A Review Paper on CMOS, SOI and FinFET Technology, Industry article.
12. Geim, A.K., and Novoselov, K.S., 2007. The rise of grapheme. *Nature Materials*, 6, pp: 183–191.
13. Chang, J., Lin, Z., Zhang C., and Hao Y., 2017. Organic field-effect transistor: Device physics, materials, and process. *INTECH*, 125–145.
14. Ostroverkhova, O., 2018. *Handbook of Organic Materials for Electronic and Photonic Devices*. Woodhead Publishing Series in Electronic and Optical Materials. Elsevier, Sawston, Cambridge, UK, pp: 875–891.
15. Joshi, V.K., 2016. *Spintronics: A Contemporary Review of Emerging Electronic Devices*. Elsevier, Sawston, Cambridge, UK, pp: 1503–1513.
16. Haron, N.Z., and Hamdioui, S., December 2008. Why is CMOS scaling coming to an END? *3rd International Design and Test Workshop (IEEE)*, pp: 98–103.
17. Stillmaker, A., and Baas, B., 2017. Scaling equations for the accurate prediction of CMOS device performance from 180 nm to 7 nm. *Integration (ELSEVIER)* 58, pp: 74–81.
18. Bohr, N., 1935. Can quantum-mechanical description of physical reality be considered complete? *Physical Review*, 48(8), pp: 696–702.
19. Jacob, A.P., Xie, R., Sung, M.G., Liebmann, L., Lee, R.T., and Taylor, B., 2017. Scaling challenges for advanced CMOS devices. *International Journal of High Speed Electronics and Systems*, 26(01n02), pp: 1–76.
20. Varshney, G., Gotra, S., Pandey, V.S., and Yaduvanshi, R.S., 2019. A proximity coupled two-port MIMO graphene antenna with pattern diversity for THz applications. *Nano Communication Networks*, 21, pp: 456–463.
21. Liu, C., Zhang, X., Zhang, J., Muruganathan, M., and Mizuta, H. 2020. Origin of non-linear current-voltage curves for suspended zigzag edge graphene nanoribbon. *Carbon*. https://doi.org/10.1016/j.carbon.2020.05.010

8 Analytical Design of FET-Based Biosensors

Khuraijam Nelson Singh and Pranab Kishore Dutta
NERIST

CONTENTS

8.1   Introduction 147
8.2   Types of Biosensors 149
    8.2.1   Electrochemical Biosensor 149
    8.2.2   Optical Biosensor... 150
    8.2.3   Piezoelectric Biosensor... 150
    8.2.4   Calorimetric Biosensor 150
8.3   Field-Effect Transistor–Based Biosensors... 150
    8.3.1   Working of Field-Effect Transistor–Based Biosensor 151
    8.3.2   Some Common Types of Field-Effect Transistor–Based
        Biosensors 152
        8.3.2.1   Ion-Sensitive Field-Effect Transistor Biosensor 152
        8.3.2.2   Nanowire Field-Effect Transistor Biosensor..................... 153
        8.3.2.3   Carbon Nanotube Biosensor 153
        8.3.2.4   Dielectrically Modulated Field-Effect
                Transistor Biosensor... 154
        8.3.2.5   Tunnel Field-Effect Transistor Biosensor 155
        8.3.2.6   Junctionless Field-Effect Transistor Biosensor................. 155
8.4   Modeling of Field-Effect Transistor–Based Biosensors.......................... 156
    8.4.1   Modeling of Dielectrically Modulated Field-Effect
        Transistor–Based Biosensors 158
        8.4.1.1   Surface Potential 158
        8.4.1.2   Electric Field.. 160
        8.4.1.3   Threshold Voltage 162
        8.4.1.4   Sensitivity.. 163
8.5   Summary 165
References.. 165

8.1   INTRODUCTION

Research on biosensors has seized the interested researchers over the past few decades due to their various advantages and applications. They are used in the discovery of drugs, monitoring of diseases, agriculture, food quality control, industrial wastage monitoring, military, etc. [1]. The sensing analyte is the main element that differentiates a biosensor from the other physical/chemical sensors. In general, the

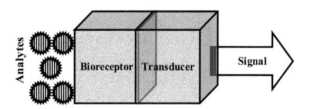

FIGURE 8.1   Schematic representation of a simple biosensor.

biosensor is a device that is used to detect an analyte using a biosensitive receptor [2]. Figure 8.1 shows a schematic representation of a simple biosensor. Its main components are as follows:

i. **Analytes**: The substance that is intended to be detected, such as glucose in a glucose sensor, ammonia in ammonia sensor, and so on.
ii. **Bioreceptors**: The bioreceptors are biosensitive elements used to detect target analyte/biomolecule. They are sensitive to the analytes of interest. Some examples of bioreceptors are antigen, DNA, enzyme, and so on.
iii. **Transducers**: The elements that are used to convert energy from one form to another are called transducers. In a biosensor, the interaction of analytes and bioreceptors produces changes in the form of heat, gas, light, ions, or electrons. These changes are then converted into a quantifiable form by the transducer. Usually, the output of the transducer is in the form of electrical or optical signals, and the generated signal is proportional to the interaction between the analyte and the biosensor.

The invention of the first biosensor by Clark and Lyons in 1962 marked the start of a new era of biomolecules detection [3]. The biosensor was used for the detection of glucose. Diabetes is a widespread chronic disease that not only affects adults but children too. In 2016, the World Health Organization (WHO) estimated that 1.6 million people died due to diabetes [4]. Such severeness has led to the rise in the demand for blood glucose detector globally. The increasing demand, in turn, increases the research of glucose detection using biosensors.

Similarly, the demand for a faster, cheaper, and easier method for detection of biomolecules in various fields has led to the advancement of biosensor research. With the recent achievements, detection of a disease-specific biomarker in vitro as well as in vivo has achieved high accuracy [5]. The biosensors are proven to be highly sensitive in the detection of disease markers such as lactate, cytokines, proteins, and antibodies. These biosensors can also detect the presence of the target biomolecules even when the bonded amount of biomolecules is meagre.

Some of the advantages of using biosensors are as follows [6]:

i. They can be used for the detection of both ionic and nonionic biomolecules.
ii. They can be used to detect a minimal amount of target biomolecules. Thus, their sensitivity is very high.
iii. They can be used to monitor any specific biomolecules continuously.

iv. They have very fast response timing.
v. Field-effect transistor (FET)–based biosensors have a minimal size. Thus, their power consumption is also minimal.
vi. They can be used to detect biomolecules even without the need for labeling.

8.2   TYPES OF BIOSENSORS

Biosensors can broadly be divided into four categories [14]:

i. Electrochemical biosensor
ii. Optical biosensor
iii. Piezoelectric biosensor
iv. Calorimetric biosensor

8.2.1   ELECTROCHEMICAL BIOSENSOR

Interaction of the biomolecules and the biosensor electrodes produces electrochemical changes on the sensing surface of the biosensor. An electrochemical biosensor measures these changes to identify the analytes embedded on its surface. Utilization of simple electrodes makes these biosensors cost-effective and straightforward and can also be miniaturized to make implantable ones. They are used in the detection of glucose, nucleic acids, peptides, etc.

They are usually made up of three main components [6]:

i. **Working electrode**: It is an electrode in which the electrochemical reaction takes place.
ii. **Reference electrode**: It is a stable electrode which potential is used as a reference for potential measurement.
iii. **Auxiliary electrode**: It is an electrode that is used for completion of the circuit of the electrochemical cell.

In a nucleic acid biosensor, DNA is bonded onto the surface of the electrode. The interaction of the DNA and the working electrode causes a hybridization reaction, resulting in a change of conductivity of the device. In a label electrochemical biosensor, the analyte is bonded in between the label and the working electrode. A label is a substance that on interaction with the analyte produces a change, which can be used to detect the presence of the analyte. Although label biosensor provides accurate results, it needs a solid binding of the labeling substance as the detection process depends on the specific reaction that takes place between the label and the biomolecule of interest, whereas a label-free biosensor detects the changes when the target biomolecules interact with the sensing surface of the biosensor scrapping the sandwiching process of biomolecules in between the label and the working electrode. Label-free detection process provides faster analysis, lower reagent cost, and continuous monitoring capability in real time. It also allows detection of the analyte without any alteration to its natural form. Popular examples of the commercially available electrochemical biosensor are pregnancy test strips and urine analysis strips.

The interaction of the electrochemical biosensor and the analytes produces various types of signals. Based on these signals, the electrochemical biosensor can be further classified as amperometric, potentiometric, conductometric, and FET-based biosensors [2].

8.2.2 OPTICAL BIOSENSOR

It is a biosensing device in which the transducer is based on the optical principle. Light from an optical source is directed toward the sensing surface on which the analyte is bonded. The light is then reflected and is detected by using a detector such as a photodiode, which estimated the physical changes occurring in the sensing surface. Interaction of the light and the analyte produces a minute change in the refractive index of the system or change in thickness. These changes are captured and translated to estimate the analyte. They are generally used to detect bacteria, DNA, protein, cell, antibodies of antigens, etc. Some of the bioreceptors that are used in these biosensors are antibodies, enzymes, and nucleic acid. The signals that are measured in these biosensors include chemiluminescence, absorbance, reflectivity, refractive index, and surface plasmon resonance [7]. The optical biosensor has the advantage of the free reference electrode, less temperature dependence, detection of multiple analytes, etc. Its main drawback is the lack of miniaturizing capability, slow response, and requirement of spectrophotometer to measure the wavelength of the reflected light.

8.2.3 PIEZOELECTRIC BIOSENSOR

It is a biosensor that is based on the piezoelectric (PE) effect, i.e., production of voltage when mechanically stressed. In this class of biosensors, binding of analytes to the PE sensing material causes a vibration with a characteristic frequency. The change in the characteristic frequency of the PE material is proportional to the mass of the analyte. The variation in the frequency of the PE crystal produces an oscillating voltage. The PE platforms currently available are compatible for biosensor construction without any addition of a reagent. However, the sensitivity of the biosensor needs to be taken care of, as the changes produced by very small microorganism should be able to produce measurable voltage [8].

8.2.4 CALORIMETRIC BIOSENSOR

Enzyme-catalyzed reactions are generally exothermic. The generated heat can be used to measure the reaction rate; thus, the concentration of the reactants can be estimated. The calorimetric biosensor takes advantage of such type of reaction to detect the analytes. Binding of the analyte on the sensing surface causes a reaction, which changes the enthalpy of the system. The change in the enthalpy of the system gives the concentration of the analyte. They are used in the detection of pathogens [9], cancer cell [10], excess enantiomeric [11], and estimation of pesticides.

8.3 FIELD-EFFECT TRANSISTOR–BASED BIOSENSORS

FET-based biosensor is a subclass of the electrochemical biosensor. It employs the change in surface charge and dielectrics of the device to detect the analytes. It provides the detection of analytes without the labeling process. Thus, the detection process becomes easier. The advancement in the metal oxide semiconductor FET (MOSFET) study also provides high miniaturization capability, which allows in vivo detection of the analytes. Its compatibility to integrate with various IC technologies, fast processing speed, ability to sense in parallel, and lower cost further expand its scope of application.

8.3.1 Working of field-effect transistor–Based Biosensor

To understand the working of FET-based biosensors, understanding the functioning of FET beforehand is preeminent. Generally, FET device is considered to have only three terminals, namely, source, gate, and drain, excluding the substrate terminal. Based on the type of the FET (n-channel or p-channel), a voltage is applied on the gate to create a channel between the source and the drain. For n-channel (p-channel), application of a positive (negative) voltage in the gate results in the repulsion of the holes (electrons) underneath the interface of gate oxide and the channel region leaving immobile negative ions (positive ions). If the gate potential is increased further, the minority charge carrier, i.e., electron (hole) gets pulled toward the interface region, and a channel of the electrons (holes) is formed, bridging the source and the drain electrically. The applied gate voltage at which a conducting channel is induced between drain and source is known as threshold voltage (V_{th}). The potential applied in the gate creates a vertical electric field that controls the channel and its conductivity. When a potential is applied between drain and source, due to the horizontal electric field, drain current starts to flow in the induced channel.

FET biosensor utilizes the channel conductance depending on the vertical electric field. In these biosensors, the gate terminal is modified by using an ion-selective membrane or molecular receptor or creation of sensing site depending on the analyte to be sensed. Binding of the analyte in the sensing site results in the change of gate charge or dielectric value, which gets reflected as a change in the electrical parameters of the FET such as threshold voltage, drain current, or transconductance [12,13]. These parameters that get changed due to the interaction of the analyte and the biosensor are known as sensing parameter.

Figure 8.2 shows a schematic diagram of a nanogap FET-based biosensor [14]. Part of the gate oxide has been etched to create the sensing site. Filling the sensing

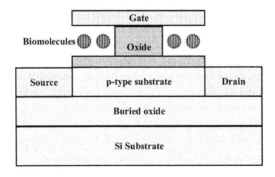

FIGURE 8.2 Schematic diagram of a nanogap FET biosensor. FET, field-effect transistor.

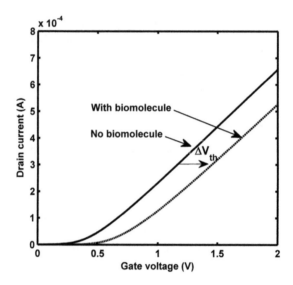

FIGURE 8.3 Variation of drain current upon binding of biomolecule.

site with the biomolecules changes the dielectric, and charge of the region changes, which changes the sensing parameter of the biosensor (threshold voltage) as shown in Figure 8.3. Due to the maturity of FET technology, it gives the advantage of the relative ease of theoretical study and fabrication.

8.3.2 SOME COMMON TYPES OF FIELD-EFFECT TRANSISTOR–BASED BIOSENSORS

8.3.2.1 Ion-Sensitive Field-Effect Transistor Biosensor

Ion-sensitive FET (ISFET) is based on the conversion of the ionic concentration of an electrolyte into an electrical signal. The detection of the analyte does not affect the chemical composition of the electrolyte, as there is not an actual transfer of charge in the whole process. The change due to the electrolyte only appears in the form of change in gate capacitance. In the beginning, ISFET was generally used for pH sensing; later, it was used for glucose monitoring, penicillin G determination, etc. [15,16].

The working of ISFET can be explained by comparing it with a MOSFET. Both of them have the same structure except for the modification in the gate of the ISFET. The metal of the gate is replaced by a reference electrode, which is connected by an electrolyte through an ion-sensitive membrane, as shown in Figure 8.4 [17]. Interaction of the charge of the biomolecules present in the electrolyte and the gate insulator changes the conductance of the channel, resulting in the change of the device current. The variation of current is used as a sensing matrix of the biosensor. Apart from the charge of the biomolecules and the gate dielectrics, the interaction also depends on the pH of the ions, enzyme reaction products, sensitivity to pH, etc. Although the threshold voltage of MOSFET is a constant quantity, it varies for ISFET depending on the pH of the electrolyte.

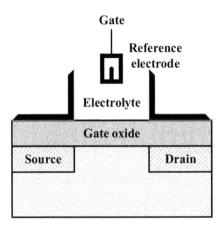

FIGURE 8.4 Schematic diagram of ISFET. ISFET, ion-sensitive field-effect transistor.

8.3.2.2 Nanowire Field-Effect Transistor Biosensor

Due to the high sensitivity of nanowires, application of nanowire as a biosensor is a lucrative choice to make. The charged molecules that get attached on the surface of the nanowire produce a gating effect, which changes the current or the conductance of the device [18]. Just like a MOSFET, a silicon nanowire (SiNW) biosensor also has a source, gate, and drain. A schematic diagram of SiNW is shown in Figure 8.5 [19]. To detect a specific analyte, a bioreceptor that recognizes the analyte is bonded on the surface of the nanowire (NW). When the analyte interacts with the surfaces of the nanowires, it changes the charge density at the surface of the NW, thereby changing the electric field. The changing electric field causes the conductance of the device to change [19]. For instance, for binding of negatively charged biomolecules on the surface of the receptor of an n-type SiNW biosensor, the conductance of the device gets reduced. For binding of positively charged biomolecules on the surface of the receptor, a p-type SiNW biosensor also produces the same result.

8.3.2.3 Carbon Nanotube Biosensor

Carbon nanotubes (CNT) described initially in 1991 by Sumio Iijima [20] can be considered as a rolled cylindrical tube of graphene. Based on the number of graphene

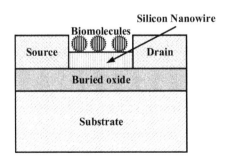

FIGURE 8.5 Schematic diagram of SiNW biosensor. SiNW, silicon nanowire.

layer, it can be classified as single-walled CNT (SWCNT) and multiwalled CNT (MWCNT). Their diameter can vary in the range of 0.75–3 and 2–30 nm, respectively, with their lengths varying from nanometer scale to micrometer scale [21]. Based on the atomic arrangement and diameter of the nanotube, SWCNT can either have metallic or semiconducting properties, whereas MWCNT has only metallic property [21,22]. Both of the CNTs have good electrical conductivity, great bioconsistency, and high adsorptive strength, which give them the ability to carry high current with relatively small heating.

In CNTFET biosensor, the channel region of the FET is modified by using CNTs instead of conventional silicon (Si) as shown in Figure 8.6 [23]. The CNT is directly exposed to the environment so that the analyte gets interacted with the nanotube surface, which changes the electrical property of the CNT [23]. The change in the electrical property of the biosensor is due to one or more of the following reasons: (i) change in electrostatic potential of the gate, (ii) scattering of charge across the CNT surface, (iii) transfer of charge between the analyte and the CNT, and (iv) modification of the Schottky barrier between the CNT and the metal electrode. One of the major problems in the usage of CNT as a biosensor is the solubility of the nanotubes. Functionalization of the CNTs can solve this problem by using materials such as enzyme, protein, DNA, and so on.

8.3.2.4   Dielectrically Modulated Field-Effect Transistor Biosensor

Dielectrically modulated FET (DMFET) biosensor is designed by modifying the gate of a MOSFET [12,14]. A nanogap is created in the gate by etching a part of the gate metal and the gate oxide, as shown in Figure 8.2. This nanogap serves as the location for the binding of the biomolecules. Binding of the biomolecules on the nanogap changes the dielectric of the gate, which changes the capacitance of the gate, the vertical electric field in the channel, and ultimately the drain current. The variation of the drain current causes the threshold voltage of the sensor to change. The change/modulation of the threshold voltage based on the properties of the biomolecule is used as a sensing matrix to determine the biomolecule.

For an n-channel DMFET, the threshold voltage is decided by the height of the nanogap. The effective dielectric of the sensing site decreases when the nanogap is empty. Due to the decreasing gate dielectric constant of the gate, the threshold

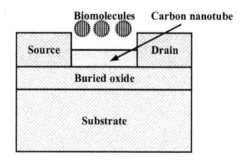

FIGURE 8.6   Schematic diagram of carbon nanotube biosensor.

voltage changes [24]. The threshold voltage also depends on the charge and dielectric constant of the bonded biomolecules.

8.3.2.5 Tunnel Field-Effect Transistor Biosensor

T. Baba first developed tunnel FET (TFET) in 1992 as a better alternative to the conventional MOSFET. It also has source, drain, and gate but unlike MOSFET, the doping of the source and drain are of opposite type [25]. The source, gate, and drain configurations are in the form of PIN junction (p-type, intrinsic, and n-type), in which the gate potential controls the potential of the intrinsic region. It works on the principle of band-to-band tunneling due to the gating effect. When the potential of the gate is high enough, the conduction band of the intrinsic region bends to the level of the valence band of the source, thereby decreasing the barrier for the electrons in the intrinsic region to tunnel toward the source region [25,26].

For biosensing, a part of the gate is etched to set as a location for binding the biomolecules, as shown in Figure 8.7 [27,28]. The gate oxide in the nanogap region is functionalized by coating it with a particular bioreceptor, which is used to detect a specific biomolecule. When the biomolecules get bonded in the nanogap region, due to the gating effect, the barrier between the different areas of the TFET gets modulated, changing the drain current of the device. This change in drain current is captured as a sensing parameter for the biosensor. Apart from the drain current, the transconductance-to-current ratio has also been used as a sensing parameter to detect the biomolecule [29].

8.3.2.6 Junctionless Field-Effect Transistor Biosensor

A junctionless transistor (JLT) is an accumulation mode (AM) transistor in which the doping types of the source, drain, and the channel region are all the same. The doping of a p-channel JLT has P+-P-P+ type doping, whereas an n-channel JLT has N+-N-N+ type doping [30]. Due to the absence of abrupt junction in the source–channel–drain region, the diffusion of impurity during the thermal process is low. It mitigates the need for annealing during the fabrication, thereby minimizing the thermal budget. Comparing with inversion mode (IM) transistor, AM transistor has been shown to have better drain-induced barrier lowering, subthreshold slope degradation, and less sensitive to doping fluctuation. Thus JLT that is in the same class as AM transistor also enjoys all the advantages of AM transistor.

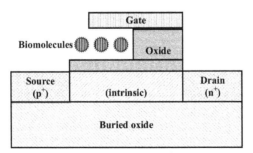

FIGURE 8.7 Schematic diagram of TFET biosensor. TFET, tunnel field-effect transistor.

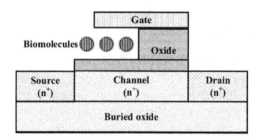

FIGURE 8.8   Schematic diagram of junctionless FET biosensor. FET, field-effect transistor.

When using JLT as a biosensor, a part of the gate oxide is etched in which the bio-molecules are bonded, as shown in Figure 8.8 [31]. Upon interaction of the biomol-ecules and the gate oxide of the sensor, the effective dielectric of the nanogap region changes, resulting in the shift of the gate capacitance. These changes are captured as a change in the electrical properties of the biosensor, such as drain current, threshold voltage, and so on.

8.4   MODELING OF FIELD-EFFECT TRANSISTOR–BASED BIOSENSORS

Modeling of any devices such as MOSFET, TFET, CNT, and so on helps to under-stand the device's properties before the fabrication begins. Thus, it is essential to develop proper models that can predict the properties of the devices, so that tweak-ing could be done to improve their performances.

There are various ways to model semiconductor devices. Some of them are physi-cal models, compact models, empirical models, black-box models, and lookup table models. Physical models are heavily based on physics. They are highly accurate but very inefficient. On the other hand, empirical models, black-box models, and lookup table models are very efficient, but their accuracy of physics is very less. The com-pact model provides an outstanding balance between the accuracy of physics and computational efficiency. Thus, this chapter provides a compact modeling of FET-based biosensor to understand its efficacy in the detection of biomolecules.

Although there are various types of FET-based biosensors, this chapter presents the analytical modeling of dielectric and charge-modulated FET-based biosen-sor. The same process can also be applied to the other types of biosensors, such as nanogap biosensor [12–14], TFET biosensor [26–29], and junctionless biosensor [31,32].

Figure 8.9 shows the schematic diagram of an underlap dielectrically modulated FET-based biosensor. Part of the gate electrode and the oxide layer are etched to cre-ate an underlap region. Binding of the biomolecules on the binding site changes the dielectric and surface charge in the sensing site. The variation in electrical proper-ties of the device is captured as variation in capacitance and flatband voltage in the modeling process.

The modeling of various parameters such as surface potential, threshold voltage, and the electric field starts with solving the Poisson's equation in all the regions of

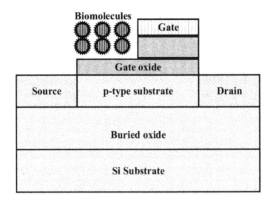

FIGURE 8.9 Schematic structure of underlap FET biosensor. FET, field-effect transistor.

the device channel. Equation 8.1 represents the Poisson's equation, upon solving of which the electric potential of the device is determined [13].

$$\frac{\partial^2 \phi_j(x,\,y)}{\partial x^2} + \frac{\partial^2 \phi_j(x,\,y)}{\partial y^2} = \frac{qNa}{\epsilon_{Si}} \tag{8.1}$$

where $j=1$ or 2 and represents the two regions of the device channel. ϕ_j $(x,\,y)$ is the potential in the device channel, q is the electronic charge, N_a is the channel doping concentration, and ϵ_{Si} is the silicon dielectric value.

The potential in the channel is then appoximated by using a parabolic approximation given by

$$\phi_j(x,\,y) = K_{j0}(x) + K_{j1}(x)\,y + K_{j2}(x)\,y^2 \tag{8.2}$$

where K_{j0} (x), K_{j1} (x), and K_{j2} (x) are all constants that depend only on x-coordinate, and their values are determined by solving the equations that exist in the boundary of the channel.

The Poisson's equation is then simplified into a 1D nonhomogeneous partial differential equation:

$$\frac{\partial^2 \phi_{fs,\,j}(x)}{\partial x^2} - A_j \phi_{fs,\,j}(x) = B_j \tag{8.3}$$

The solutions of the simplified Poisson's equation give the electric potential profile in the channel:

$$\phi_{fs1}(x) = P_1 e^{\lambda_1 x} + P_2 e^{-\lambda_1 x} + \gamma_1, \text{ for R-I} \tag{8.4}$$

$$\phi_{fs2}(x) = P_3 e^{\lambda_1 x} + P_4 e^{-\lambda_1 x} + \gamma_2, \text{ for R-II} \tag{8.5}$$

The electric potential can be used to find the threshold voltage of the device. The value of the gate voltage where the potential becomes lowest gives the threshold voltage of the device.

8.4.1 MODELING OF DIELECTRICALLY MODULATED FIELD-EFFECT TRANSISTOR–BASED BIOSENSORS

Figure 8.10 shows the structure of an underlap double-gate dielectrically modulated FET (UDG DMFET)-based biosensor. Part of the gate dielectric and electrodes are etched to create space for binding the biomolecules. Binding of the biomolecules in these underlap regions changes the dielectric and surface charge of the device, which in turn changes the electrical characteristics of the device. The modification of the gate resulted in the parting of the channel into two regions: R-I and R-II. The gate oxide is also staked with a high-k material (HfO$_2$) so that the oxide is thick enough to lower the gate leakage current [33,34]. Usage of high-k also has an additional benefit of creating enough space for the biomolecules binding, as space must be large enough to hold the biomolecules physically.

The device has been simulated by using ATLAS (SILVACO). Models used in the simulation are CONMOB, FLDMOB, SRH, AUGER, and BOLTZMANN. Binding of biomolecules in the sensing site is realized by inserting a dielectric with a value larger than unity. The charge of the biomolecules is realized by inserting a surface charge in between the oxide and the channel. The antibody of avian influenza (anti-AI) and DNA are considered as a test subject to check the performance of the UDG DMFET as a biosensor. Anti-AI is simulated by considering a dielectric value of 3 and charge of -6×10^{11} C/cm^2 [35], whereas DNA is simulated by considering a dielectric value of 8 and charge of -2×10^{12} C/cm^2 [36,37].

8.4.1.1 Surface Potential

The electric potential $\phi_j(x, y)$ existing in the channel of UDG DMFET can be realized by solving the 2D Poisson's equation:

$$\frac{\partial^2 \phi_j(x, y)}{\partial x^2} + \frac{\partial^2 \phi_j(x, y)}{\partial y^2} = \frac{qNa}{\epsilon_{Si}} \tag{8.6}$$

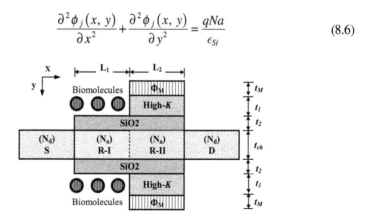

FIGURE 8.10 Schematic diagram of UDG DMFET. UDG DMFET, underlap double-gate dielectrically modulated field-effect transistor.

where $j=1$ or 2 and it corresponds to R-I or R-II of the device as shown in Figure 8.5, q is the electronic charge, N_a is channel doping concentration, and ϵ_{Si} is dielectric value of silicon.

Since the potential in the channel has parabolic nature, it can be approximated as [32]

$$\phi_j(x, y) = K_{j0}(x) + K_{j1}(x)y + K_{j2}(x)y^2 \tag{8.7}$$

where the x-dependent contants K_{j0}, K_{j1}, and K_{j2} are solved by using the boundary conditions shown in the following:

$$\phi_j(x, 0) = \phi_{fsj}(x) \tag{8.8}$$

$$\phi_j(x, t_f) = \phi_{bsj}(x) \tag{8.9}$$

$$\left. \frac{\partial \phi_j(x, y)}{\partial y} \right|_{y=0} = \frac{C_j}{\epsilon_{Si}} \left[\phi_{fsj}(x) - V'_{gsj} \right] \tag{8.10}$$

$$\left. \frac{\partial \phi_j(x, y)}{\partial y} \right|_{y=t_{ch}} = -\frac{C_j}{\epsilon_{Si}} \left[\phi_{bsj}(x) - V'_{gsj} \right] \tag{8.11}$$

where $\phi_{fsj}(x)$ and $\phi_{bsj}(x)$ are the front and back surface potential of the jth region. The binding of the biomolecules causes the capacitance in R-I (C_1) to change, whereas the capacitance in R-II (C_2) remains the same which values are given in the following:

$$C_1 = \frac{C_{ox2}C_{bio}}{C_{ox2} + C_{bio}}, \quad C_{ox2} = \frac{\epsilon_{ox}}{t_2}, \quad C_{bio} = \frac{2\epsilon_{bio}}{n\pi L_1}\sinh\left[\cosh^{-1}\left(\frac{t_1 + t_M}{t_1}\right)\right]$$

$$C_2 = \frac{C_{ox2}C_{high}}{C_{ox2} + C_{high}}, \quad C_{high} = \frac{\epsilon_{high-K}}{t_1}$$

where ϵ_{ox}, ϵ_{high}, and ϵ_{bio} are dielectric constants of SiO$_2$, high-k, and biomolecule, respectively; C_{bio} is the fringing capacitance due to binding of biomolecules [33], and n must satiate $\left|\sin\left\{\frac{n\pi}{2}\right\}\right| = 1$. Also, $V'_{gsj} = V_{gs} - V_{FBj}$, where V_{gs} is the gate voltage and V_{FBj} is the jth region flatband voltage given by

$$V_{FB1} = V_{FB2} - \frac{qQ_{bio}}{C_1}, \quad V_{FB2} = \phi_M - \phi_{Si}$$

where Q_{bio}, ϕ_M, and ϕ_{Si} are charge density of biomolecules, work functions of gate electrode, and work function of silicon, respectively.

Now, substituting the values of K_{j0}, K_{j1}, and K_{j2} in Equation (8.7), the Poisson's equation can be simplified as

$$\frac{\partial^2 \phi_{fsj}(x)}{\partial x^2} - A_j \phi_{fsj}(x) = B_j \tag{8.12}$$

where

$$A_j = \frac{2C_j}{\epsilon_{Si} t_{Si}} \phi_{fsj}(x), \ B_j = \frac{qN_a}{\epsilon_{Si}} - \frac{2C_j}{\epsilon_{Si} t_{Si}} V'_{gsj}$$

The solutions of the simplified Poisson's equation can be written as

$$\phi_{fs1}(x) = P_1 e^{\lambda_1 x} + P_2 e^{-\lambda_1 x} + \gamma_1, \quad \text{for R-I} \tag{8.13}$$

$$\phi_{fs2}(x) = P_3 e^{\lambda_2 (x-L_1)} + P_4 e^{-\lambda_2 (x-L_1)} + \gamma_2, \quad \text{for R-II} \tag{8.14}$$

where $\lambda_j = \sqrt{A_j}$, and $\gamma_j = -\frac{B_j}{A_j}$. The values of P_1, P_2, P_3, and P_4 can be solved by using the boundary conditions given in the following:

$$\phi_{fs1}(0) = V_{bi} \tag{8.15}$$

$$\phi_{fs1}(L_1) = \phi_{fs2}(L_1) \tag{8.16}$$

$$\phi_{fs2}(L_1 + L_2) = V_{bi} + V_{ds} \tag{8.17}$$

$$\left.\frac{\partial \phi_1(x, y)}{\partial x}\right|_{x=L_1} = \left.\frac{\partial \phi_2(x, y)}{\partial x}\right|_{x=L_1} \tag{8.18}$$

where V_{bi} is the built-in potential and V_{ds} is the drain terminal voltage.

Figure 8.11 shows the effect of the binding of anti-AI and DNA on the surface potential profile of UDG DMFET. It shows that binding of the biomolecules shifts the surface potential plot from its original position before the binding happens. And since DNA has a larger value of dielectrics and charge value, the change it brings is more in comparison with that of anti-AI. If the sensing site length increases, the area for loading biomolecules increases. The increase in biomolecules binding causes a greater shift in the surface potential, as shown in Figure 8.12.

8.4.1.2 Electric Field

The electric field in the channel of the device is determined by differentiating equations (8.13) and (8.14) with respect to the horizontal distance.

$$E_1 = \frac{d\phi_{fs1}(x)}{dx} = \lambda_1 \left(P_1 e^{\lambda_1 x} - P_2 e^{-\lambda_1 x} \right), \quad \text{for R-I} \tag{8.19}$$

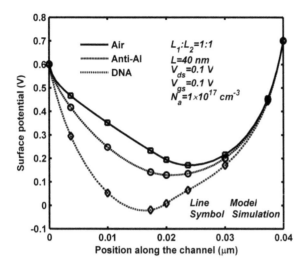

FIGURE 8.11 Variation of surface potential along UDG DMFET channel when $L_1:L_2=1:1$. UDG DMFET, underlap double-gate dielectrically modulated field-effect transistor.

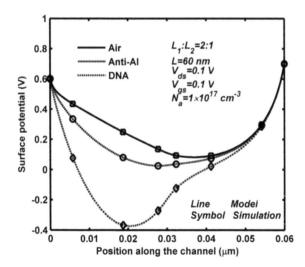

FIGURE 8.12 Variation of surface potential along UDG DMFET channel when $L_1:L_2=2:1$. UDG DMFET, underlap double-gate dielectrically modulated field-effect transistor.

$$E_2 = \frac{d\phi_{fs1}(x)}{dx} = \lambda_1 \left(P_3 e^{\lambda_2(x-L_1)} - P_4 e^{-\lambda_2(x-L_1)} \right), \quad \text{for R-II} \qquad (8.20)$$

Figure 8.13 shows the variation in the electric field of the UDG DMFET along its channel. It is observed that binding of the biomolecules does not change the electric field in the drain side of the UDG DMFET. Thus, no hot electron effect happens due

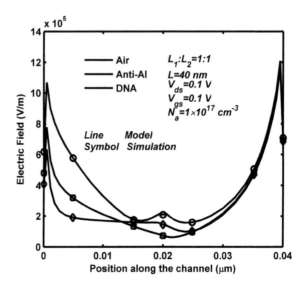

FIGURE 8.13 Effect on the electric field along the channel of UDG DMFET when biomolecules bind when $L_1:L_2=1:1$. UDG DMFET, underlap double-gate dielectrically modulated field-effect transistor.

to the binding of the biomolecules on the device. The figure also shows that binding of the biomolecules changes the electric field in the R–I of the device. The high electric field in the source side also causes the injection of the electrons from the source to channel to increase. Thus, the mobility of the device also increases. The effect of increasing the sensing site to the electric field is shown in Figure 8.14. As more biomolecules are bonded, a greater shift in the electric field is produced. But increasing the sensing site still does not cause any hot electron effect.

8.4.1.3 Threshold Voltage

The threshold voltage (V_{th}) is defined as the value of V_{gs} where the minimum surface potential is equal to twice the fermi potential [38]. Its value can be determined by sloving the following equation:

$$\phi_{fsj}\left(x_{min}\right)\Big|_{V_{gs}=V_{th}} = 2\psi_F \qquad (8.21)$$

where x_{min} and ψ_F are the minimum potential location and fermi potential.

In a MOSFET with a multiple-gate electrode, the higher work function determines the location of the minimum potential [39,40]. But, in this study, there is a variation in the electrical parameters due to the biomolecules binding. Thus, threshold voltage has been a model for the two regions.

The minimum potential location is determined by solving the following equation:

$$\frac{d\phi_{fs,i}\left(x\right)}{dx}\Bigg|_{x=x_{min}} = 0 \qquad (8.22)$$

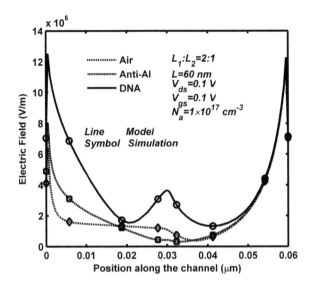

FIGURE 8.14 Effect on the electric field along the channel of UDG DMFET when biomolecules bind when $L_1 : L_2 = 2:1$. UDG DMFET, underlap double-gate dielectrically modulated field-effect transistor.

Substituting x_{min} value in Equation (8.21) and then solving the equation, V_{th} is found as

$$V_{th} = \frac{C_2 + \sqrt{C_2^2 - 4C_1 C_3}}{2C_1}, \quad \text{for R-I} \tag{8.23}$$

$$V_{th} = \frac{C_5 + \sqrt{C_5^2 - 4C_4 C_6}}{2C_4}, \quad \text{for R-II, for R-II} \tag{8.24}$$

Figure 8.15 shows the effect of binding the anti-AI and DNA on the threshold voltage of the UDG DMFET. It is observed that binding of biomolecules anti-AI and DNA changes the threshold voltage. Both anti-AI and DNA have negative charges. The negative charge of the biomolecules hinders the charge inversion in the channel; thus the threshold voltage increases. Their dielectric values are higher than unity (air) which helps the molecules to oppose the external electric filed, which tries to align the molecules. Thus, the higher the negative charge and dielectric value, the larger the threshold voltage.

8.4.1.4 Sensitivity

The sensitivity of FET-based biosensor is defined as the change in the threshold voltage of the device due to the biomolecules binding [41]. It is represented as

$$\Delta\Delta V_{th} = V_{th}|_{\text{biomolecule}} - V_{th}|_{\text{no biomolecule}} \tag{8.25}$$

Figure 8.16 presents the sensitivity of the UDG DMFET as a biosensor. UDG DMFET has a sensitivity of 200 mV when bonded with anti-AI and of 440 mV when

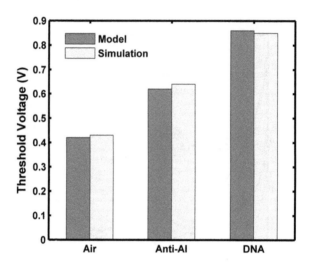

FIGURE 8.15   Threshold voltage of UDG DMFET with and without binding of biomolecules. UDG DMFET, underlap double-gate dielectrically modulated field-effect transistor.

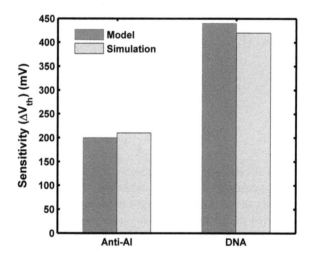

FIGURE 8.16   Sensitivity of UDG DMFET in detecting anti-AI and DNA. UDG DMFET, underlap double-gate dielectrically modulated field-effect transistor; anti-AI, antibody of avian influenza.

bonded with DNA. Thus, enough shift in threshold voltage is produced when biomolecules are bonded to the UDG DMFET. It shows that UDG DMFET is a very option to choose for the detection of such biomolecules. On the top of having high sensitivity or massive threshold voltage shift, binding of biomolecules produces extra short-channel effect (SCE). Thus, such FET-based biosensors provide outstanding performance for detection of biomolecules, which provide high sensitivity, lower cost, label-free detection, and faster processing.

8.5 SUMMARY

The application of the FET device as a biosensor is presented in this chapter. Different types of FET-based biosensor are reviewed to ingrain with the basic and the available structures. This chapter focuses on UDG DMFET, as it has a simple structure, good electrical characteristics, and high structural stability. The study has been made based on the modeling of the device. All the model data have been verified by the simulation results of ATLAS (SILVACO). Electrical parameters such as surface potential, electric field, threshold voltage, and sensitivity are discussed in detail to understand the effect of biomolecules binding on the electrical characteristics of the device. The higher the value of dielectric and charge of the device, the higher the change. The binding of the biomolecules causes no hot electron effect, as the electric field toward the drain is not affected by the binding. Thus, UDG DMFET is an excellent choice in the detection process of biomolecules, as it has high sensitivity as well as low SCEs. The low supply voltage and size is a factor that contributes immensely in lowering the power consumption of the device.

REFERENCES

1. Koyun, A., Ahlatcolu, E., Koca, Y., & Kara, S. (2012). Biosensors and their principles. A Roadmap of Biomedical Engineers and Milestones, IntechOpen, London, pp. 117–142.
2. Thévenot, D. R., Toth, K., Durst, R. A., & Wilson, G. S. (2001). Electrochemical biosensors: recommended definitions and classification. *Analytical Letters*, *34*(5), 635–659.
3. Clark Jr, L. C., & Lyons, C. (1962). Electrode systems for continuous monitoring in cardiovascular surgery. *Annals of the New York Academy of Sciences*, *102*(1), 29–45.
4. Roglic, G., & World Health Organization (Eds.). (2016). *Global Report on Diabetes*. Geneva: World Health Organization.
5. Malima, A., Siavoshi, S., Musacchio, T., Upponi, J., Yilmaz, C., Somu, S., & Busnaina, A. (2012). Highly sensitive microscale in vivo sensor enabled by electrophoretic assembly of nanoparticles for multiple biomarker detection. *Lab on a Chip*, *12*(22), 4748–4754.
6. Koyun, A., Ahlatcolu, E., Koca, Y., & Kara, S. (2012). Biosensors and their principles. A Roadmap of Biomedical Engineers and Milestones, IntechOpen, London, pp. 117–142.
7. Patel, S., Nanda, R., Sahoo, S., & Mohapatra, E. (2016). Biosensors in health care: the milestones achieved in their development towards lab-on-chip-analysis. *Biochemistry Research International*, 1, 1–12.
8. Pohanka, M. (2018). Overview of piezoelectric biosensors, immunosensors and DNA sensors and their applications. *Materials*, *11*(3), 448.
9. Ivnitski, D., Abdel-Hamid, I., Atanasov, P., & Wilkins, E. (1999). Biosensors for detection of pathogenic bacteria. *Biosensors and Bioelectronics*, *14*(7), 599–624.
10. Park, S. C., Cho, E. J., Moon, S. Y., Yoon, S. I., Kim, Y. J., Kim, D. H., & Suh, J. S. (2007). A calorimetric biosensor and its application for detecting a cancer cell with optical imaging. In *World Congress on Medical Physics and Biomedical Engineering* 2006 (pp. 637–640). Springer, Heidelberg.
11. Hundeck, H. G., Weiss, M., Scheper, T., & Schubert, F. (1993). Calorimetric biosensor for the detection and determination of enantiomeric excesses in aqueous and organic phases. *Biosensors and Bioelectronics*, *8*(3–4), 205–208.
12. Choi, J. M., Han, J. W., Choi, S. J., & Choi, Y. K. (2010). Analytical modeling of a nanogap-embedded FET for application as a biosensor. *IEEE Transactions on Electron Devices*, *57*(12), 3477–3484.

13. Chakraborty, A., & Sarkar, A. (2017). Analytical modeling and sensitivity analysis of dielectric-modulated junctionless gate stack surrounding gate MOSFET (JLGSSRG) for application as biosensor. *Journal of Computational Electronics*, *16*(3), 556–567.

14. Im, H., Huang, X. J., Gu, B., & Choi, Y. K. (2007). A dielectric-modulated field-effect transistor for biosensing. *Nature Nanotechnology*, *2*(7), 430–434.

15. Kimura, J., Ito, N., Kuriyama, T., Kikuchi, M., Arai, T., Negishi, N., & Tomita, Y. (1989). A novel blood glucose monitoring method an ISFET biosensor applied to trans-cutaneous effusion fluid. *Journal of the Electrochemical Society*, *136*(6), 1744–1747.

16. Palan, B., Santos, F. V., Karam, J. M., Courtois, B., & Husak, M. (1999). New ISFET sensor interface circuit for biomedical applications. *Sensors and Actuators B: Chemical*, *57*(1–3), 63–68.

17. Bergveld, P. (1986). The development and application of FET-based biosensors. *Biosensors*, *2*(1), 15–33.

18. Ahn, J. H., Choi, S. J., Han, J. W., Park, T. J., Lee, S. Y., & Choi, Y. K. (2010). Double-gate nanowire field effect transistor for a biosensor. *Nano letters*, *10*(8), 2934–2938.

19. Zhang, G. J., & Ning, Y. (2012). Silicon nanowire biosensor and its applications in disease diagnostics: a review. *Analytica Chimica Acta*, *749*, 1–15.

20. Iijima, S. (1991). Helical microtubules of graphitic carbon. *Nature*, *354*(6348), 56.

21. Yang, N., Chen, X., Ren, T., Zhang, P., & Yang, D. (2015). Carbon nanotube based biosensors. *Sensors and Actuators B: Chemical*, *207*, 690–715.

22. Gupta, S., Murthy, C. N., & Prabha, C. R. (2018). Recent advances in carbon nanotube based electrochemical biosensors. *International Journal of Biological Macromolecules*, *108*, 687–703.

23. Tran, T. T., & Mulchandani, A. (2016). Carbon nanotubes and graphene nano field-effect transistor-based biosensors. *TrAC Trends in Analytical Chemistry*, *79*, 222–232.

24. Gu. B., Park, T. J., Ahn, J. H., Huang, X. J., Lee, S. Y., & Choi, Y. K. (2009). Nanogap field-effect transistor biosensors for electrical detection of avian influenza. *Small*, *5*(21), 2407–2412.

25. Baba, T. (1992). Proposal for surface tunnel transistors. *Japanese Journal of Applied Physics*, *31*(4B), L455.

26. Sarkar, D., & Banerjee, K. (2012). Proposal for tunnel-field-effect-transistor as ultra-sensitive and label-free biosensors. *Applied Physics Letters*, *100*(14), 143108.

27. Narang, R., Saxena, M., & Gupta, M. (2015). Comparative analysis of dielectric-modulated FET and TFET-based biosensor. *IEEE Transactions on Nanotechnology*, *14*(3), 427–435.

28. Xu, H. F., Dai, Y. H., Gui Guan, B., & Zhang, Y. F. (2017). Two-dimensional analytical model for asymmetric dual-gate tunnel FETs. *Japanese Journal of Applied Physics*, *56*(1), 014301.

29. Dwivedi, P., & Kranti, A. (2016). Applicability of transconductance-to current ratio (gm/Ids) as a sensing metric for tunnel FET biosensors. *IEEE Sensors Journal*, *17*(4), 1030–1036.

30. Lee, C. W., Afzalian, A., Akhavan, N. D., Yan, R., Ferain, I., & Colinge, J. P. (2009). Junctionless multigate field-effect transistor. *Applied Physics Letters*, *94*(5), 053511.

31. Ajay, Narang, R., Saxena, M., & Gupta, M. (2015). Investigation of dielectric modulated (DM) double gate (DG) junctionless MOSFETs for application as a biosensors. *Superlattices and Microstructures*, *85*, 557–572.

32. Wagaj, S. C., Patil, S., & Chavan, Y. V. (2018). Performance analysis of shielded channel double-gate junctionless and junction MOS transistor. *International Journal of Electronics Letters*, *6*(2), 192–203.

33. Tripathi, S. L., Mishra, R., & Mishra, R. A. (2012). Multi-gate MOSFET structures with high-k dielectric materials. *Journal of Electron Devices*, *16*, 1388–1394.

34. Zhang, X. Y., Hsu, C. H., Cho, Y. S., Lien, S. Y., Zhu, W. Z., Chen, S. Y., ... & Huang, S. X. (2017). Simulation and Fabrication of HfO2 Thin Films Passivating Si from a Numerical Computer and Remote Plasma ALD. *Applied Sciences*, *7*(12), 1244.

35. Ahn, J. H., Choi, S. J., Im, M., Kim, S., Kim, C. H., Kim, J. Y.,& Choi, Y. K. (2017). Charge and dielectric effects of biomolecules on electrical characteristics of nanowire FET biosensors. *Applied Physics Letters*, *111*(11), 113701.

36. Cuervo, A., Dans, P. D., Carrascosa, J. L., Orozco, M., Gomila, G., & Fumagalli, L. (2014). Direct measurement of the dielectric polarization properties of DNA. *Proceedings of the National Academy of Sciences*, *111*(35), E3624–E3630.

37. Narang, R., Saxena, M., Gupta, R. S., & Gupta, M. (2011). Dielectric modulated tunnel field-effect transistor—A biomolecule sensor. *IEEE Electron Device Letters*, *33*(2), 266–268.

38. Tiwari, P. K., Dubey, S., Singh, M., & Jit, S. (2010). A two-dimensional analytical model for threshold voltage of short-channel triple-material double-gate metal-oxide-semiconductor field-effect transistors. *Journal of Applied Physics*, *108*(7), 074508.

39. Kumar, M. J., & Chaudhry, A. (2004). Two-dimensional analytical modeling of fully depleted DMG SOI MOSFET and evidence for diminished SCEs. *IEEE Transactions on Electron Devices*, *51*(4), 569–574.

40. Goel, E., Kumar, S., Singh, K., Singh, B., Kumar, M., & Jit, S. (2016). 2-D analytical modeling of threshold voltage for graded-channel dual-material double-gate MOSFETs. *IEEE Transactions on Electron Devices*, *63*(3), 966–973.

41. Singh, K. N., & Dutta, P. K. (2019, March). Comparative analysis of underlapped silicon on insulator and underlapped silicon on nothing dielectric and charge modulated FET based biosensors. In 2019 Devices for Integrated Circuit (DevIC), Kalyani, India (pp. 231–235). IEEE Conference Proceedings.

9 Low-Power FET-Based Biosensors

Prasantha R. Mudimela and Rekha Chaudhary
Lovely Professional University

CONTENTS

9.1 Introduction 169
9.2 Principle of Operation 170
9.3 Silicon Nanowire Biosensor 173
9.4 Organic Field-Effect Transistor 174
9.5 Classification and Advances in Bio-FETs 176
9.6 ImmunoFET 179
9.7 Cell-Based Bio-FET 180
9.8 Conclusions 183
References 184

9.1 INTRODUCTION

A biosensor, defined by the International Union of Pure and Applied Chemistry (IUPAC), is a device that uses specific biochemical reactions mediated by isolated enzymes, immune systems, tissues, organelles, or whole cells to detect chemical compounds by using electrical, thermal, or optical signals [1]. FET (field-effect transistor)-based biosensor (Bio-FET) is an electrically and chemically insulating layer that separates the analyte solution from the semiconducting device. The first Bio-FET developed by Piet Bergveld was the ion-sensitive field-effect transistor (ISFET) used for electrochemical and biological applications in 1970. The Bio-FET is a field-effect transistor (metal oxide semiconductor field-effect transistor [MOSFET] based) that is gated by variations in the surface potential induced by molecules binding. When charged molecules, such as biomolecules, bind to the FET gate, which is usually a dielectric material, they can change the charge distribution of the underlying semiconductor material resulting in a change in conductance of the FET channel. As shown in Figure 9.1, Bio-FET consists of two main compartments: (i) the biological recognition element and (ii) the FET. Its construction is basically centered on the ISFET, a type of MOSFET where the metal gate is substituted by a membrane, solution of electrolyte, and reference electrode.

In other words, biosensors are defined as an operative tool that is used to study the biomolecular interactions such as DNA hybridization, antibody–antigen interactions, protein–protein interactions, receptor–ligand binding, DNA–protein binding,

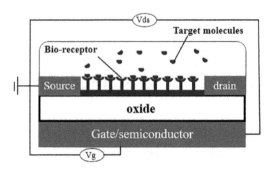

FIGURE 9.1   Side view of Bio-FETs. Bio-FET, field-effect transistor–based biosensor.

and other types of interaction [2–6]. A basic biosensor structure consists of three segments: biorecognition element, a transducer, and a signal-processing unit. The biorecognition element senses the biological interactions; the transducer converts the response into electrical signal; and the signal-processing unit processes and amplifies the signals. Clark and Lyon reported the very first biochemical-based biosensors in 1962 [7]. After that, a lot of different biosensors have been reported. As the biosensor's development has emerged, an interesting sensing approach has been proposed, i.e., Bio-FETs. Due the fast development in the solid-state technologies, Bio-FETs have come into the picture and have drawn the attention of researchers in sensing mechanism. Bio-FETs are most suitable candidates for applications that demand very high sensitivity, very fast response time, and mass production with low cost.

9.2   PRINCIPLE OF OPERATION

The basic principle of operation of Bio-FET is the detection of changes in potential due to binding of analyte. To detect nucleic acids and proteins, the Bio-FETs couple a transistor device with a biosensitive layer. The Bio-FET system comprises a FET that acts as a transducer separated by an insulator layer of SiO_2 from the biological recognition element which is selective to the target molecule called analyte. Once the analyte binds to the recognition element, the charge dissemination at the surface changes relative to the change in the electrostatic surface potential of the semiconductor. This change in the surface potential of the semiconductor acts like a gate voltage, which would result in a traditional MOSFET. The change in current can be measured; thus, the binding of the analyte can be sensed.

> **Applications:** Bio-FETs are important for detection in the field of medical diagnostics. It has been utilized in the various real-life applications of biology research, environmental safety, and food safety analysis. Bio-FETs have competitive advantages over optical methods as they consume less time and are less expensive and companionable to real-time monitoring.
>
> **FET, a Bio-FET's Basic Structure:** A FET is a three-terminal device consisting of source, gate, and drain terminals. The current conduction occurs

from source to drain through a channel when gate voltage is applied at gate terminal. The polarity gate voltage switches the FET between on and off state. The current conduction from source to drain occurs either due to electrons or due to holes. In n-type FET, the applied gate voltage will drive the flow of electrons to pass from source to drain. If applied positive gate voltage is increased, the current conduction from source to drain increases significantly [8]. But if negative gate voltage is applied, then channel formation will pinch off. On the other hand, in p-type FET, when negative voltage is applied at gate terminal, channel formation takes place from source to drain. But positive gate voltage will turn off the device [9].

FET-based sensors consist of a MOS (metal oxide semiconductor) structure. The change in the metal potential (ψ_m) induces the band bending of the semiconductor channel accordingly. It results in three different kinds of phenomena such as accumulation, depletion, and inversion. In MOS, the applied gate voltage turns on or off the device. When the applied voltage is greater than threshold voltage $(V_g > V_{th})$, the device turns on. This gate voltage can be applied either directly or by different factors such as solution potentials or charge of biomolecules. The concentration of carriers is affected by these factors, and according to that, the current–voltage characteristics shift positively and negatively. The difference in the threshold voltage presents the change in characteristics and can be defined as indicator of sensitivity.

The electrolyte–insulator–semiconductor (EIS) model is same as the MOS model; the only difference is that gate electrode is replaced by electrolyte. The biomolecular reactions in the electrolyte change the charge concentration on surface; thus, a potential difference is created. Thus, the inversion phenomenon occurs. The solid–liquid interface behavior is explained by electrical double-layer model [10]. In this model, ions present in the electrolyte/solution are divided into three different layers: stern layer, diffuse layer, and bulk solution. The diffuse layer is also known as Gouy–Chapman double layer or electric double layer. The Gouy–Chapman model states that in electrolyte, charges are distributed toward the bulk electrolyte. The electrostatic potential drop shown in Figure 9.2 can be categorized as follows:
1. Potential drop electrode–electrolyte interface
2. Potential drop at electrolyte–insulator interface
3. Potential drop within insulator
4. Potential drop within depletion charges in semiconductor

The operation of Bio-FETs can be divided into three stages: (i) the change in charge on the interface of sensor is due to change in concentration in electrolyte; (ii) this change in charge leads to the variation in effective gate voltage; and (iii) the variation in effective gate voltage results in change in drain current. The overall sensitivity depends upon all the three factors mentioned earlier.

ISFET, a pH Sensor: ISFET (Ion sensitive field effect transistor) was first proposed by Bergveld in 1970 to detect pH values of solutions [11]. ISFET

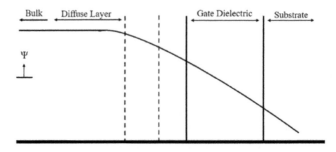

FIGURE 9.2   Potential distribution at interface of electrolyte–insulator semiconductor.

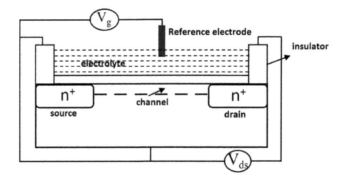

FIGURE 9.3   Schematic diagram of ISFET. ISFET, ion-sensitive field-effect transistor.

has same structure as MOSFET, but in ISFET, the metal gate is replaced by dielectric layer, electrolyte, and reference electrode. The dielectric layer acts as a sensing membrane. The schematic diagram of ISFET is shown in Figure 9.3. The capacitance of dielectric defines the sensitivity of sensor. The dangling bonding sites available on dielectric surface determine the charge concentration on the surface. The use of high-k dielectric has been done to study the sensing mechanism [12, 13]. The high-k material improves sensitivity and prevents drifting. The voltage is applied at reference electrode. The potential at reference electrode will break the electrolyte/pH into ions (H^+ or OH^-). The dangling bonds or hydroxyl groups present at dielectric will attract the ions due to which charge is formed at dielectric surface and current conduction occurs in the channel. The hydroxyl groups can donate a proton or accept a proton. The schematic of site-binding model of SiO_2 is shown in Figure 9.4. The surface hydrolysis bond of Si–OH differs in various solutions because of different pH values. Nernst limit is defined as the maximum sensitivity of ISFET, i.e., 59.2 mV/pH.

Dual-Gate ISFET (DG-ISFET): DG-ISFET consists of have a back gate on the other side with respect to active channel. The literature shows that the sensitivity of DG-ISFET can go beyond Nernst limit [14–18]. The

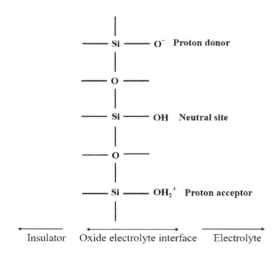

FIGURE 9.4   Site-binding model for SiO_2.

sensitivity is defined as the shift in the threshold voltage due to change in pH ($\Delta Vt/\Delta pH$). Using silicon-on-insulator substrate having high-k dielectric, DG-ISFET gives very high sensitivity.

ISFET Bioapplications: ISFET is used to detect the protein concentration [19–23], DNA detection [13,24–26], genetic diagnosis, DNA sequencing [27], and DNA amplification test [28].

9.3   SILICON NANOWIRE BIOSENSOR

Nanowire FETs are a kind of nanowire sensor. In standard FET, the variation in applied external voltage causes the variation in electric potential. In the same way, chemical/biological species can also change the potential, which in turn alter the conductivity by binding the species at gate terminal. In nanowire FETs, nanowires are used instead of doped channel. Cui et al. reported the first silicon nanowire–based (SiNW) biosensor [29]. A variety of research studies on SiNW have been done that report the usage of sensors in applications such as detection of nucleic acid [30–34], proteins [9,35–37], and virus [2]. In SiNW, a wire-like structure is taken as channel based on polysilicon. In Figure 9.5, the schematic diagram is given which shows the source and drain electrodes [4]. Resistor type and transistor type are the two main classes of SiNW sensor. When the doping concentration is high, SiNW behaves as a resistor. The change in conductance of resistor depends upon change in the number of charged analytes on sensor surface. The electrostatic forces are responsible for change in conductivity. Figure 9.5 shows the working principle of SiNW sensor. The high doping reduces the resistance, which in turns lowers the sensitivity [38]. In SiNW, the surface-to-volume (S/V) ratio is high when compared with other planar structure. The working principle of nanowire FET sensor resembles that of the typical FET-based sensor. In Figure 9.6, working of n-type SiNW-based sensor is shown. The middle one is the actual state of SiNW-based biosensor. Whereas the first state

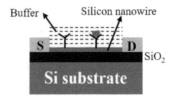

FIGURE 9.5   Silicon nanowire biosensor with applilied bias voltages.

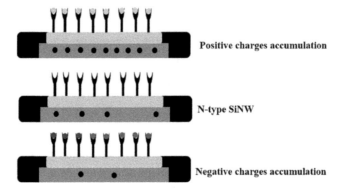

FIGURE 9.6   n-type silicon nanowire biosensor.

shows the accumulation of positive charges, the last state shows the accumulation of negative charges. The sensitivity of nanowire FET sensor is high because of channel confinement effect [39]. If the nanowire diameter decreases in nanoscale, the S/V ratio increases rapidly. The sensitivity of device is dominated by high S/V ratio.

> **Fabrication Process**: Two different methods are used for the fabrication of SiNW biosensor, i.e., bottom-up and top-down. In bottom-up method [29], chemical vapor deposition was used to grow SiNW on substrate. But in top-down method [40], SiNW is formed by lithography process. For nanopatterning and reactive ion etching, e-beam lithography is used to create a three-dimensional wire structure. In the created wire structure, light doping is done in the sensing region, and heavy doping is done in the source/drain contact. Top-down method is mostly used due to its compatibility with standard CMOS process.
>
> **Applications**: This CMOS SiNW has been used to detect hepatitis B virus DNA [41], cardiac troponin I [42], and NT-proBNP [43] in clinical sample successfully.

9.4   ORGANIC FIELD-EFFECT TRANSISTOR

Organic material has been used since the past few decades and has emerged as a powerful material in microelectronics applications. Organic FET (OFET) is same as that of thin film transistor in which organic semiconductor is used as a semiconductor. This

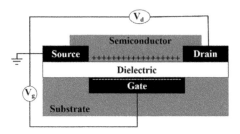

FIGURE 9.7   Silicon nanowire biosensor.

organic semiconductor layer is used as a receptor in sensing as shown in Figure 9.7. The receptor layer has its own electronics and mechanical properties. The material selection categorizes it into p-type or n-type. OFET works in accumulation mode [44]. The channel current is controlled by gate voltage. The mobility of OFET (10^{-1} to $10^{-2} cm^2/Vs$) is lesser when compared with the mobility of silicon. But the organic semiconductor is vastly well suited with flexible substrate. OFET has been used for various sensing applications such as chemical sensing, biological sensing, and humidity sensing. On the basis of structure of the device, OFET can be categorized as follows:

1. Bottom gate/bottom contact (BGBC)
2. Bottom gate/top contact (BGTC)
3. Top gate/bottom contact (TGBC)
4. Top gate/top contact (TGTC)

Applications: OFET sensors can be used to detect dangerous and explosive gases [45], for volatile organic compounds [46], and for ionizing radiation dosimetry [47] [48].

Graphene FET (GFET) Biosensors: Graphene has attracted the attention of researchers since 2000 due to its various electrical and mechanical properties. Since graphene is a zero-bandgap material, electrons and holes can easily be generated in the electric field. Such an effect is known as ambipolar effect in which drain current is generated by both positive and negative voltages. When minimum drain current value is achieved, the corresponding gate voltage is called charge neutral point. In this case, undoped graphene is taken. Single gate or double gate can be used for operation of GFET. In bottom-gate GFET, voltage applied at the back side controls the channel. As the bottom-gate GFET is dipped in electrolyte, the electric double layer appears at top gate. When the voltage is changed in electrolyte, the top gate shows the ambipolar behavior. The disturbance in electrolyte varies the surface potential and makes GFET a promising device for biosensing. The schematic diagram of GFET biosensor is shown in Figure 9.8.

Applications: GFET biosensors have been used for various applications such as detection of DNA [49–53], protein specific recognition [54–56], glucose [57], and living cells [58].

Depending upon the biorecognition element, Bio-FETs can be categorized as shown in Figure 9.9

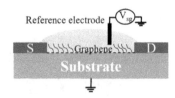

FIGURE 9.8 Graphene FET biosensor. FET, field-effect transistor.

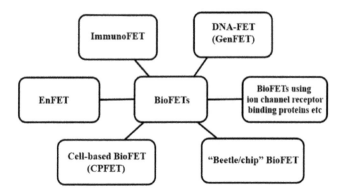

FIGURE 9.9 Bio-FET classification. Bio-FET, field-effect transistor–based biosensor.

9.5 CLASSIFICATION AND ADVANCES IN BIO-FETs

Enzyme FET (EnFET): Janata and Caras reported the first penicillin-sensitive biosensor [59] in 1980. In EnFET, ISFET structure is used in which enzymes are immobilized over gate surface. The deposition of enzyme layer on the gate insulator is very crucial. The immobilization of enzymes can be done by several methods such as adsorption, entrapment, covalent binding, cross-linking, and so on [60–63]. The techniques named drop-on, spin coating, and dip coating are the simplest methods that have been used for deposition of enzyme membrane. A very high adhesion is needed between transducer and enzyme membrane. To improve the adhesion, a process called surface salinization is usually done.

Working: In EnFET, when the analyte is in contact with the enzyme layer, two possibilities are there: either the products are generated or the reactants are consumed. The underlying ISFET structure monitors the variation in concentration. The obtained ISFET signal is compared with original concentration. In Figure 9.10, a penicillin-sensitive EnFET with Ta_2O_5 gate is demonstrated. In EnFET, the variation in pH is caused by the enzymatic reactions. The hydrolysis of penicillin is catalyzed by enzyme penicillinase, which changes the pH over the gate terminal of ISFET. The variation in output is calculated by the amount of penicillin present in the solution. For any kind of enzyme, such an EnFET structure can be created. But actually, only a few enzymes are present that are capable to produce or devour the electrochemically active species.

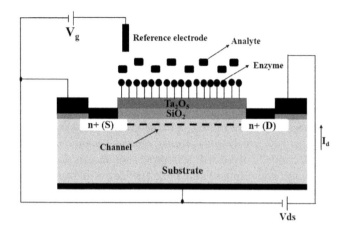

FIGURE 9.10 Schematic diagram of EnFET. EnFET, enzyme field-effect transistor.

TABLE 9.1
Major Benchmarks in Bio-FET Device

Year	Concept Given/Structure Proposed
1970	ISFET
1976	Bio-FET
1980	EnFET
1980	ImmunoFET
1981	Coupling of cells with MOS
1991	CPFET
1997	Beetle/chip Bio-FET
1997	DNA hybridization

ISFET, ion-sensitive field-effector transistor; Bio-FET, field-effect transistor–based biosensor; EnFET, enzyme field-effect transistor; MOS, metal oxide semiconductor.

With different sensors design, a plethora of EnFETs are available nowadays for the detection of analytes such as glucose, urea, penicillin, ethanol, lactose, sucrose, maltose, ascorbic acid, lactate, and so on [64,65]. In Table 9.1, developed EnFETs are summarized.

Different Improvement Techniques: To improve the characteristics of EnFETs, immense efforts have been done. But still the practical use of EnFET is restricted. The restrictions are mainly faced in the operating principle. Also, there are some other factors such as (i) the sensor output dependency on buffer capacity, available sites, and pH, (ii) high detection limit value, (iii) less response time, and (iv) light sensitivity. To solve the issues in the working of EnFETs, a lot of solutions have been proposed. If an additional charged polymeric membrane like Nafion is deposited over enzyme membrane, some of the problems can be removed. The charged polymeric membrane controls the diffusion between substrate and product [66–68]. These methods minimize the effect of salt concentration on sensor output and increase the sensitivity of glucose- [66] and urea-sensitive EnFETs [67]. But this approach has not been used universally. In urea-sensitive EnFET, a structure was proposed where the pH was kept at a predetermined constant value. Thus, the linearity of EnFET is improved [69]. To solve the issues in case of glucose-sensitive EnFET, a Pt electrode is deposited over gate terminal. In this case, electrolysis process generates additional hydrogen ions when Pt electrode is deposited. It improves the factors such as sensitivity, dynamic range, and response time [70]. The idea of in situ electrochemical generation of ions was came into the picture for the improvement of recovery time of glucose-sensitive EnFET [71]. The use of monolayer/multilayer enzyme in Bio-FETs was developed to minimize the response time. The thin-layered enzyme sensors have fast response time. But in such cases, they have low lifetime and less stability.

To create a miniaturized array of EnFETs, a lot of efforts have been done by the researchers. An array of EnFETs helps in the measurements of different electrolyte at the same time. An array of EnFETs has developed to measure glucose, ascorbic acid, and citric acid simultaneously [72]. Another array of EnFETs was developed to measure four different kinds of analytes [73]. The main hurdle in the array concept is cross-sensitivity. Since the between the different sensors is less. So the electrochemical reaction on one enzyme membrane can affect the reactions on other membrane. So the total output signal can vary. This problem of cross-sensitivity can be removed by optimizing the pH value for all the enzyme membranes [74].

The deposition of enzyme layer is manually done. Although this process is simple, it consumes a lot of time. Some industries have launched EnFET (in the form of pH ISFET) using silicon-integrated technology. Different masks are used to implement multi biosensor. Although the fabrication is quite simple, the most important factor is encapsulation and packaging of device. As the device works in electrolyte, it needs encapsulation on selected portion of the device while the sensor should be kept exposed. To get the desired performance of the device, the full automatization of fabrication should be done. In device fabrication process, although encapsulation and packaging are the main crucial steps, still these processes are done manually.

Applications: EnFETs can be used in applications such as determination of glucose in blood serum [75], urine [76], urea [77], and so on.

DNA-modified FET or GenFET: For DNA detection, hybridization process is used in which the probe molecule is used to identify the single-stranded DNA (ssDNA). When ssDNA comes in contact with probe molecule, it forms double-stranded DNA (dsDNA) structure. In the past, various methods using different types of transducers have been reported. In Figure 9.11, a GenFET is shown. In GenFET, the sequences of ssDNA are immobilized onto a transducer, which converts them into an output signal. The

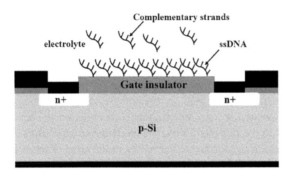

FIGURE 9.11 Schematic diagram of GenFET. GenFET, DNA-modified FET; ssDNA, single-stranded DNA.

miniaturization of FET devices and their advanced fabrication technologies makes the FET devices suitable for DNA detection. Such a device is called GenFET. But only limited work about ISFET for DNA sensing has been done. A fast, cheap, and disposable GenFET is always required. The first GenFET was proposed in 1997 by Souteyrand et al. [62]. A small amount of complementary DNA solution was added to buffer electrolyte for the detection of real-time hybridization process. During hybridization process, at gate voltage, some changes are observed due to change in surface charge. The practical realization of GenFET is actually problematic because the process of binding is affinity binding and charged molecules (DNA) are used. The ions present in the electrolyte screen the charge of macromolecules, i.e., DNA. But still there are points that lead to the working of GenFET. Some of them are mentioned in the following:

1. In DNA structure, the charges are evenly distributed on the surface. The inorganic counterions of electrolyte neutralize the negative charges.
2. In double layer, the charge density distribution is affected by hybridization process.
3. Both DNA immobilization and hybridization are affected when the surface is charged.

Applications: GenFET can be used in the biorecognition of DNA and RNA [65, 78].

9.6 IMMUNOFET

The enhanced proclivity to utilize the high specificity and sensitivity of biomolecules and living biological systems for sensor function has been the major force for the increased use of the biosensor in numerous real-life situations. The collaboration of antibody–antigen results in the ability to recognize the biological molecules. The antibody is an intricate biomolecule containing hundreds of individual amino acids. Structure of these amino acids is well organized. The antigen could be a macromolecule structure which is generated in response with the antibody. This generation of antigen against antibody structure happens due to a defensive mechanism of the organism. Proteins with molecular weights larger than 5000 Daltons are normally immunogenic that are susceptible to being recognized. The minute change in the chemical modification of the molecular structure of the antigen leads to a tremendous reduction in its affinity toward the original antibody. So immune sensors are useful for measuring the wellness of the human immune system that subsequently leads to a superb diagnostic tool in the field of medicine.

The increased possibility for monitoring the direct relationship between the antibody and antigen leads to the pursuits toward the modeling of biosensors, as the actual recognition takes place at the surface of the sensor in the layer of antibodies. The concept of the direct immunizing ISFET was developed by Schenck in 1978. In immunoFET, the gate is modified by immobilizing the antibodies or antigens mostly in the structured membrane as shown in Figure 9.12. The formation of the antibody and antigen pattern on the gate of the ISFET results in the detectable change in the charge distribution. This consequently leads to the direct modulation of the direct current of the ISFET. The unique possibility of the direct detection of immune logical reactions has been realized through the application of the immunoFET. Under the ideal conditions, the immunoFET is at least theoretically capable to measure the concentration of immune molecules with a very low detection limit. The ideal conditions required here are a truly capacitive interface that leads to immobilization of immunological binding sites, complete coverage of the antibody, highly charged antigen, and comparatively low ionic strength.

The measure problem observed in the immunoFET is to transmit or transduce the recognizable molecular action between the antibody and antigen into a signal that can be measured. It has been observed that potential charge distribution plays a significant role in the transfer of immunological signals to the ISFET.

Applications: ImmunoFET is used for the detection of antibody–antigen binding [79] and recognition of immunological molecules.

9.7   CELL-BASED BIO-FET

In cell-based Bio-FETs, cells are used as recognition element. Cells are the smallest biological entity. In this structure, the cell is in direct contact with the

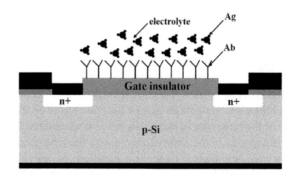

FIGURE 9.12   Schematic structure of an immunoFET with immobilized antibody (Ab) molecules. Ag, antigen molecules.

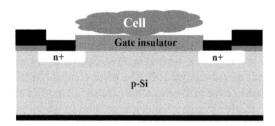

FIGURE 9.13 Schematic structure of cell-based Bio-FET. Bio-FET, field-effect transistor–based biosensor.

gate insulator of FET. Figure 9.13 shows the side view of a cell -transistor. Cells are actually the living microstructures that consist of chemicals, ions, enzymes, different kinds of proteins, and organic molecules in high concentration. Cells are capable of processing different received signals via activation in a parallel manner of various signal pathways. Although there are some problems with the life span of cells, the motivation behind developing a Bio-FET is that living components can directly respond to received information [80, 81]. Such information along with some additional information is very useful in clinical diagnosis and many other fields. Using cell-based Bio-FETs, one can easily study the effects of different compounds, pollutants, toxic materials, and so on on cellular metabolism.

Working: Two different methods are used to find out the state of cells. The first method uses the energy of cell metabolism. The ISFET-based chemical sensors are used to measure the variation in extracellular pH, ion concentration, and so on. In cell-based Bio-FETs, Si_3N_4 or Al_2O_3 is used as a gate insulator for extracellular acidification measurements [82–84]. In the measurement of extracellular acidification, the working of cell-based Bio-FET is explained as follows: In a chamber having an array of cell-based Bio-FETs, measurements of rate of acidification are done generally. A proton amount of 10^8 protons/s is produced by a single cell in steady-state conditions. This quantity can be increased to high value if external receptor stimulation is done. But that also depends upon type of cell and the receptor used. When the pump gets stopped, the protons that are generated get accumulated and the extracellular medium gets acidified. The rate of change of extracellular acidification is detected by the ISFET. The high sensitivity can be achieved if the volume of chamber is small. But the confinement of produced protons to small volume–sized chamber is still challenging. Different signaling pathways are parallelly activated. The integral of acidification rate is not enough to get the information about the effects of different parameters on cells. So a method has been reported in which both acidification and respiration in a cell has been done by a cell-based Bio-FET. In this, a noble metal electrode that performs reduction of oxygen molecules has been used [85]. The ISFET can be integrated with other microsensors on a single chip.

Other method involves the features of different cells such as muscle cells or cell network. It involves the potential measurement of extracellular and intracellular voltages. To record the extracellular voltage of muscle tissues, ISFET-based sensor was developed in 1970. Such a sensor was used for neurophysiological measurements [18]. The point contact model is the common method that has been used to demonstrate the signal transfer from cells to FET structure and further explains the characteristics of output signal [81]. Some other methods are also reported [86]. But all the methods do not include the effects of ion-sensitive property or pH, but these can play an important role in changing the interfacial potential over the gate surface. Another important parameter is "adhesion." A strong adhesion between neuron and transducer surface is desired. So the specific resistance in the adhesion region was measured [87]. The advancement in the area depends on getting the positive results, repeatability, and making a disposable device structure.

Applications: Cell-based Bio-FETs are used for the detection of cell metabolism and extracellular potential measurement [88, 89].

Beetle/Chip FET: The idea of bioelectronic sensor was proposed by Rechnitz in 1986 in which the complete sensory organ of an insect is taken [90]. A beetle/chip sensor is a kind of biosensor in which an insect antenna is directly coupled to a FET [75]. This approach was investigated and then optimized [91–93]. In this method, when the antenna detects a particular odor, then a voltage is generated in the antenna. Because of the potential developed, the drain current is changed in the transistor. The insect antenna is made up of different parts such as small hairs called sensilla. The sensilla contains neurons in which the recognition of odor is processed. Because of the sensed odor, a potential is developed in the insect antenna. Thus, drain current flows in FET [94–96].

A bioelectronic interface is set up between insect antenna and FET so that the signal generated by the antenna can be measured. There are two methods to couple the antenna and FET [93]. In the first method, the whole beetle is fixed in the cell, and antenna tip is dipped on the electrolyte. So in this way, the antenna is in direct contact with insulator of FET through electrolyte as shown in Figure 9.14. In between the head and neck, the reference electrode like platinum wire is placed.

In the second method, the antenna is separated from the insect and then placed on the electrolyte as shown in Figure 9.15. When the odor is sensed by the antenna, the air current over antenna depolarizes the electrolyte, which in turn changes the conductance of FET between source and drain. The change in conductance changes the drain current in the FET. Practically, a beetle/chip-based FET was developed in which antenna of Colorado potato beetle was taken [94]. The more intense the odor, the stronger the signal was obtained. The changes in drain current define the sensor signal.

Application: Beetle/chip FET is used to find out the right time for pesticide application by the detection of plant damage in potato field [65]; smoldering fires such as burning coal, paper, and wood can be detected by means of monitoring fire-specific odors [95–98].

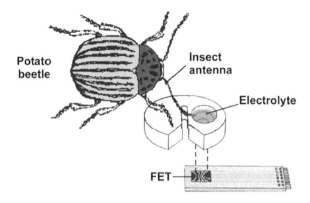

FIGURE 9.14   Whole beetle Bio-FET. Bio-FET, field-effect transistor–based biosensor.

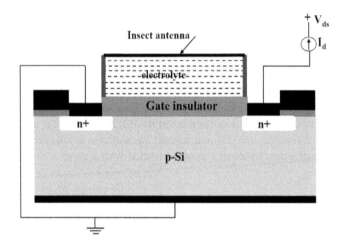

FIGURE 9.15   Isolated antenna Bio-FET. Bio-FET, field-effect transistor–based biosensor.

Table 9.2 summarizes the types of FETs and their applications.

9.8   CONCLUSIONS

In this chapter, the operating principle and application of low-power FET-based bio-sensor is explained. The working of the biosensor is discussed in detail. In future, the Bio-FETs can be developed in areas such as follows: (i) Bio-FETs use biological receptor to bind with particular drug, which leads to change in protein structure; (ii) an array of ISFET can be used to analyze the nucleic acids and can be used in various applications such as clinical diagnosis, environment monitoring, and food industry; the DNA hybridization can be done using ISFET structure; (iii) the formation of large neural network for drug detection using conductive polymer field-effect transistors (CPFET) structure; (iv) the bioanalytical microsystems integrate the Bio-FETs with

TABLE 9.2
Types of Bio-FETs

Type of Bio-FET	Application
EnFET	EnFETs can be used in applications such as determination of glucose in blood serum, urine, urea, and so on
GenFET	Biorecognition of DNA and RNA
ImmunoFET	Detection of antibody–antigen binding, recognition of immunological molecules
Cell based Bio-FET	Detection of cell metabolism and extracellular potential measurement
Beetle/chip FET	To find out the right time for pesticide application by the detection of plant damage in potato field, smoldering fires such as burning coal, paper, and wood can be detected by means of monitoring fire-specific odors

Bio-FET, field-effect transistor–based biosensor; EnFET, enzyme field-effect transistor; GenFET, DNA-modified FET.

some other kinds of sensors; such products are cost-effective; and (v) antennas of insect are very useful when combined with the FET structure. It can be used for applications such as finding out the leakage of gas in a laboratory or smoke. Although a lot of research has been done in low-power-based Bio-FETs, still there is a huge scope for different applications that are still untouched. Remarkable results are obtained with FET-based miniaturized structures when combined with sensor.

REFERENCES

1. PAC, *IUPAC Recommendations* **64**, 143 (1992).
2. F. Patolsky, G. Zheng, O. Hayden, M. Lakadamyali, X. Zhuang, and C. Lieber, *Proc. Natl. Acad. Sci. USA* **101**, 14017 (2004).
3. D. Kim, Y. Jeong, H. Park, J. Shin, P. Choi, J. Lee, and G. Lim, *Biosens. Bioelectron.* **20**, 69 (2004).
4. E. Stern, J. F. Klemic, D. A. Routenberg, P. N. Wyrembak, D. B. Turner-Evans, A. D. Hamilton, D. A. LaVan, T. M. Fahmy, and M. A. Reed, *Nature* **445**, 519 (2007).
5. E. Stern, E. Steenblock, M. Reed, and T. Fahmy, *Nano Lett.* **8**, 3310 (2008).
6. S. D. Caras, D. Petelenz, and J. Janata, *Anal. Chem.* **57**, 1920 (1985).
7. L. C. Clark Jr. and C. Lyons, Ann. N. Y. *Acad. Sci.* 102, 29 (1962).
8. A. Sedra and K. Smith, *Microelectronic Circuits*, Oxford University Press, USA (2004).
9. G. Zheng, F. Patolsky, Y. Cui, W. U Wang, and C. M. Lieber, *Nat. Biotechnol.* **23**, 1294 (2005).
10. A. J. Bard and L. R. Faulkner, *Electrochemical Methods, Fundamentals and Applications*, 2nd ed., John Wiley & Sons, Inc., New York (2001).
11. P. Bergveld, *IEEE Trans. Biomed. Eng.*, 17, 70 (1970).
12. S. Chen, J. G. Bomer, E. T. Carlen, and A. van den Berg, *Nano Lett.*, **11**, 2334 (2011).
13. I.-K. Lee, M. Jeun, H.-J. Jang, W.-J. Chob, and K. H. Lee, *Nanoscale*, **7**, 16789 (2015).
14. O. Knopfmacher, A. Tarasov, W. Fu, M. Wipf, B. Niesen, M. Calame, and C. Schönenberger, *Nano Lett.*, **10**, 2268 (2010).

15. M.-J. Spijkman, J. J. Brondijk, T. C. T. Geuns, E. C. P. Smits, T. Cramer, F. Zerbetto, P. Stoliar, F. Biscarini, P. W. M. Blom, and D. M. de Leeuw, *Adv. Funct. Mater.*, **20**, 898 (2010).

16. B. Khamaisi, O. Vaknin, O. Shaya, and N. Ashkenasy, *ACS Nano*, **4**, 4601 (2010).

17. C. Duarte-Guevara, F.-L. Lai, C.-W. Cheng, B. Reddy Jr., E. Salm, V. Swaminathan, Y.-K. Tsui, H. C. Tuan, A. Kalnitsky, Y.-S. Liu, and R. Bashir, *Anal. Chem.*, **86**, 8359 (2014).

18. J. Go, P. R. Nair, and M. A. Alam, *J. of Appl. Phys.*, **112**, 034516 (2012).

19. C. Duarte-Guevara, V. V. Swaminathan, B. Reddy Jr., J.-C. Huang, Y.-S. Liu, and R. Bashir, *RSC Adv.*, **6**, 103872 (2016).

20. D. Goncalves, D. M. F. Prazeres, V. Chua, and J. P. Conde, *Biosens. Bioelectron.*, **24**, 545, (2008).

21. F. Uslu, S. Ingebrandt, D. Mayer, S. B¨ocker-Meffert, M. Odenthal, and A. Offenh¨ausser, *Biosens. Bioelectron.*, **19**, 1723 (2004).

22. J. Go, P. R. Nair, B. Reddy Jr., B. Dorvel, R. Bashir, and M. A. Alam, *IEDM*, **10**, 202 (2010).

23. D. M. Garner, H. Bai, P. Georgiou, T. G. Constandinou, S. L. M. Shepherd, W. Wong Jr, K. T. Lim, and C. Toumazou, *ISSCC*, **27**(4), 492 (2010).

24. M. Kamahori, Y. Ishige, and M. Shimoda, *Biosens. Bioelectron.*, **22**, 3080 (2007).

25. M. Yano, K. Koike, K. Mukai, T. Onaka, Y. Hirofuji, K.-I. Ogata, S. Omatu, T. Maemoto, and S. Sasa, *Phys. Status Solidi A*, **211**, 2098 (2014).

26. N. Bhalla, M. D. Lorenzo, P. Estrela, and G. Pula, *Drug Disc. Today*, **22**, 204 (2017).

27. J. M. Rothberg, W. Hinz, T. M. Rearick, J. Schultz, W. Mileski, M. l Davey, J. H. Leamon, K. Johnson, M. J. Milgrew, M. Edwards, J. Hoon, J. F. Simons, D. Marran, J. W. Myers, J. F. Davidson, A. Branting, J. R. Nobile, B. P. Puc, D. Light, T. A. Clark, M. Huber, J. T. Branciforte, I. B. Stoner, S. E. Cawley, M. Lyons, Y. Fu, N. Homer, M. Sedova, X. Miao, B. Reed, J. Sabina, E. Feierstein, M. Schorn, M. Alanjary, E. Dimalanta, D. Dressman, R. Kasinskas, T. Sokolsky, J. A. Fidanza, E. Namsaraev, K. J. McKernan, A. Williams, G. T. Roth, and J. Bustillo, *Nature*, **475**, 348 (2011).

28. C. Guiducci and F. M. Spiga, *Nature Methods*, **10**, 617 (2013).

29. Y. Cui and C. M. Lieber, *Science*, **291**, 851 (2001).

30. J.-I. Hahm and C. M. Lieber, *Nano Lett.*, **4**, 51 (2004).

31. G.-J. Zhang, J. H. Chua, R.-E. Chee, A. Agarwal, and S. M. Wong, *Biosens. Bioelectron.*, **24**, 2504 (2009).

32. A. Agarwal, K. Buddharaju, I. K. Lao, N. Singh, N. Balasubramanian, and D. L. Kwong, *Sens and Actu. A*, **145–146**, 207 (2008).

33. Z. Gao, A. Agarwal, A. D. Trigg, N. Singh, C. Fang, C.-H. Tung, Y. Fan, K. D. Buddharaju, and J. Kong, *Anal. Chem.*, **79**, 3291 (2007).

34. A. Gao, N. Lu, P. Dai, T. Li, H. Pei, X. Gao, Y. Gong, Y. Wang, and C. Fan, *Nano Lett.*, **11**, 3974 (2011).

35. N. N. Mishra, W. C. Maki, E. Cameron, R. Nelson, P. Winterrowd, S. K. Rastogi, B. Filanoski, and G. K. Maki, *Lab Chip*, **8**, 868 (2008).

36. J. H. Chua, R.-E. Chee, A. Agarwal, S. M. Wong, and G.-J. Zhang, *Anal. Chem.*, **81**, 6266 (2009).

37. M. M. A. Hakim, M. Lombardini, K. Sun, F. Giustiniano, P. L. Roach, D. E. Davies, P. H. Howarth, M. R. R. de Planque, H. Morgan, and P. Ashburn, *Nano Lett.*, **12**, 1868 (2012).

38. Y. Cui, X. Duan, J. Hu, and C. M. Lieber, *J. Phys. Chem. B*, **104**, 5213 (2000).

39. E. Stern, R. Wagner, F. Sigworth, R. Breaker, T. Fahmy, and M. Reed, *Nano Lett.* **7**, 3405 (2007).

40. O. H. Elibol, D. Morisette, D. Akin, J. P. Denton, and R. Bashira, *App. Phys. Lett.*, **83**, 4613 (2003).

41. C.-W. Huang, H.-T. Hsueh, Y.-J. Huang, H.-H. Liao, H.-H. Tsai, Y.-Z. Juang, T.-H. Lin, S.-S. Lu, and C.-T. Lin, *Sens. & Actu. B*, 181, 867 (2013).

42. P.-W. Yen, C.-W. Huang, Y.-J. Huang, M.-C. Chen, H.-H. Liao, S.-S. Lu, and C.-T. Lin, *Biosens. Bioelectron.*, **61**, 112 (2014).

43. J.-K. Lee, I.-S. Wang, C.-H. Huang, Y.-F. Chen, N.-T. Huang, and C.-T. Lin, *Sensors*, **17**, 2733 (2017).

44. L. Torsi, M. Magliulo, K. Manoli, and G. Palazzo, *Chem. Soc. Rev.*, **42**, 8612 (2013).

45. S. H. Jeong, J. Y. Lee, B. Lim, J. Lee, Y. Y. Noh, *Dyes Pigm.*, **140** 244–249 (2017).

46. D. Khim, G. Ryu, W. Park, H. Kim, M. Lee, and Y. Noh, *Adv. Mater.*, **28** 2752–2759 (2016).

47. S. S. Prasad and D. R. Nair, *Int. J. Sci. Res.*, **6** 1258–1261 (2017).

48. D. K. Nair and K. S. Kumar, Development of gamma radiation dosimeter using copper phthalocyanine & zinc phthalocyanine based OFET, *IEEE International Conference on Circuits and Systems (ICCS)*, 17662327, (2017).

49. B. Cai, S. Wang, Y. Ning, L. Huang, Z. Zhang, and G.-J. Zhang, *ACS Nano*, **8**, 2632 (2014).

50. Z. Wang and Y. Jia, *Carbon*, **130**, 758 (2018).

51. X. Dong, Y. Shi, W. Huang, P. Chen, and L.-J. Li, *Adv. Mater.*, **22**, 1649 (2010).

52. B. Cai, L. Huang, H. Zhang, Z. Sun, Z. Zhang, and G.-J. Zhang, *Biosens. Bioelectron.*, **74**, 329 (2015).

53. T.-Y. Chen, P. T. K. Loan, C.-L. Hsu, Y.-H. Lee, J. T.-W. Wang, K.-H. Wei, C.-T. Lin, and L.-J. Li, *Biosens. Bioelectron.*, **41**, 103 (2013).

54. D. J. Kim, I. Y. Sohn, J. H. Jung, O. J. Yoon, N. E. Lee, and J. S. Park, *Biosens. Bioelectron.*, **41**, 621 (2013).

55. D.-J. Kim, H.-C. Park, Il Y. Sohn, J.-H. Jung, O. J. Yoon, J.-S. Park, M.-Y. Yoon, and N.-E. Lee, *Small*, **9**, 3352 (2013).

56. S. Mao, G. Lu, K. Yu, Z. Bo, and J. Chen, *Adv. Mater.*, **22**, 3521 (2010).

57. Y. X. Huang, X. C. Dong, Y. M. Shi, C. M. Li, L. J. Li, and P. Chen, *Nanoscale*, **2**, 1485 (2010).

58. P. K. Ang, A. Li, M. Jaiswal, Y. Wang, H. W. Hou, J. T. L. Thong, C. T. Lim, and K. P. Loh, *Nano Lett.*, **11**, 5240 (2011).

59. S. Caras and J. Janata, *Anal. Chem.*, **1980**, 52, (1935–1937).

60. D. R. Thevenot, K. Toth, R. A. Durst and G. S. Wilson, *Biosens. Bioelectron.*, **16**, 121–131 (2001).

61. E. A. H. Hall, *Biosensors*, Open University Press, Milton Keynes, 1990.

62. R. Ulber and T. Scheper, Disposable Bioprocessing Systems, in *Enzyme and Microbial Biosensors*, ed. A. Mulchandani and K. R. Rogers, Humana Press, Totowa, NJ, 1998, 35–50.

63. K. Wan, J. M. Chovelon, N. Jaffrezic–Renault and A. P. Soldatkin, *Sens. Actuators, B*, **58**, 399–408 (1999).

64. J. Hu, *Sens. Mater.*, **8**, 477–484 (1996).

65. M. J. Schöning, and A. Poghossian, *Analyst*, **127**(9), 1137–1151 (2002).

66. V. Volotovsky, A. P. Soldatkin, A. A. Shul'ga, V. K. Rossokhaty, V. I. Strikha and A. V. El'skaya, *Anal. Chim. Acta*, **322**, 77–81 (1996).

67. D. V. Gorchkov, A. P. Soldatkin, S. Poyard, N. Jaffrezic-Renault and C. Martelet, *Mater. Sci. Eng.*, **C 5**, 23–28 (1997).

68. A. P. Soldatkin, D. V. Gorchkov, C. Martelet and N. Jaffrezic–Renault, *Mater. Sci. Eng.*, **C 5**, 35–40 (1997).

69. B. H. van der Schoot, H. Voorthuyzen and P. Bergveld, *Sens. Actuators, B*, **1**, 546–549 (1990).

70. H. I. Seo, C. S. Kim, B. K. Sohn, T. Yeow, M. T. Son and M. Haskard, *Sens. Actuators, B*, **40**, 1–5 (1997).

71. K. Y. Park, S. B. Choi, M. Lee, B. K. Sohn and S. Y. Choi, *Sens. Actuators, B*, **83**, 90–97 (2002).
72. V. Volotovsky and N. Kim, *Sens. Actuators, B*, **49**, 253–257 (1998).
73. A. B. Kharitonov, M. Zayats, A. Lichtenstein, E. Katz and I. Willner, *Sens. Actuators, B*, **70**, 222–231 (2000).
74. E. Tobias-Katona and M. Pecs, *Sens. Actuators, B*, **28**, 17–20 (1995).
75. S. V. Dzyadevich, Y. I. Korpan, V. M. Arkhipova, M. Yu. Alesina, C. Martelet, A. V. El'skaya and A. P. Soldatkin, *Biosens. Bioelectron.*, **14**, 283–287 (1999).
76. A. Poghossian, *Sens. Actuators, B*, **44**, 361–364 (1997).
77. D. G. Pijanowska and W. Torbicz, *Sens. Actuators, B*, **44**, 370–376 (1997).
78. J. P. Cloarec, N. Deligianis, J. R. Martin, I. Lawrence, E. Souteyrand, C. Polychronakos and M. F. Lawrence, *Biosens. Bioelectron.*, **17**, 405–412 (2002).
79. A. B. Kharitonov, J. Wasserman, E. Katz and I. Willmer, J. Phys. Chem. B, **105**, 4205–4213 (2001).
80. L. Bousse, *Sens. Actuators, B*, **34**, 270–275 (1996).
81. A. Offenhäusser and W. Knoll, *Trends Biotechnol.*, **19**, 62–66 (2001).
82. A. Fanigliulo, P. Accossato, M. Adami, M. Lanzi, S. Martinoia, M. Grattarola and C. Nicolini, *Sens. Actuators, B*, **32**, 41–48 (1996).
83. M. Lehmann, W. Baumann, M. Brischwein, R. Ehret, M. Kraus, A. Schwinde, M. Bitzenhofer, I. Freund and B. Wolf, *Biosens. Bioelectron.*, **15**, 117–124 (2000).
84. S. Martinoia, N. Rosso, M. Grattarola, L. Lorenzelli, B. Margesin and M. Zen, *Biosens. Bioelectron.*, **16**, 1043–1050 (2001).
85. M. Lehmann, W. Baumann, M. Brischwein, H. J. Gahle, I. Freund, R. Ehret, S. Drechsler, H. Palzer, M. Kleintges, U. Sieben and B. Wolf, *Biosens. Bioelectron.*, **16**, 195–203 (2001).
86. C. Sprössler, M. Denyer, S. Britland, W. Knoll and A. Offenhäusser, *Phys. Rev. E*, **60**, 2171–2176 (1999).
87. V. Kiessling, B. Müller and P. Fromherz, *Langmuir*, **16**, 3517–3521 (2000).
88. B. Wolf, M. Brischwein, W. Baumann, R. Ehret and M. Kraus, *Biosens. Bioelectron.*, **13**, 501–509 (1998).
89. W. H. Baumann, M. Lehmann, A. Schwinde, R. Ehret, M. Brischwein and B. Wolf, *Sens. Actuators, B*, **55**, 77–89. (1999).
90. S. Belli and G. Rechnitz, *Anal. Lett.*, **19**, 403–405 (1986).
91. M. J. Schöning, S. Schütz, P. Schroth, B. Weissbecker, A. Steffen, P. Kordos, H. E. Hummel and H. Lüth, *Sens. Actuators, B*, **47**, 235–238 (1998).
92. P. Schroth, M. J. Schöning, S. Schütz, Ü. Malkoc, A. Steffen, M. Marso, H. E. Hummel, P. Kordos and H. Lüth, *Electrochim. Acta*, **44**, 3821–3826 (1999).
93. P. Schroth, M. J. Schöning, P. Kordos, H. Lüth, S. Schütz, B. Weissbecker and H. E. Hummel, *Biosens. Bioelectron.*, **14**, 303–308 (1999).
94. M. J. Schöning, P. Schroth and S. Schütz, *Electroanalysis*, **12**, 645–652 (2000).
95. P. Schroth, H. Lüth, H. E. Hummel, S. Schütz and M. J. Schöning, *Electrochim. Acta*, **47**, 293–297 (2001).
96. M. J. Huotari, *Sens. Actuators, B*, **71**, 212–222 (2000).
97. M. J. Schöning, P. Schroth and S. Schütz, *Electroanalysis*, **12**, 645–652 (2000).
98. P. Schroth, M. J. Schöning, H. Lüth, B. Weissbecker, H. E. Hummel and S. Schütz, *Sens. Actuators, B*, **78**, 1–5 (2001).

10 Nanowire Array–Based Gate-All-Around MOSFET for Next-Generation Memory Devices

Krutideepa Bhol
VIT-AP University

Biswajit Jena
Koneru Lakshmaiah Education Foundation

Umakanta Nanda
VIT-AP University

CONTENTS

10.1 Introduction 189
10.2 Brief Review on Sentaurus TCAD 193
    10.2.1 Sentaurus TCAD Codes 194
10.3 Device Design and Simulation 195
10.4 Results and Discussions... 195
10.5 Conclusion 200
References..200

10.1 INTRODUCTION

Planar transistors have been considered as the heart of integrated circuits for more than a few decades; during this period, the size of the individual transistors has regularly decreased. As the size decreases, planar transistors progressively suffer from the abominable short-channel effect (SCE), especially "off-state" leakage current, which increases the idle power required by the device.

Several gates on multiple surface surround the channel in case of a multigate device. It thus provides a better electrical control over the channel, allowing more effective suppression of "off-state" leakage current. Multiple gates also allow enhanced current in the "on" state, also known as drive current. Among different multigate devices, cylindrical surrounding-gate MOSFET (metal oxide semiconductor field-effect transistor) provides better gate control over the channel. Also saturated device dimensions produce high gate leakage current and several SCEs such as threshold voltage roll-off, drain-induced barrier lowering, corner effect, and so on. [1,2]. However, many device geometries have already been developed and investigated in detail to improve the performance. Cylindrical gate-all-around (GAA) MOSFET provides superior gate controllability compared with single-gate and other multigate structures [3]. It is considered as one of the better models that has the capability to overcome the physical scaling limit of conventional CMOS (complementary metal oxide semiconductor) technology. The GAA MOSFET in the fully depleted regime shows improved robustness against SCEs and also reduces the threshold voltage and subthreshold swing (SS) [4]. The SCEs in the small-dimension devices result in high off-state leakage current (I_{off}) and high SS. So I_{on}/I_{off} ratio of the device decreases significantly, which affects the circuit speed and dissipation power. However, the impact of SCEs can be reduced by scaling the gate oxide thickness and increasing the channel doping concentration for channel length beyond 100 nm [5].

The objective of device scaling is to shrink transistor dimension so that more number of transistors can be placed on a chip. Typically, the scaling factor "k" is $\sqrt{2}$, so that the area is reduced to one-half and the number per area increases by a factor of 2. Integration of billions of transistors on a single chip has been feasible due to the possibility to pattern every minor feature on silicon through optical lithography. As optical lithography enters the subwavelength regime, light diffraction and interference from subwavelength pattern feature causes image disorder. Therefore, patterning becomes difficult without adopting resolution enhancement techniques. As the traditional planar MOSFET design slowed down due to scaling problems, new material or gate engineering was necessary in order to match current requirement of semiconductor industries. Evolution of MOSFET technology started and achieved milestone both in device design and performance. Gate engineering technology changed traditional MOSFET to GAA MOSFET as shown in Figure 10.3. In fully depleted double-gate (DG) and cylindrical surrounding-gate MOSFET, the SCEs are governed by the electrostatic potential, and in these structures, the channel is confined by the gate metal. So gate provides superior controllability to reduce these types of effects (Figure 10.1).

Just after the invention of transistor, researchers realize that by reducing the size of the transistor, its performance can be improved. After that, MOSFET scaling played an important role in device architecture. Basically, there are various methods of scaling, but constant electric field scaling proposed by Dennard et al. played a vital role [7]. Since the early 1990s, semiconductor manufacturing companies and researchers have come to the front in order to define the future semiconductor industry. This initiative gave birth to ITRS (International Technology Roadmaps for Semiconductors) [6]. Every year, ITRS publishes an issue that helps to set a benchmark for future device improvement. The CMOS technology is one of the frontline developments achieved by semiconductor industries from a long time. And the basic component behind the CMOS technology is the MOS transistor. At the end of the 1990s, a new device structure was developed which was efficient to increase the

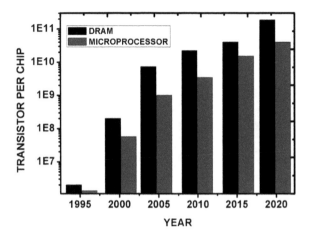

FIGURE 10.1 Evolution of the number of transistors per chip predicted [6]. DRAM, dynamic random access memory.

performance of conventional MOSFET. And the new device was named as silicon-on-insulator (SOI) in which the transistors are made in a thin silicon layer sitting on the top of a SiO$_2$ layer. SOI device has the ability to reduce parasitic capacitances. But these devices were not that much efficient to reduce SCEs. SCE occurs when the dimensions of a transistor shrunk down and the closeness between source and drain hampers the controllability of the gate to control the potential distribution and flow of current through the channel. So for future device purpose, it was impossible to reduce the length anymore. Research to develop this device came to an end with multigate device structure.

With the continuous development for improved drain current and SCE, a classical planar device is transformed into multigate device structures such as double-/triple-/quadruple-gate devices. The first article on the double-gate MOSFET was given by T. Sekigawa and Y. Hayashi in 1984 [8]. The triple-gate MOSFET is an improved model of the double-gate MOSFET where the gates are surrounded to three sides of the channel. Π-gate and Ω-gates are some developed models of triple gate for improved electrostatic characteristics. The structure that theoretically offers the best possible control of the channel region with the help of gate is surrounding-gate MOSFET. Surrounding-gate device with square or circular cross section was given by N. Singh et al. in 2006 [9]. But the analytical model for cylindrical surrounding-gate MOSFET was developed by several researchers [10–26]. Some GAA structures are given in Figures 10.2–10.4.

Continuous downscaling in device dimension to fulfill the demands of present semiconductor technology by introducing high-speed and low-power devices is pushing the CMOS technology to the ultimate nanoscale dimension. Out of different nanoscale structures, cylindrical GAA MOSFET is acting as a workhorse for the scaled device family. Due to cylindrical geometry, the device has the ability to preserve the device ability toward SCEs. Apart from this, a tight capacitive coupling with higher electrostatic controllability is obtained in case of a cylindrical GAA

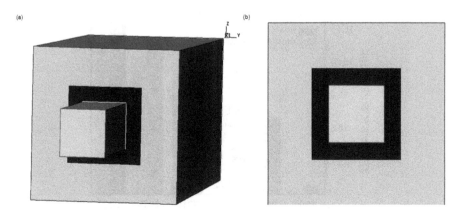

FIGURE 10.2   GAA with cubical structure. GAA, gate-all-around.

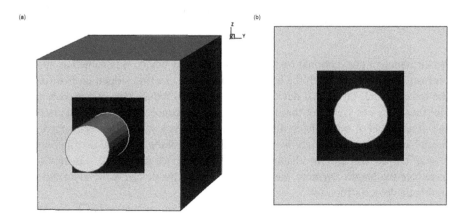

FIGURE 10.3   GAA structure with cubical gate and oxide but cylindrical channel. GAA, gate-all-around.

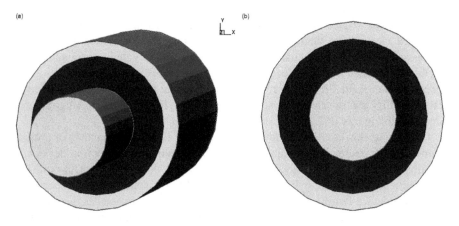

FIGURE 10.4   GAA cylindrical structure. GAA, gate-all-around.

geometry. Introduction of multiple numbers of cylindrical geometry with channel thickness below 10 nm creates nanowire array. Though GAA MOSFET is the leader in present CMOS technology, the current driving capability of the device can be improved by introducing an array of nanowires, which resembles to the small forest with tall trees. Introduction of vertical nanowires not only improves the device characteristics but also provides extra mechanical strength to the proposed device. The advantages that will provide add-on to the device performance is that the silicon arrays as channel are controlled by single gate. Arrangement of silicon pillars as channel affects the device performance extensively. In this chapter, we have arranged 13 silicon pillars adjacent to each other to form a square structure.

10.2 BRIEF REVIEW ON SENTAURUS TCAD

Sentaurus Structure Editor is a structure editor for 2D and 3D device structures. It has three specific operational modes: 2D structure editing, 3D structure editing, and 3D process emulation. From the graphical user interface (GUI), 2D and 3D device models are created geometrically, using 2D or 3D structures, such as rectangles, polygons, cuboids, cylinders, and spheres. The GUI of Sentaurus Structure Editor features a command line window, in which Sentaurus Structure Editor prints script commands corresponding to the GUI operations (Figure 10.5).

There are various carrier transport models used in Sentaurus Structure Editor for different applications based on different situations. The drift–diffusion phenomenon has been used to study the electrostatic characteristic of the proposed model. The drift–diffusion transport model solves self-consistently the Poisson and carrier continuity equations in the selected device regions with specified boundary conditions. For MOSFETs, with an inversion condition operating under low-drain bias, it is often sufficient to solve the Poisson equation and the continuity equations with

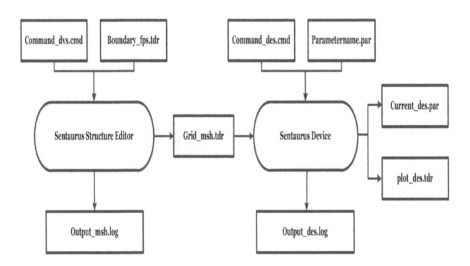

FIGURE 10.5 Software process flow.

only one type of carrier. However, generation–recombination processes such as thermal Shockley–Red–Hall (SRH) generation, avalanche generation, and band-to-band tunneling may become significant for larger drain voltage. As a result, the continuity equations for both electrons and holes are necessary to achieve the necessary simulation accuracy [27].

10.2.1 SENTAURUS TCAD CODES

The design of the device is started with selecting the material and proper dimensions. As the device structure is cylindrical, the dimensions should be chosen in X, Y, and Z directions. The coordinate should have starting and ending values during the design of the structure. All the codes for the design purpose should be written in Sentaurus Structure Editor.

```
............................ . .
(sde:set-default-material "Metal/Oxide/Silicon")
(sdegeo:create-cylinder (position 0 0 0.002) (position 0 0
0.088) 0.023 "Metal/Oxide/Silicon" "Gate1/Dielectric/Channel")
............................ . .
```

For multiple nanowires, which is called nanowire array, different silicon nanowires should be taken for different coordinates. Similarly, all other regions associated with MOSFET are defined with proper coordinates and accurate materials. In order to provide the contact points for the device, the device code can be written as follows:

```
(sdegeo:define-contact-set "source/drain/gate" 4 (color:rgb 1
0 1) "==")
(sdegeo:set-current-contact-set "source/drain/gate")
```

After complete declaration of the electrode, the corresponding region is doped, and mesh is created by giving minimum and maximum values. In order to define and execute the physics behind the device, drift–diffusion/hydrodynamic/quantum potential–based models are added in the SDEVICE section. The doping at any point between nodes (or any physical quantity calculated by Sentaurus Device) can be obtained by interpolation. Each virtual device structure is described in the Synopsys TCAD tool suite by a TDR file containing the following information:

  I. The grid (or geometry) of the device contains a description of the various regions, that is, boundaries, material types, and the locations of any electrical contacts.
  II. The data fields contain the properties of the device, such as the doping profiles, in the form of data associated with the discrete nodes.

```
File
 {
* input files:
* output files:
```

```
}
Electrode {
{ Name="source/drain/gate" Voltage=0.0/0.0/0.0 }
 }
Physics {
}
Plot {
}
```

10.3   DEVICE DESIGN AND SIMULATION

The 3D view with clear vision toward silicon pillars of the proposed nanowire array–based MOSFET is shown in Figure 10.6a and b, respectively. All the pillars present in the proposed model are isolated from each other using individual SiO_2 layer as dielectric. The work function of the metal gate used in this proposed model is varied from 4.2 to 5.0 eV. The gate length of the device is kept 90 nm initially and varied up to 16 nm to study the recent technology node behavior. Individual silicon pillar has a thickness of 6 nm with channel length of 90 nm. The oxide thickness (t_{ox}) is taken as 2 nm for each silicon pillar. The doping concentration of the channels is kept to be $10^{16} cm^{-3}$. Similarly the doping concentration of source/drain is kept to be $10^{20} cm^{-3}$. The design and simulation of the proposed model is carried out using gold standard semiconductor simulator from Synopsys. This simulator includes all the possibilities, not only for simple structures but also for more complex design and simulation to study the characteristics extensively. This simulator includes different carrier transport models such as drift–diffusion model, hydrodynamic model, thermodynamic model, and quantum model to study the movement of electron inside the channel. Apart from electron movement, the movement of holes inside the device can be studied easily by taking appropriate cut-set of the corresponding devices. All the simulations are carried out using Sentaurus TCAD from Synopsys. This simulator is a state of the art simulator to design and study the advanced semiconductor materials and their behavior. This is an electronic design automation tool with GUI, which supports both 2D and 3D device designing (Figure 10.6).

10.4   RESULTS AND DISCUSSIONS

A nanowire MOSFET with cylindrical structure has the ability to increase the total W_{eff} without increasing layout area penalty, as a result of which the transistor's on current (I_{on}) increases due to strong electrostatic controllability. At the same time, introducing multiple nanowires increases the device drive current more effectively. In this simulation, we have used quantum–based model as the channel length is below 10 nm. The drift–diffusion model also considers the field-dependent mobility, concentration-dependent model, and velocity saturation model. The BGN model is considered to determine the carrier concentration. Also the SRH model is used to design the short-channel devices. The device designs with proper dimensions were included in the SDE section, and the corresponding device physics was included in the SDEVICE section. The results thus obtained are visualized using INSPECT and SVISUAL tools.

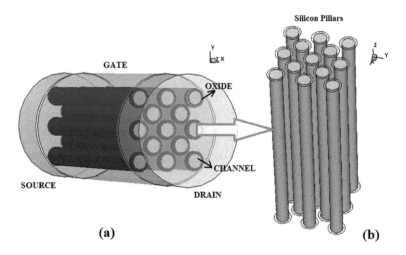

FIGURE 10.6 Three-dimensional device structure of the proposed model with nanowire array.

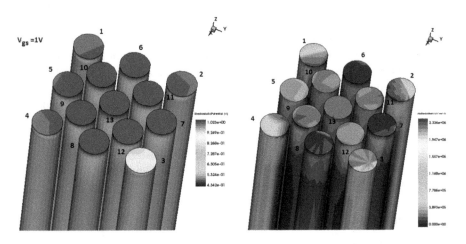

FIGURE 10.7 (a) and (b) The electrostatic potential distribution of the proposed nanowire array–based MOSFET model along the channel for two different gate biases. MOSFET, metal oxide semiconductor field-effect transistor.

With a lower gate bias of 0.2 V, the corresponding potential distribution in each nanowire is shown in Figure 10.7a. From the figure, it can be observed that the potential is lower at the source side and comparatively higher at the drain side. The source side has a potential nearly equal to the built-in potential, whereas for drain side, the built-in potential is added up with the applied drain voltage. This is the reason that we always get asymmetric potential curve. However, for a gate bias of nearly 0.7 V, the potential is high throughout the channel, but the starting potential at the source side and end potential at the drain side remain similar to the lower gate bias. Again it can be observed that, for each and every nanowire, the potential distribution is not uniform. This is because of the fact that the arrangement of nanowires to form the array and number of nanowires plays a vital role in potential distribution. Apart from this, due to the channel and drain junction with different drain bias, the potential curve is affected a little. Each nanowire has a different potential profile and hence a different potential minima. The average of these potential minima is used to calculate overall threshold voltage of the device. It is also observed that the nanowires placed at the center have uniform potential curve with lower potential minima, as a result of which their contribution in deciding threshold voltage of the device is more compared with other peripheral nanowires.

The analysis is further extended to examine the effect of higher gate bias on the potential distribution as well as electric field distribution of individual nanowires as shown in Figure 10.8a and b, respectively. The nanowires are individually numbered from 1 to 13 as shown in the figure. Except pillar number 1, 2, 3, and 4, almost all the pillars exhibit similar potential distribution throughout the pillar. But there is a little deviation in the peripheral nanowires due to their positioning. However, the electric field distribution for the silicon pillars is different for each pillar due to the different potential distribution and depletion lines due to drain bias. The results can be well observed from Figure 10.8b. The analysis is further extended toward the switching ratio calculation of the device by illustrating the on current and off current.

Figure 10.9 shows the gate voltage versus drain current analysis of the proposed device for different drain current. An inset is also given for proper analysis of on and off current in logarithmic form. From the figure, the dependency of drain voltage on drain current is observed. With the increase in drain voltage, the corresponding on current is increasing by keeping the off current constant. An acceptable range of switching ratio is observed for the proposed device, and the dependency of drain current on drain voltage clearly indicates toward the stability of the device in high electric field.

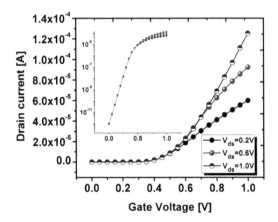

FIGURE 10.8 (a) Higher gate bias on the potential distribution. (b) Electric field distribution of individual nanowires.

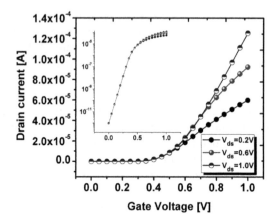

FIGURE 10.9   Gate voltage versus drain current analysis of the proposed device for different drain voltages.

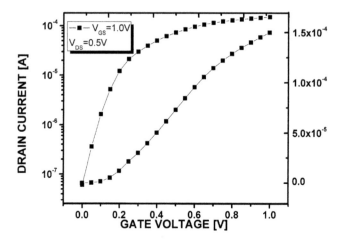

FIGURE 10.10   Transfer characteristic analysis of the proposed device in linear and logarithm scales.

Figure 10.10 shows the transfer characteristics curve of the proposed model for a gate voltage of 1 V with a drain voltage of 0.5 V. The gate voltage versus drain current curve is represented in both linear and logarithmic forms in order to ensure effective I_{On}/I_{Off} ratio. As this ratio plays a vital role during memory design, so with an improved switching ratio, this device seems to be a suitable candidate in logic applications. The threshold voltage of the proposed device is extracted by taking the derivative of transconductance as shown in Figure 10.11. A lower value of threshold voltage for the proposed model indicates toward better memory device with fast switching also.

The variation in switching ratio with different threshold voltage is shown in Figure 10.12. From the figure, it can be observed that the device with higher threshold

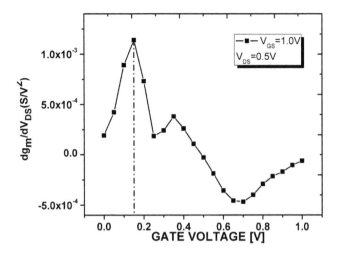

FIGURE 10.11 Threshold voltage extraction from the transconductance of the proposed device.

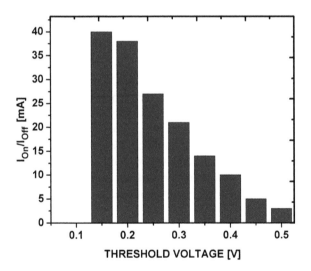

FIGURE 10.12 Threshold voltage versus I_{On}/I_{Off} ratio of the proposed device.

voltage is having lower switching ratio and vice versa. However, the proposed device exhibits a threshold voltage of 0.1 V with a switching ratio about 40 mA, which indicates toward next-generation memory device.

The variation in drain current with respect to the drain voltage for different gate voltage is illustrated in Figure 10.13. From the figure, the improved output characteristic of the proposed model is observed, and the variation with different gate voltage is also observed. Higher drain current is observed due to higher electron mobility in

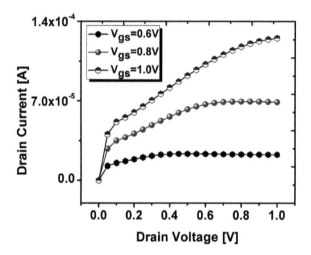

FIGURE 10.13 Gate voltage versus drain current analysis of the proposed device for different drain voltage.

the silicon pillars. Array of nanowires that act as channels are responsible for high mobility of the electrons.

10.5 CONCLUSION

Introducing multiple nanowires in place of single silicon pillar increased the drain current with reduced threshold voltage. The potential distribution in individual nanowire results in improved electron mobility by enhancing the electrostatic controllability. However, arrangement of nanowires to form the channel plays a vital role during device performance metrics calculation. Distribution of nanowires in circular form exhibits better performance compared with other geometrical arrangements of the nanowires. With improved characteristics, the proposed device can be a fruitful member in GAA family for next-generation memory–based devices.

REFERENCES

1. K. Suzuki, T. Tanaka, Y. Tosaka, H. Horie, and Y. Arimoto, "Scaling theory for double-gate SOI MOSFETs", *IEEE Tran. Electron Dev.*, vol. 40, (1993) 2326.
2. A. Chaudhry and M. J. Kumar, "Controlling short-channel effects in deep submicron SOI MOSFETs for improved reliability: A review", *IEEE Trans. Dev. Mater. Rel.*, vol. 4, (2004) 99.
3. A. A Orouji, M. Jagadesh Kumar, "A new symmetrical double gate nano scale MOSFET with asymmetrical side gates for electrically induced source/drain", *Microelectron. Eng.*, vol. 83, (2006) 409.
4. T. K. Chiang, "A scaling theory for fully-depleted, surrounding-gate MOSFET's: Including effective conducting path effect", *Microelectron. Eng.* vol. 77, (2005) 175.
5. J. P. Colinge, "An SOI voltage-controlled bipolar-MOS device", *IEEE Trans. Electron Devices*, vol. 34 (1987) 845.

6. B. Davari, H. Dennard, AND G. G. Shahidi, "CMOS Scaling for high performance and low power-the next ten years", *Proc. IEEE*, vol. 83, (1995) 595.
7. B. Jena, S. Dash, S. Routray, G. P Mishra, "Inner-gate-engineered GAA MOSFET to enhance the electrostatic integrity", *NANO: Brief Rep. Rev.*, vol. 14, (2019), 1950128.
8. T. Sekigawa and Y. Hayashi, "Calculated threshold-voltage characteristics of an XMOS transistor having an additional bottom gate", *Solid-State Electron.*, vol. 27, (1984) 827.
9. N. Singh, A. Agarwal, L. K. Bera, T. Y. Liow, R. Yang, S. C. Rustagi, C. H. Tung, R. Kumar, G. Q. Lo, N. Balasubramanian, D. L. Kwong, "High- performance fully depleted silicon nanowire (diameter<5 nm) gate-all- around CMOS devices", *IEEE Electron Dev. Lett.*, vol. 27,(2006) 383.
10. C. P. Auth, J. D. Plummer, "Scaling theory for cylindrical fully-depleted surrounding gate MOSFETs", *IEEE Electron Dev. Lett.*, vol. 18, (1997) 74.
11. S. L. Jang, S. S Liu, "An analytical surrounding gate MOSFET model", *Solid State Electron.*, vol. 42, (1998) 721.
12. A. Kranti, S. Haldar, R.S. Gupta, "Analytical model for threshold voltage and I-V characteristics of fully depleted short channel cylindrical/surrounding gate MOSFET", *Microelectron Eng.*, vol. 56, (2001) 241.
13. S. Hyun Oh, D. Monroe, J. M. Hergenrother, "Analytic description of short channel effects in fully depleted double gate and cylindrical surrounding gate MOSFET", *IEEE Electron Dev. Lett*, vol. 21, (2000) 445.
14. D. Jimenez et al., "Continuous analytic I-V model for surrounding gate MOSFETs", *IEEE Electron Dev. Lett.*, vol. 25, (2004) 571.
15. T. K. Chiang, "A new two dimensional analytical model for threshold voltage in undoped surrounding gate MOSFETs", *Solid-State Int. Circ. Technol.*, (2006) 1234.
16. L. Zhang, J. He, J. Zhang, J. Feng, "A continuous yet Explicit In 2006, a carrier based core model for the long channel undoped cylindrical surrounding gate MOSFETs", *NSTI-Nanotech.*, vol. 3, (2008) 590.
17. J. He, Y. Tao, F. Liu, J. Feng, "Analytical channel potential solution to the undoped surrounding gate MOSFETs", *Solid State Electron.*, vol. 51, (2007) 802.
18. H. Kaur, S. Kabra, S. Bindra, S. Haldar, R. S. Gupta, "Impact of graded channel (GC) design in fully depletedcylindrical/surrounding gate MOSFET (FD CGT/SGT) forimproved short channel immunity and hot carrier reliability", *Solid-State Electron.*, vol. 51, (2007) 398.
19. F. Liu, J. He, L. Zhang, J. Zhang, J. Hu, C. Ma, and M. Chan, "A charge-based model for long-channel cylindrical surrounding-gate MOSFETs from intrinsic channel to heavily doped body", *IEEE Trans. Electron Dev.*, vol. 55, (2008) 2187.
20. T. K. Chiang, "A new compact subthreshold behavior model for dual-material surrounding gate (DMSG) MOSFETs", *Solid-State Electron.*, vol. 53, (2009) 490.
21. C. Li, Y. Zhuang, R. Han, "Cylindrical surrounding-gate MOSFETs with electrically induced source/drain extension", *Microelectron. J.*, vol. 42, (2011) 341.
22. V. M. Srivastava, K. S. Yadav, G. Singh, "Design and performance analysis of cylindrical surrounding double-gate MOSFET for RF switch", *Microelectron. J.*, vol. 42, (2011) 1124.
23. P. Ghosh, S. Haldar, R. S. Gupta, M. Gupta, "An analytical drain current model for dual material engineered cylindrical/surrounded gate MOSFET", *Microelectron. J.*, vol. 43, (2012) 17.
24. B. Jena, S. Dash, K. P Pradhan, S. K Mohapatra, P. K Sahu, G. P Mishra, "Performance analysis of undoped cylindrical gate all around (GAA) MOSFET at subthreshold regime", *Adv. IN Nat. Sci.: Nanosci. Nanotechnol.*, vol. 6, (2015), 035010.
25. B. Jena, S. Dash, G. P Mishra, "Effect of underlap length variation on DC/RF performance of dual material cylindrical MOS", *Int. J. Num. Model: Electron Network, Dev. Fields*, vol. 30, (2016), e2175.

26. L. Zhang, C. Ma, J. He, X. Lin, M. Chan, "Analytical solution of subthreshold channel potential of gate underlap cylindrical gate-all-around MOSFET", *Solid-State Electron.*, vol. 54, (2010) 806.
27. Guide, Sentaurus Device User, and N. Version. "Synopsys TCAD Sentaurus." San Jose, CA, USA (2017).

11 Design of 7T SRAM Cell Using FinFET Technology

T. Santosh Kumar and Suman Lata Tripathi
Lovely Professional University

CONTENTS

11.1   Introduction 203
11.2   SRAM Cell Architectures Based on CMOS Technology 204
    11.2.1   6T SRAM Cell...204
    11.2.2   7T SRAM Cell...205
    11.2.3   8T SRAM Cell...205
    11.2.4   10T SRAM Cell...206
    11.2.5   12T SRAM Cell...206
11.3   SRAM Cell Architectures Based on FinFET Technology 206
    11.3.1   Proposed Design 7T SRAM Cell 207
    11.3.2   Operations of 7T SRAM 208
11.4   Result Analysis 209
11.5   Different Types of Leakage Current in SRAM 210
    11.5.1   Subthreshold Leakage Current 210
    11.5.2   Gate Leakage 210
    11.5.3   Junction Tunneling Leakage..211
    11.5.4   Different Leakage Reduction Techniques 211
        11.5.4.1   Self-controllable Voltage Level Technique 211
        11.5.4.2   Lower Self-controllable Voltage Level 211
        11.5.4.3   Upper Self-controllable Voltage Level........................211
11.6   Conclusion 213
References...213

11.1   INTRODUCTION

In the current chip technology, the capability of silicon-on-chip (SOC) memory is quickly developing to increment worldwide execution. As a more prominent reserve memory is required, SRAM (static random access memory) assumes a continuously critical job in current chip frameworks, compact gadgets, and cell phones [1]. To achieve more noteworthy speed, SRAM-dependent reserve memories and SOC memories are typically utilized [2]. The gadget scaling, SRAM cell configuration has a few difficulties such as power utilization issues, steadiness, and area. Several kinds of research have been carried out for low-power stable SRAM cell operation using many circuit design techniques 6T, 7T, 9T, and 11T bit cell using CMOS (complementary metal oxide semiconductor) technology [3–7]. 6T SRAM cell is minimally utilized as a memory cell [8,9]. A 9T SRAM cell was designed using CMOS in 32-nm technology which shows improvements in power dissipation and stability in comparison with earlier designs for low-power memory processes [10]. A 6T SRAM cell was described showing reduction of 12%–38% of total power dissipation related to conventional SRAM bit cell [11]. FinFET technology has enormous potential to replace CMOS technology in most recent SRAM because of its greater channel flexibility with increased control of gate over the channel without conceding performance [12]. Also, the leakage current is usually less in FinFET as compared with MOSFET (metal oxide semiconductor field-effect transistor) at subthreshold system. Therefore, a FinFET-based efficient smart embedded memory incorporated with new circuit topologies is explored for future industry trends [13].

11.2   SRAM CELL ARCHITECTURES BASED ON CMOS TECHNOLOGY

This section describes the design of static random access memory (SRAM) using complementary metal oxide field effect transistor (CMOS) technology.

11.2.1   6T SRAM CELL

The fundamental cell of customary SRAM comprises 6T among them, two are data-passing gates and two are inverters that store the information [14]. Though the data-passing transistor's reason for existing is for choosing the cell that is initiated with WoL passes the information inputs that are to be read or write in accordance with the inverters [15–18]. A conventional 6T SRAM bit cell is presented in Figure 11.1. The SRAM core circuit is shown in Figure 11.1. The middle four transistors form a pair of inverters (P1, P2, N1, and N3) and are used to store the values. The other two transistors (N2 and N4) mainly control access to the memory cell by the bit lines. A read or write is performed through the selected cell. In read operation, bit and bit_b are precharged to supply (V_{dd}) before the select line is allowed for high. In write operation, the bit and bit_b lines are set to the required values, and then select is kept.

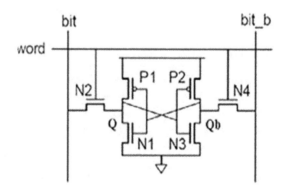

FIGURE 11.1   Conventional 6T SC.

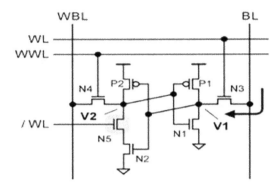

FIGURE 11.2   7T SRAM cell.

11.2.2   7T SRAM Cell

The 7T cell design [19] also employs dispersed read and write signal lines but uses only one extra NMOS (n-type MOS) transistor to achieve read-disturb-free operation, thus increasing the cell area by 13%. NMOS transistor, whose gate is controlled by signal WL, is additional between node V2 and NMOS transistor, to the 7T cell design as seen in Figure 11.2. Whereas the cell is being accessed, WL is set to "0" to turn off NMOS transistor. In the case of a "0" read, even if the voltage at node V1 reaches the Vth of NMOS transistor N2, node V2 cannot be pulled down to "0," thus preserving the stored data. During data retention period, WL is set to "1," and the cell operates in the similar process as the 6T cell circuit.

11.2.3   8T SRAM Cell

To achieve better stability, 8T bit cells have been proposed to isolate the read from the write enabling low-voltage operations. Figure 11.3 exhibits an 8T bit-cell configuration. Adding two FETs to a 6T bit cell provides a read mechanism that does not disturb the internal nodes of the cell [20]. This requires separate read and write word lines (RWL, WWL) and can accommodate dual-port operation with separate read and write bit lines as shown in Figure 11.3. The read operation is carried out by

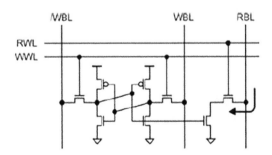

FIGURE 11.3   8T SRAM cell.

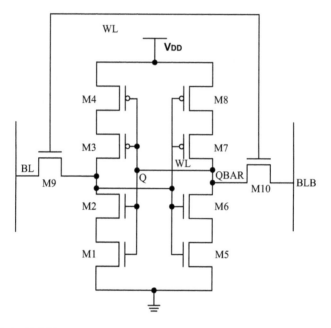

FIGURE 11.4　10T SRAM cell.

precharging the RBL and activating RWL. If 1 is stored at node Q, then M6 turns
ON and makes a low resistance path for the flow of cell current through RBL to
ground, which can be sensed by the sense amplifier [21].

11.2.4　10T SRAM Cell

A 10T SRAM cell in Figure 11.4 has a read buffer on each side to improve the read
performance and also write buffer on each side to progress the write performance;
apart from that, it has six main body transistors that make its functionality similar
to that of a 6T SRAM cell.

11.2.5　12T SRAM Cell

A 12T transistor SRAM is proposed to further improve the steadiness of the SRAM
cell as shown in Figure 11.5 with one transistor on the top and the other in the bottom
connected to V_{dd} and G_{nd}, respectively.

11.3　SRAM CELL ARCHITECTURES BASED
ON FINFET TECHNOLOGY

FinFETs are proved to be suitable for memory design because of their low static
power leakage compared with CMOS-based memory cells. SRAM cell can be
designed with FinFET in all available topologies of CMOS-based SRAM cell.

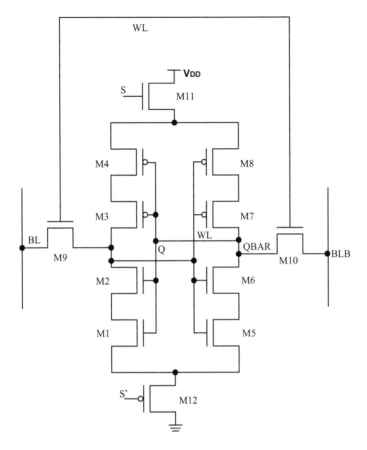

FIGURE 11.5 12T SRAM cell.

11.3.1 Proposed Design 7T SRAM Cell

The fundamental goal of proposing the 7T SRAM cell is to device abundant read steadiness and a good value of signal to noise ratio (SNMs). The proposed 7T SRAM cell is presented in Figure 11.6. This new SRAM cell is comprised of seven transistors and utilizes one BL, WL, and RL. However, writing into the cell, BL and WL are utilized, and RL is not utilized, whereas during reading process, BL and RL are utilized, and WL is not utilized. For any activity suggested, 7T SRAM cell utilizes single BL and WL or RL. But a 6T SRAM cell requires all the three lines for any activity, henceforth consuming more power contrasted with the 7T SC.

As 7T SRAM cell utilizes just a single BL, there will be a reduction in the control required for other operations. Consequently, utilization of just a single BL decreases the voltage necessary to ON and OFF the BL transistor to around 50%, on grounds that just a single BL is charged amid a read activity rather than two. The BL is charged 50% of the time rather than each time during a write activity and is necessary, in this situation an equivalent likelihood of writing a 1 or 0. The proposed 7T SC utilizes two transistors NMOS4 and NMOS5 with RL for read task.

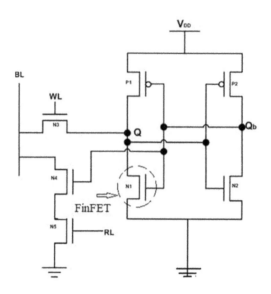

FIGURE 11.6   Proposed 7T FinFET SRAM cell.

11.3.2 OPERATIONS OF 7T SRAM

A. Write:

During the write process, the information to be written is stacked with BL, and then WL is enabled. The transistor N3 permits BL to suppress the SC, so that the cell is written with essential information. BL should be charged to V_{DD} for writing "1" into the SC. In event that "o" has to be written, BL is low, and WL will be level of power supply. RL is not active in write mode.

B. Read:

The information in the cell can be read with the BL which is already-charged to V_{DD}. Once initially charged, then bit line RL is triggered. Relying on whether or not the BL discharges or holds the command charge, information stored within the 7T SC may be set. If BL discharges once by escalating the read line to V_{DD}, it specifies 7T SC is holding "0." If bit line holds the command charge, then the stored data are "1." WL will be inactive in read process.

Consider the 7T SC holds a "0" primarily. The two transistors NMOS4 and NMOS5 will be ON by precharging the BL to V_{DD} and pulling RL to V_{DD}. As the data stored is "0," the transistor NMOS4 will be ON, and as RL is at V_{DD}, the transistor NMOS5 is ON. Currently, BL has a connection to ground via NMOS4 and NMOS5 by discharging to logic "0" specifying data stored is "0." Consider that 7T SC is storing "1." RL is increased to V_{DD} by precharging the BL, NMOS4 will be OFF, and NMOS5 will be ON. Because Q_b is at "0," NMOS4 is OFF, and because RL is at V_{DD}, NMOS5 is ON. The data stored is shown as "1" as there is no path for BL to decrease to 0 levels.

C. Hold:
   In this state, the SRAM maintains the data until the power supply is ON. Depending on the data stored, i.e., if data contained in the cell is "1," then Q will be at V_{DD} and Q_b will be at "0 V" and vice versa if the stored data is "0."

11.4   RESULT ANALYSIS

Here, comparison of proposed 7T CMOS SRAM cell and FinFET SRAM cell is shown using power calculations and read, write, and hold state SNMs.

A. Power:
   The complete power consumed by SRAM cell is the sum of power drawn from the source, sources accustomed to charge and discharge the BLs, and sources used for WL and RL. Power consumed by explicit supply is considered a product of the average source current and source voltage. The total power mainly has three components such as (1) static or leakage power, (2) dynamic or charging–discharging power, and (3) short circuit power consumption during switching (Figure 11.7).

$$P_{\text{total}} = P_{\text{static}} + P_{\text{dynamic}} + P_{\text{short circuit}} \tag{11.1}$$

$$P_{\text{total}} = V_{dd} I_{\text{static}} + \alpha V_{dd}^{2} C_L f_{clk} + t_{sc} V_{dd} I_{\text{peak}} f_{clk} \tag{11.2}$$

Here, α=activity factor; C_L=load capacitance; f_{clk}=clock frequency; t_{sc}=switching time

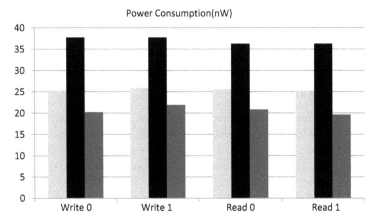

FIGURE 11.7   Power comparisons of CMOS SRAM and FinFET SRAM cells.

11.5 DIFFERENT TYPES OF LEAKAGE CURRENT IN SRAM

In deep submicron CMOS technology, leakage current becomes an enormous problem. In semiconductor devices, leakage is a quantum occurrence where mobile charge carriers (electrons or holes) tunnel through an insulating region. As the thickness of the insulating region decreases, leakage increases exponentially. Junction tunneling leakage can also take place across between heavily doped P-type and N-type semiconductor junctions. Carriers can also leak between source and drain terminals of a metal oxide semiconductor (MOS) transistor, other than tunneling via the gate insulators or junctions. This is identified as subthreshold conduction. Although the primary source of leakage occurs inside transistors, electrons can also leak between the interconnects. Leakage increases power consumption and, if sufficiently large, can cause complete circuit failure. The three major components of leakage current such as subthreshold leakage, gate leakage, and junction tunneling leakage have been explained as follows.

11.5.1 SUBTHRESHOLD LEAKAGE CURRENT

For FinFET transistors, the key component of the static leakage current is the subthreshold current. The front-gate leakage is nearly independent of the back-gate in DG transistors. By and large, subthreshold leakage is the drain source current of the transistor when the gate source voltage is less than the threshold voltage $V_{gs} < V_{th}$. In a double-gate FinFET, the subthreshold current can be illustrated as [22].

$$I_{subthreshold} = \mu C_{dep} \frac{W}{L} V_T^2 e^{\frac{V_{gs} - V_T}{V_T}} (1 - e^{\frac{-V_{ds}}{nV_T}}) \qquad (11.3)$$

where K stands for Boltzmann constant, T denotes temperature in Kelvin, A and b are the fitting parameters, T_{si} is silicon body thickness, L_{eff} is the length of gate, H_{fin} is the height of fin, V_{gs} depicts gate-to-source voltage, V_t: means the threshold voltage, q stands for charge of electron (1.6×10^{-19}), V_{ds} is the drain-to-source voltage, and I_{sub} shows the subthreshold current.

11.5.2 GATE LEAKAGE

Oxide scaling increases the field across the oxide. The high electric field along with the low oxide thickness causes gate tunneling leakage current from the gate to the channel and source/drain overlap region or from the source/drain overlap region to the gate. Gate leakage chiefly bears three components: gate-to-source/drain overlap current, gate-to-channel current, and gate-to-substrate current. Gate-to-source/drain overlap current exists when the transistor is in OFF state, and gate-to-channel leakage current takes place when the transistor is in ON state. Gate-to-substrate leakage is less than the gate-to-channel and gate-to-source/drain overlap current so that gate leakage current in OFF state is less than gate leakage current in ON state [23].

11.5.3   Junction Tunneling Leakage

Normally, junction tunneling leakage takes place in reverse-biased PN junction which consists of two constituents. One is generation of electron–hole pairs in the depletion region of reverse-biased junction, and other is diffusion of minority carriers (either electrons or holes) near the edge of depletion region. It is an exponential function of reverse bias voltage and junction doping [24]. Junction leakage does not add more in the total leakage current.

11.5.4   Different Leakage Reduction Techniques

11.5.4.1   Self-controllable Voltage Level Technique

Several methods are used to reduce the power and maintain the high speed performance. Self-controllable voltage level (SVL) circuit is the important type that not only decreases power but also improves the high speed performance. The operating speed is made maximum by generating the maximum or minimum supply voltage and the ground level voltage to minimum or maximum during the switching states between active mode and standby mode. The circuit includes switches that operate in the weak inversion region when their gate voltages are just below their threshold voltages. This makes the drain source voltages (V_{ds}) of the "off FinFETs" in the standby load circuits to decrease and increases the substrate p-type voltage. This increases the threshold voltage; consequently, subthreshold current I_{sub} of the "off FinFETs" decreases, so power is reduced, while data are retained. The SVL circuit consists of three types, namely (i) the upper SVL (USVL) circuit, (ii) the lower SVL (LSVL) circuit, and (iii) combination of the USVL and LSVL circuits [28].

11.5.4.2   Lower Self-controllable Voltage Level

The LSVL circuit comprises of a one n-channel FinFET transistor and two p-channel FinFET transistors associated in arrangement, put between a ground-level power supply V_{ss} and FinFET-based 7T SRAM cell [25,26]. The LSVL circuit supplies V_{ss} to the 7T SRAM cell through the M8 transistor and furthermore supplies V_{ss} to the remaining by 7T SRAM cell using M9 and M10 transistors. On account of LSVL circuit (Figure 11.8), 0 V is connected on the control signal (CSB) of the 7T SRAM cell; hence, M10 transistor is turned ON and M8 is off so that V_{ss} is associated with the 7T SRAM cell. During dynamic mode, the CSB switch gives 0 V at ground level and a raised virtual ground during the idle mode. This technique is same as the diode footed storage structure strategy proposed to subedge leakages and control signal in FinFET-based 7T SRAM cell. [27]

As result, LSVL method decreases leakage flows through M4, M5, and M1. The outcome of the method can be expressed in such a way that all subthreshold currents are reduced utilizing LSVL technique, by reducing gate leakage.

11.5.4.3   Upper Self-controllable Voltage Level

The USVL circuit comprises of a single p-channel FinFET transistor and two n-channel FinFET transistors associated in arrangement [28]. The ON p-channel FinFET transistor associates a power supply V_{DD} and FinFET-based 7T SRAM cell

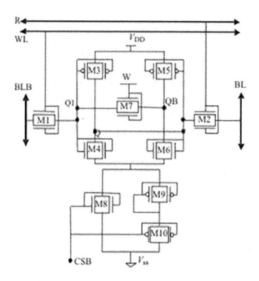

FIGURE 11.8   7T SRAM LSVL. LSVL, lower self-controllable voltage level.

in the dynamic mode on solicitation, and ON n-channel FinFET transistor associates with V_{DD} and 7T SRAM cell in sleep mode. Figure 11.3 demonstrates the
FinFET-based 7T SRAM cell with SVL (USVL) circuit. In this technique, a full
supply voltage (V_{DD}=1 V) is connected to the FinFET-based 7T SRAM cell in
dynamic mode, whereas the supply voltage V_{DD} level to FinFET-based 7T SRAM is
reduced to voltage level VD in sleep mode. According to simulation, gate voltage of

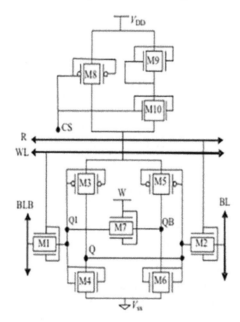

FIGURE 11.9   7T SRAM USVL. USVL, upper self-controllable voltage level.

transistor M6 is decreased, and gate leakage current through transistor M6 is also reduced. Because of decrease in channel voltage of transistor M4 than gate leakage current, a lower gate channel voltage crosswise over transistor M4 additionally reduces. The gate leakage current of transistor M1 stays unaltered. The p-channel FinFET transistor M8 does not include huge leakage current utilizing USVL circuit as shown in Figure 11.9. USVL is a superior method to reduce leakage current. In any case, the sublimit leakage current is better with difference than this method. The subedge leakage current is decreased through transistors M4 and M5, and leakage current stays unaltered crosswise over transistor M1.

11.6   CONCLUSION

Different SRAM cell design topologies of 7T, 8T, 10T, and 12T performances have been compared in terms of leakage power. The speed of SRAM cell is increased, as the process technologies continue to advance, but devices will be more vulnerable to gaps, which damage the static noise margin of SRAM cells. An 18-nm FinFET-based SRAM cell circuit is implemented to reduce the leakage power. It is found that by applying SVL technique, there was a reduction of leakage current in 18-nm FinFET-based SRAM cell design.

REFERENCES

1. N. Verma, A. P. Chandrakasan 2008. A 256 kb 65 nm 8T subthreshold SRAM employing sense-amplifier redundancy. *IEEE Journal of Solid-State Circuits* 43(1): 141–149.
2. T.-H. Kim, J. Liu, J. Keane, C. H. Kim 2008. A 0.2 V, 480 kb subthreshold SRAM with 1 k cells per bitline for ultra-low-voltage computing. *IEEE Journal of Solid-State Circuits* 43: 518–529.
3. B. H. Calhoun, A. P. Chandrakasan 2007. A 256-kb 65-nm sub-threshold SRAM design for ultra-low-voltage operation. *IEEE Journal Solid-State Circuits*, 42: 680–688.
4. M.-H. Tu, J.-Y. Lin, M.-C. Tsai, S.-J. Jou 2010. Single-ended sub-threshold SRAM with asymmetrical write/read-assist. *IEEE Transactions on Circuits and Systems*, 57(12): 3039–3047.
5. M. Sharifkhani, M. Sachdev 2009. An energy efficient 40 Kb SRAM module with extended read/write noise margin in 0.13 um CMOS. *IEEE Journal Solid-State Circuits*, 44: 620–630.
6. S. K. Jain, P. Agarwal 2006. A low leakage SNM free SRAM cell design in deep sub micron CMOS technology. *19th International Conference on VLSI Design (VLSID'06)*, IEEE: 1063–9667.
7. I. J. Chang, J. J. Kim, S. P. Park, K. Roy 2008. A 32 kb 10 T subthreshold SRAM array with bit-interleaving and differential read scheme in 90 nm CMOS. *ISSCC Digest of Technical Papers*, 3–7: 388–622.
8. P. Athe, S. Dasgupta 2009. A comparative study of 6T, 8T and 9T Decanano SRAM cell. *IEEE Symposium on Industrial Electronics & Applications*: 889–894.
9. A. A. Mazreah, M. R. Sahebi, M. T. Manzuri, S. J. Hosseini 2008. A novel zero-aware four-transistor SRAM cell for high density and low power cache application. *International Conference on Advanced Computer Theory and Engineering*, IEEE: 571–575.
10. S. Lin, Y.-B. Kim and F. Lombardi 2008. A 32 nm SRAM design for low power and high stability. *51st Midwest Symposium on Circuits and Systems*, IEEE: 422–425.

11. P. Upadhyay, R. Mehra, N. Thakur 2010. Low power design of an SRAM cell for portable devices. *International Conference on Computer and Communication Technology (ICCCT)*, IEEE: 255–259.

12. S. S. Rathod, A. K. Saxena, S. Dasgupta 2010. A proposed DG-FinFET based SRAM cell design with RadHard capabilities. *Microelectronics Reliability*, 50(8): 1039–1190.

13. C. B. Kushwah, S.K. Vishvakarma, D. Dwivedi 2016. A 20 nm robust single-ended boostless 7T FinFET sub-threshold SRAM cell under process–voltage–temperature variation. *Microelectronics Journal* 51: 75–88.

14. Kumar T. S., Tripathi S. L. 2019. Implementation of CMOS SRAM cells in 7, 8, 10 and 12-transistor topologies and their performance comparison. *International Journal of Engineering and Advanced Technology*, 8(2S2).

15. G. Chen, D. Sylvester, D. Blaauw, T. Mudge 2010. Yield-driven near-threshold SRAM design. *IEEE Transaction on VLSI System*, 18: 1590–1598.

16. U. R. Karpuzcu, A. Sinkar, N.S. Kim, J. Torrellas 2013. Energy smart: Toward energy efficient many cores for near-threshold computing, *Proceedings of IEEE HPCA*: 542–553.

17. R. G. Dreslinski, M. Wieckowski, D. Blaauw, D. Sylvester, T. Mudge 2010. Nearthreshold computing: Reclaiming Moore's law through energy efficient integrated circuits. *Proceedings of the IEEE*, 98: 253–266.

18. B. H. Calhoun, D. Brooks 2010. Can subthreshold and near-threshold circuits go mainstream?. *IEEE Micro*, 30: 80–85.

19. K. Takeda 2006. Low-Vdd static-noise-margin-free for read SRAM cell in high-speed applications. *Journal of Solid-State Circuits*: 113–121.

20. L. Chang, D. M. Fried, J. Hergenrother, J. W. Sleight, R. H. Dennard, R. K. Montoye, L. Sekaric, S. J. McNab, A. W. Topol, A. D. Adams, K. W. Guarini, and W. Haensch 2005. Stable SRAM cell design for the 32 nm node and beyond. *Proceedings of the IEEE on Very Large Scale Integration (VLSI) Technology*: 128–129.

21. C. Visweswariah 2003. Death, taxes and failing chips. *Proceedings of the IEEE on Design Automation Conference*: 343–347.

22. Y. Taur, X. Liang, W. Wang, and H. Lu 2004. A continuous, analytic drain current model for DG MOSFETs. *IEEE Transactions on Electron Device Letters*. 25(2): 107–109.

23. V. Sikarwar, S. Khandelwal and S. Akashe 2013. Analysis of leakage reduction techniques in independent gate DG FinFET SRAM Cell. *Chinese Journal of Engineering*: 2013: 1–8. https://doi.org/10.1155/2013/738358

24. S. Birla, N. Kr. Shukla, R. K. Singh and M. Pattanaik 2010, June. Leakage current reduction in 6T single cell SRAM at 90 nm technology. *Proceedings of the IEEE International Conference on Advances in Computer Engineering*: 292–294.

25. C. Duari and S. Birla 2018, October. Leakage power improvement in SRAM Cell with clamping diode using reverse body bias technique. *Proceedings of the 2nd International Conference on Data Engineering and Communication Technology, Advances in Intelligent Systems and Computing*: 828.

26. M. Manorama, S. Khandelwal, S. Akashe 2013, 12–14 April. Design of a FinFET based inverter using MTCMOS and SVL leakage reduction technique, *Students Conference on Engineering and Systems (SCES)*.

27. S. Akashe, S. Sharma 2013. Leakage current reduction techniques for 7T SRAM Cell in 45 nm technology. *Wireless Personal Communication* 71: 123–136.

28. K. Endo, S-I. O'inchi, Y. Ishikawa, E. Suzuki 2009. Independent-double-gate FinFET SRAM for leakage current reduction. *IEEE Electron Device letters* 30(7): 757–759.

12 Performance Analysis of AlGaN/GaN Heterostructure Field-Effect Transistor (HFET)

Yogesh Kumar Verma
Lovely Professional University

Santosh Kumar Gupta
Motilal Nehru National Institute of Technology Allahabad

CONTENTS

12.1 Introduction 215
12.2 Model Description 216
12.3 Results and Discussions... 216
12.4 Conclusion 221
References.. 221

12.1 INTRODUCTION

AlGaN and GaN materials are being exploited in the semiconductor electronics market from the past decade. The materials are popularly used in heterojunction bipolar transistors and heterojunction field-effect transistors (HFETs) [1–9]. Because of the merits of FETs over bipolar transistors, the HFETs are still under research. The difference between the electron affinities of both the materials causes high inversion charge at the heterointerface. The AlGaAs/GaAs HFETs were extensively a main area of research before the past decade. However, the magnitude of 2-DEG density was limited only to 10^{16}m^{-2}, which is 10 order of magnitude lesser than the AlGaN/GaN HFET [2–19]. The formation of 2-DEG density in AlGaAs/GaAs HFET depends on the doping concentration in the barrier layer. There is AlN buffer layer to reduce the leakage current, thereby minimizing the leakage issue in the device. The geometry of the device is also responsible for enhancing the performance of the device. The spacing between source–gate and gate–drain is deterministic to improve the performance of the device [20–34]. In this chapter, the spacing between source–gate and gate–drain is varied, and its effect on the current–voltage characteristics of the device is analyzed. This chapter

215

is divided into four sections: first section is introduction, second section is model description, third section represents the results and discussions, and fourth section represents the results.

12.2   MODEL DESCRIPTION

Figure 12.1 represents the stucture of AlGaN/GaN HFET. The structure comprises of different layers including AlGaN and GaN.

12.3   RESULTS AND DISCUSSIONS

Figure 12.2 represents the output current with respect to different source–gate spacing. The characteristics are calculated for different values of gate voltage (= 0 and −1 V).

Figure 12.3 represents the output characteristics compared for 1-1, 1-2, 1-3, 1-4, 1-5, and 1-6 at gate voltage of 0 and −1 V, respectively.

The transfer characteristics are compared for different values of spacing between source and gate. The drain voltage is kept constant at 1 and 2 V as represented by Figure 12.4a and b, respectively. The aluminum composition in the barrier layer is 0.30. The complete analysis is performed for identical conditions and device dimensions to maintain uniformity.

The transfer characteristics are compared for 1-1, 1-2, 1-3, 1-4, 1-5, and 1-6 at drain voltages of 1 and 2 V represented by Figure 12.5a and b, respectively.

Figure 12.6 represents the output voltage characteristics of the present device for different values of the load resistance.

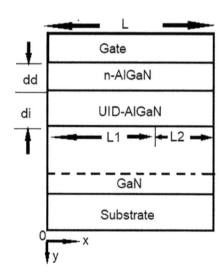

FIGURE 12.1   Structure of AlGaN/GaN HFET [1]. HFET, heterostructure field-effect transistor.

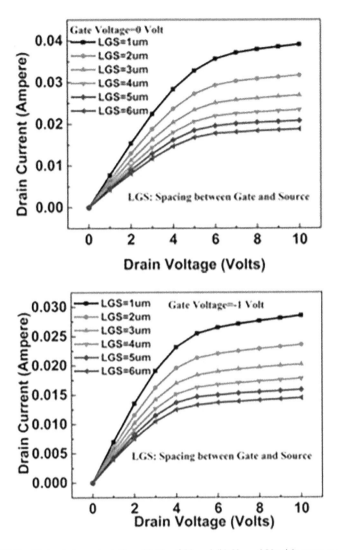

FIGURE 12.2 Output characteristics: (a) $V_G=0\,\mathrm{V}$ and (b) $V_G=-1\,\mathrm{V}$ with respect to different source spacing.

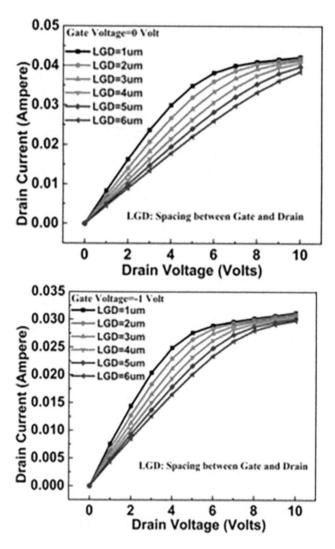

FIGURE 12.3 Output characteristics of AlGaN/GaN HFET for 1-1, 1-2, 1-3, 1-4, 1-5, and 1-6 at gate voltage of (a) 0 V and (b) −1 V. HFET, heterostructure field-effect transistor.

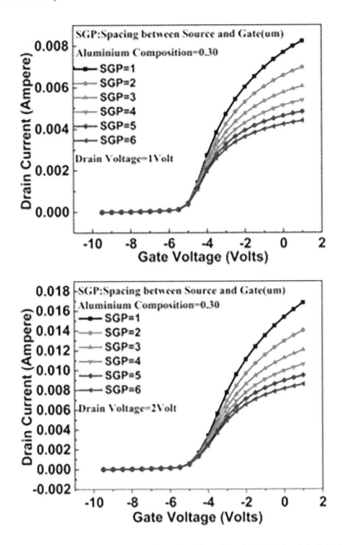

FIGURE 12.4   (a, b) Transfer characteristics of AlGaN/GaN HFET for 1-1, 2-1, 3-1, 4-1, 5-1, and 6-1 at drain voltage of (a) 1 V and (b) 2 V. HFET, heterostructure field-effect transistor.

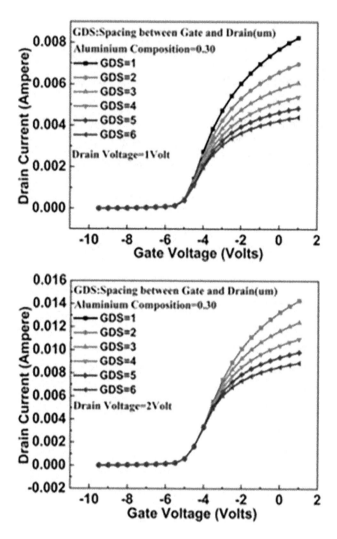

FIGURE 12.5 Transfer characteristics for 1-1, 1-2, 1-3, 1-4, 1-5, and 1-6 at (a) $V_D = 1$ V and (b) $V_D = 2$ V.

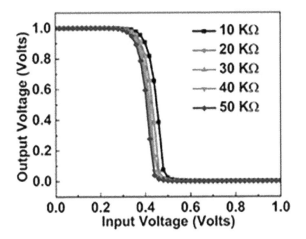

FIGURE 12.6 VTC curve for different values of load resistance. VTC, voltage transfer characteristics.

12.4 CONCLUSION

In this chapter, the analysis of AlGaN/GaN HFET for different device geometries is performed. The invertor circuit is also designed to utilize the present analysis. It has been revealed that the drain current reduces significantly, when the spacing between source and gate is reduced. However, the effects of reducing the spacing between gate and drain are less significant on the current–voltage characteristics.

REFERENCES

1. S. Khandelwal, Y. S. Chauhan, and T. A. Fjeldly. Analytical modeling of surface-potential and intrinsic charges in AlGaN/GaN HEMT devices. *IEEE Transactions on Electron Devices*. 2012, vol. 59, No. 10, pp. 2856–2860. ISSN: 0018-9383. DOI: 10.1109/TED.2012.2209654.
2. Y. K. Verma, V. Mishra, P. K. Verma, and S. K. Gupta. Analytical modelling and electrical characterisation of ZnO based HEMTs. *International Journal of Electronics (IJE, Taylor and Francis Group)*. 2019, November, vol. 106, No. 5, pp. 707–720. DOI: 10.1080/00207217.2018.1545931.
3. L. Wang, J. Liu, W. Zhou, Z. Xu, Y. Wu, and H. Tao. A novel method to dynamic thermal impedance and channel temperature extraction of GaN HEMTs. *International Journal of Numerical Modeling: Electron Networks, Devices, and Fields*. 2019, p. 2599. ISSN: 1099-1204, DOI: 10.1002/jnm.2599.
4. N. Mendiratta, and S. L. Tripathi. A review on performance comparison of advanced MOSFET structures below 45 nm technology node. *Journal of Semiconductor, IOP science*. 2020, vol. 41, pp. 1–10.
5. Y. K. Verma, V. Mishra, and S. K. Gupta. Analog/RF and linearity distortion analysis of MgZnO/CdZnO quadruple-gate field effect transistor (QG-FET). *Silicon*, Springer. 2020. DOI: 10.1007/s12633-020-00406-4.
6. S. L. Tripathi, R. Patel, and V. K. Agrawal. Low leakage pocket junction-less DGTFET with bio sensing cavity region. *Turkish Journal of Electrical Engineering and Computer Sciences*. 2019, vol. 27, No. 4, pp. 2466–2474.

7. S. K. Gupta, A. S. Rawat, Y. K. Verma, and V. Mishra. Linearity distortion analysis of junctionless quadruple gate MOSFETs for analog applications. *Silicon*, Springer. 2019 February, vol. 11, No. 1, pp. 257–265. DOI: 10.1007/s12633-018-9850-z.

8. V. Mishra, Y. K. Verma, and S. K. Gupta. Surface potential based analysis of ferroelectric dual material gate all around (FE-DMGAA) TFETs. *International Journal of Numerical Modelling, Wiley Publications*. DOI: 10.1002/jnm.2726.

9. S. L. Tripathi and G. S. Patel. Design of low power $Si_{0.7}Ge_{0.3}$ pocket junction-less tunnel FET using below 5 nm technology. *Wireless Personal Communication*. 2019, pp. 1–10. DOI: 10.1007/s11277-019-06978-8.

10. S. Russo and A. Di Carlo. Influence of the source-gate distance on the AlGaN/GaN HEMT performance. *IEEE Transactions on Electron Devices*. 2007, vol. 54, No. 5, pp. 1071–1075. ISSN: 0018-9383, DOI: 10.1109/TED.2007.894614.

11. V. Mishra, Y. K. Verma, P. K. Verma, and S. K. Gupta. EMA based modeling of surface potential and drain current of dual material gate all around TFETs. *Journal of Computational Electronics*, Springer. 2018, December, vol. 17, No. 4, pp. 1596–1602, DOI: 10.1007/s10825-018-1250-5.

12. W. Dongfang, W. Ke, Y. Tingting, and L. Xinyu. High performance AlGaN/GaN HEMTs with 2.4 μm source-drain spacing. *Journal of Semiconductors*. 2010, vol. 31, No. 3, p. 34001, ISSN: 2058-6140, DOI: 10.1088/1674-4926/31/3/034001.

13. S. L. Tripathi and S. Saxena. Asymmetric gated Ge-Si0.7Ge0.3 nHTFET and pHTFET for steep subthreshold characteristics. *International Journal of Microstructure and Materials Properties*, Inderscience. 2019, vol. 14 No. 6, pp. 497–509.

14. W. Dongfang, Y. Tingting, W. Ke, C. Xiaojuan, and L. Xinyu. Gate-structure optimization for high frequency power AlGaN/GaN HEMTs. *Journal of Semiconductors*. 2010, vol. 31, No. 5, p. 54003, ISSN: 2058-6140, DOI: 10.1088/1674-4926/31/5/054003.

15. Y. K. Verma, V. Mishra, and S. K. Gupta. A physics based analytical model for MgZnO/ZnO HEMT. *Journal of Circuits, Systems, and Computers (JCSC*, World Scientific Publishers). 2020 January, vol. 29, No. 1, pp. 2050009-1. DOI: 10.1142/S0218126620500097.

16. M. T. Bin Kashem and S. Subrina. Analytical modeling of channel potential and threshold voltage of triple material gate AlGaN/GaN HEMT including trapped and polarization-induced charges. *International Journal of Numerical Modeling: Electron Networks, Devices, and Fields*. 2019, vol. 32, No. 1, p. 2476. ISSN: 1099-1204, DOI: 10.1002/jnm.2476.

17. S. K. Sinha, S. L. Tripathi, and G. S. Patel. Mole-fraction engineering in Germanium source pocket based tunnel field effect transistor. *Sensor Letters*. 2019, vol. 17, pp. 470–473.

18. N.-Q. Zhang, S. Keller, G. Parish, S. Heikman, S. P. Denbaars, and U. K. Mishra. High breakdown GaN HEMT with overlapping gate structure. *IEEE Electron Device Letters*. 2000, vol. 21, No. 9, pp. 421–423. ISSN: 0741-3106, DOI: 10.1109/55.863096.

19. Y. K. Verma, V. Mishra, and S. K. Gupta. Electrical characterization of AlGaN/GaN quadruple gate heterostructure field effect transistor for analog applications. *Journal of Nanoelectronics and Optoelectronics*, American Scientific Publishers. 2019 October, vol. 14, No.10, pp. 1491–1502(12). DOI: 10.1166/jno.2019.2655.

20. T. S. Kumar and S. L. Tripathi. Implementation of CMOS SRAM cells in 7, 8, 10 and 12-transistor topologies and their performance comparison. *International Journal of Engineering and Advanced Technology*. 2019 January vol. 8, No. 2S2, pp. 227–229.

21. Y. K. Verma, V. Mishra, and S. K. Gupta. Linearity distortion analysis of III-V and Si quadruple gate field effect transistor (QG-FET) for analog applications. *Journal of Nanoelectronics and Optoelectronics*, American Scientific Publishers. 2020, vol. 15, pp. 1–18. DOI: 10.1166/jno.2020.2741.

22. B. Vandana, S. K. Mahapatra, and S. L. Tripathi. Impact of channel engineering (Si1-0.25ge0.25) technique on gm (transconductance) and its higher order derivatives of 3D conventional and wavy Junctionless FINFETS (JLT). *Facta Universitatis, Series: Electronics and Energetics.* 2018 June, vol. 31, No. 2, pp. 257–265.

23. V. Mishra, Y. K. Verma, and S. K. Gupta. Investigation of localized charges and temperature effect on device performance of ferroelectric dual material gate all around TFETs. *Journal of Nanoelectronics and Optoelectronics*, American Scientific Publishers. 2019 February, vol. 14, No. 2, pp. 161–168. DOI: 10.1166/jno.2019.2462.

24. S. L. Tripathi, S. K. Sinha, G. S. Patel, and S. Awasthi. High performance low leakage pocket SixGe1-x junction-less single-gate tunnel FET for 10 nm technology. *IEEE EDKCON.* 2019, pp. 161–165.

25. V. Mishra, Y. K. Verma, and S. K. Gupta, Investigation of localized charges on linearity and distortion performance of ferroelectric dual material gate all around TFETs. *Journal of Nano and Electronic Physics (JNEP).* 2019, vol. 11, No. 04, 04014 (6 pp).

26. O. Ambacher, B. Foutz, J. Smart, J. R. Shealy, N. G. Weimann, K. Chu, M. Murphy, A. J. Sierakowski, W. J. Schaff, L. F. Eastman, R. Dimitrov, A. Mitchell, and M. Stutzmann. Two dimensional electron gases induced by spontaneous and piezoelectric polarization in undoped and doped AlGaN/GaN heterostructures. *Journal of Applied Physics*, 2000, vol. 87, No. 1, pp. 334–344. ISSN: 0021-8979, DOI: 10.1063/1.371866.

27. Y. K. Verma, S. K. Gupta, V. Mishra, and P. K. Verma. Surface potential based analysis of MgZnO/ZnO high electron mobility transistors. *IEEE International Students' Conference on Electrical, Electronics and Computer Science (SCEECS)*, 2018, MANIT, Bhopal, pp. 1–4.

28. Y. K. Verma, S. K. Gupta, and R. K. Chauhan. Analysis of inherent properties of SiGe hetero-structure device using analytical modelling and simulation (I2CT). *IEEE, 2nd International Conference for Convergence in Technology*, 2017 April 07–09, Pune.

29. V. Mishra, S. K. Gupta, Y. K. Verma, V. Ramola, and A. Bora. A high-gain, low-power latch comparator design for oversampled ADCs. *Signal Processing and Integrated Networks (SPIN), IEEE*, 2018, February 22–23, Amity University, Noida.

30. V. Mishra, Y. K. Verma, P. K. Verma, N. Q. Singh, and S. K. Gupta. Performance of double gate tunnel FET devices with source pocket. Advances in VLSI, Communication, and Signal Processing, 2020, Springer, Singapore. pp. 387–396.

31. D. S. Mehta, V. Mishra, Y. K. Verma, and S. K. Gupta. A hardware minimized gated clock multiple output low power linear feedback shift register. Advances in VLSI, Communication, and Signal Processing, 2020, Springer, Singapore. pp. 367–376.

32. J. W. Chung, W. E. Hoke, E. M. Chumbes, T. Palacios, and A. We. AlGaN/GaN HEMT with 300 GHz fmax. *IEEE Electron Device Letters.* 2010, vol. 31, No. 3, pp. 195–197, ISSN: 0741-3106, DOI: 10.1109/LED.2009.2038935.

33. D. S. Mehta, V. Mishra, Y. K. Verma, and S. K. Gupta. A novel dual material extra insulator layer fin field effect transistor for high-performance nanoscale applications. Advances in VLSI, Communication, and Signal Processing, 2020, Springer, Singapore. pp. 377–386.

34. Y. K. Verma, V. Mishra, P. K. Verma, S. K. Gupta, and R. K. Chauhan. Impact of extrinsic reliability issues including radiation and temperature on SiGe HBT. *Computational and Characterization Techniques in Engineering and Sciences (CCTES)*, 2018 September 14–15, Integral University, Lucknow, UP, India.

13 Synthesis of Polymer-Based Composites for Application in Field-Effect Transistors

Amit Sachdeva
Lovely Professional University

Pramod K. Singh
Sharda University

CONTENTS

13.1 Introduction 225
13.2 Polymer-Based Composites 226
13.3 Methods of Synthesis... 228
    13.3.1 Solution Casting Method 228
    13.3.2 Copolymerization 228
    13.3.3 Addition of Ceramic Fillers.. 228
    13.3.4 Sol–Gel Process... 229
    13.3.5 Plasticization.. 229
    13.3.6 Nanofillers... 230
13.4 Conclusion 230
References.. 231

13.1 INTRODUCTION

From almost the past decade, polymer-based composites have gained a lot of attraction on account of its interesting properties for the scientific community. Variation in properties may be in the form of enhanced electrical properties, mechanical properties, magnetic properties, or optical properties. On account of some excellent properties, these materials have been widely used in different applications in the field of medical, food, or consumer goods, thus serving the society. In regard to a huge number of research articles, scientific reports, or journals, this chapter aims at narrating various methods of synthesis of these polymer-based composites.

Composite materials are defined as materials that are made by combination of two or more than two materials, leading to some enhanced properties. While synthesizing a composite material, one must make sure that there is no chemical reaction among components of a composite. All the components making a complete composite are superficially placed without any chemical reaction between them. Nowadays, we have a special class of composites known as nanocomposites. In nanocomposites, at least one of the components of composite must have a nanoscale dimension [1,2]. Also for application in FETs, we use a special class of composites known as polymer electrolyte [3].

13.2 POLYMER-BASED COMPOSITES

When we talk about composites, one of the main applications of polymer-based composites is in the form of electrolytes. Before discussing about polymer electrolytes, one must be clear with the term electrolyte. Electrolytes are defined as materials that conduct electricity either in their molten state or in the form of aqueous solution [4]. Generally, electrolytes are formed when a salt is dissolved in a polar solvent such as water. On dissolving in polar solvent, the salt gets dissociated into respective positively and negatively charged ions, which are responsible for its conduction. Such materials that conduct via movement of ions are known as ionic conductors.

Classification of electrolytes (Figure 13.1):

    a. Liquid electrolytes
    b. Gel electrolytes
    c. Solid electrolytes

Liquid electrolytes are ion-containing solutions that allow free movement of ions. But they suffer major setback on account of their leakage problems. On the other hand, gel electrolytes are in the form of a semisolid mass that show more stability and equivalent conduction as in liquid electrolytes. In liquid and gel electrolytes, charge carriers, i.e., ions conduct while moving in a polar solvent such as ethylene carbonate, propylene carbonate, butyrolactone, and others or their mixtures [5–7]. Value of ionic conductivity of these electrolytes is sufficient for their application in the commercial lithium battery (for laptops or and cellular phones) working at ambient temperature. Another drawback associated with these electrolytes is formation of an additional insulating layer during their application in batteries. The decomposition products of these electrolyte get deposited on the electrode, form insulating layer at electrode–electrolyte interface, and hinder the movement of ions. This ultimately reduces the life span of the product. Furthermore, it is very difficult to transport devices based on liquid electrolyte due to its bulky nature. Usually, liquid electrolytes have low boiling

liquid electrolyte   gel electrolyte   hybrid liquid/solid electrolyte

FIGURE 13.1 Types of electrolytes.

point as compared with solid electrolytes. So use of liquid electrolytes in batteries or other devices puts a limit on the temperature range within which the material can be used commercially. These drawbacks paved way for the discovery of solid electrolytes range 2×10^{-2} and 10^{-3} S/cm [8–11]. Polymer electrolytes are a special class of solid electrolytes that are synthesized by blending specialized polymers such as PEMA, PEMA, PEO, PEG, and so on, with metal salts (Choudhary and Pradhan, 2007).

Polymer electrolytes have a number of advantages over liquid electrolytes such as follows:

- Polymer electrolytes are very easy to process as compared with liquid electrolytes. This ensures high flexibility in designing these polymers.
- They have adequate mechanical properties along with good thermal and electrochemical stability. Also, they form acceptable electrode/electrolyte interfacial contact.
- They are more safe as they lack the existence of any organic solvent that is flammable.

For its use as a solid-state device for electrochemical applications, a polymer electrolyte composite must have the following properties:

a. High value of transference number, i.e., Li-ion test.
b. Must possess chemical stability, thermal stability, mechanical stability along with electrochemical stability.
   - Polymer electrolyte must be compatible with electrode materials used in the cell.
   - Activation energy must decrease with increase in temperature.
   - Polymer electrolyte must not show any change in its phase within the operating temperature region.
   - For higher value of ionic conductivity, the glass transition temperature of polymer electrolyte must be low.

On account of such outstanding properties of polymers, scientific community across the globe has accepted to use polymer electrolytes for battery applications in recent years. While fabricating a polymer electrolyte, one has to choose the appropriate salt along with polymer very carefully. Below mentioned are some of the characteristics that can be considered while choosing a polymer and a salt.

a. Polymers that possess atoms or have special groups that are basically electron donors show a strong capability of making coordinate bonds with salts (cations present in salt).
b. Host polymer must have low value of glass transition temperature.
c. Polymers that show flexibility and possess low values of cohesive energy density have higher likelihood to interact with metal salts.
d. The metal salts that have low values of lattice energy have higher tendency to form a polymer salt complex. Such salts generally have monovalent small sized alkali ions and a large size anions (e.g., NH_4I, NaI, KI, etc.).

13.3   METHODS OF SYNTHESIS

13.3.1   Solution Casting Method

Films of polymer-based composites are generally prepared by solution cast-ing method. It is one of the most easiest and cheapest methods used to obtain polymer-based composites in the form of a film. Primary requirement of this method is availability of a common solvent for all the components of a composite. Also it is advisable that solvent that is being used must be volatile. In this method, all the components of a composite are separately dissolved in the common solvent using a magnetic stirrer. All the dissolved solutions in appropriate ratio are mixed together homogeneously using a magnetic stirrer [12]. As the components mix, the volatile solution gets evaporated, and finally, we have a highly viscous solution available in the beaker. This viscous solution consists of all the components of a composite in an appropriate ratio. Homogeneously mixed viscous solution is then poured in polypro-pylene petriplate where the solvent evaporates and final composite gets deposited at the bottom of the plate in the form of a film (Figure 13.2).

13.3.2   Copolymerization

In this method, one more polymer with a low value of T_g is added in addition to the original host. This tends to produce a flexible polymer backbone and hence lowers crystallinity and T_g of original host polymer and enhances conductivity (σ) (Druger et al., 1971). PEO–PPO copolymer complex with alkali metal salts has also been reported (Figure 13.3).

13.3.3   Addition of Ceramic Fillers

In order to enhance the mechanical stability of polymer complex films that are, in gen-eral, unstable, either (i) the available host polymer electrolytes are added to polymer having high value of "T"-like polyetherene (Scrosati, 1987), polymethyl methacrylate

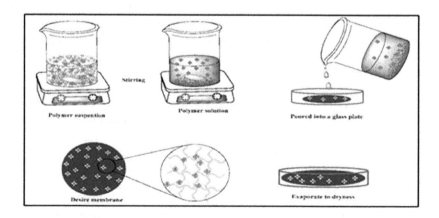

FIGURE 13.2   Solution casting method.

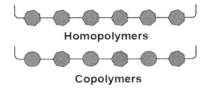

FIGURE 13.3 Homopolymers and copolymers.

(PMMA) (West et al., 1985) or (ii) some inorganic/organic or ceramic fillers (Al$_2$O$_3$, LiAlO$_2$, SiO$_2$, Li$_3$N, Nascon, PMMA, etc.) during casting process are added to the polymer solution to form composites [13–15], which have high conductivity also.

13.3.4 SOL–GEL PROCESS

Restrict the movement of liquid electrolyte in a polymeric matrix by sol–gel process to get "polymeric gel electrolytes." These gel electrolytes not only retain most of the high conductivity characteristics of the liquid phase but are also mechanically stable. The polymer simply provides the structural support (Figure 13.4).

13.3.5 PLASTICIZATION

The addition of low-molecular-weight polar molecules called plasticizers improves the segmental flexibility of original host polymer backbone chain and hence

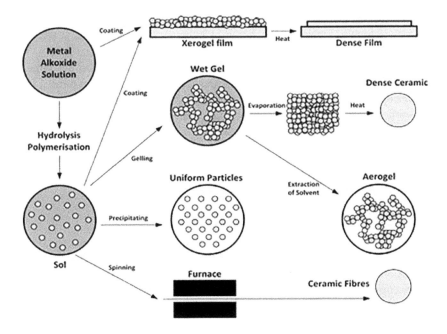

FIGURE 13.4 Sol–gel process in detail.

enhances conductivity. Some common plasticizers are propylene carbonate (PC), ethylene carbonate (EC), γ-buty-rolactone (γ-BL), *N,N*-do-methylacetamide (DMA), *N,N*-dimethylformamide (DMF), diethyl carbonate (DEC), and dimethyl carbonate (DMC). A number of reports with plasticizers are available [16].

13.3.6 NANOFILLERS

The existence of polymer nanocomposites can be dated a number of decades back. The extensive research shows that polymer nanocomposites existing in nature have nanofillers in the form of diatoms, black carbon, and silica, which had been introduced as additives in polymer matrices. The augmented properties introduced by the addition of nanofillers were not fully apprehended at that time due to lack of interest and lack of available techniques. Extensive study on the outcome of incorporation of nanofillers in polymer matrix started only after the first published research work on polyamide doped with nanoclays by Park (2014) and Torsi et al. (2003) [17,18]. Both the mentioned research publications used the term "hybrid" material instead of nanocomposite material. Swiftly, investigation in the field increased manifold and led to the appearance of introductory use of the denomination "nanocomposites." After the aforementioned voyage, a large number of young researches initiated the work on various nanosized fillers. The emergence of this field was on account of continual development in the practical development of thermoplastic as well as thermoset polymeric materials, which has further led to development of new technologies. List of various nanofillers being used has increased at an enormous rate from the past few years like we started with nanoclays, nanooxides and are moving towards carbon nanotubes, etc. With the increase in number of fillers, the matrix in which they are incorporated has also increased manifold ranging from polymers and ceramics to glasses.

As per present scenario, the evolution of polymer nanocomposites is the most active area and key aspect of progress of fields related to nanomaterial [19,20]. The existing properties introduced by the incorporation of nanoparticles are large in number and primarily focus on enhancing the electrical conduction and resist various barrier properties to effects of temperature, presence of various gases and liquids along with providing fire-retardant properties. Nanofillers significantly enhance or adjust various properties of the base materials into which they are introduced, such as optical property, electrical property, mechanical property, thermal property, or fire-retardant properties. These properties of nanocomposite materials are significantly affected by ratio of organic matrix to the nanofillers.

13.4   CONCLUSION

Organic FET is a three-terminal device consisting of drain, source, and gate. Dielectric material of gate can be synthesized using a polymer-based composite that assists in charge accumulation around gate on application of source voltage. These polymer composites are synthesized by a number of methods such as wet chemical method, ball milling, solution casting method, and so on. Out of all the methods, solution casting method is widely known for synthesis of polymer-based composites on account of ease of fabrication, low cost, and better output.

REFERENCES

1. Cavallari, M.R.; Izquierdo, J.E.E.; Braga, G.S.; Dirani, E.A.T.; Pereira-da-Silva, M.A.; Rodriguez, E.F.G.; Fonseca, F.J. Enhanced sensitivity of gas sensor based on poly(3-hexylthiophene) thin-film transistors for disease diagnosis and environment monitoring. *Sensors* **2015**, *15*, 9592–9609.

2. Klug, A.; Denk, M.; Bauer, T.; Sandholzer, M.; Scherf, U.; Slugovc, C.; List, E.J.W. Organic field-effect transistor based sensors with sensitive gate dielectrics used for low-concentration ammonia detection. *Org. Electron.* **2013**, *14*, 500–504.

3. Lu, C.; Fu, Q.; Huang, S.; Liu, J. Polymer electrolyte-gated carbon nanotube field-effect transistor. *Nano Lett.* **2004**, *4*, no. 4, 623–627.

4. Lienerth, P.; Fall, S.; Leveque, P.; Soysal, U.; Heiser, T. Improving the selectivity to polar vapors of OFET-based sensors by using the transfer characteristics hysteresis response. *Sens. Actuator B Chem.* **2016**, *225*, 90–95.

5. Sandberg, H.G.O.; Backlund, T.G.; Osterbacka, R.; Jussila, S.; Makela, T.; Stubb, H. Applications of an all-polymer solution-processed high-performance, transistor. *Synth. Met.* **2005**, *155*, 662–665.

6. Pacher, P.; Lex, A.; Proschek, V.; Etschmaier, H.; Tchernychova, E.; Sezen, M.; Scherf, U.; Grogger, W.; Trimmel, G.; Slugovc, C.; et al. Chemical control of local doping in organic thin-film transistors: From depletion to enhancement. *Adv. Mater.* **2008**, *20*, 3143–3148.

7. Ryu, G.S.; Park, K.H.; Park, W.T.; Kim, Y.H.; Noh, Y.Y. High-performance diketopyrrolopyrrole-based organic field-effect transistors for flexible gas sensors. *Org. Electron.* **2015**, *23*, 76–81.

8. Yang, Y.; Zhang, G.X.; Luo, H.W.; Yao, J.J.; Liu, Z.T.; Zhang, D.Q. Highly sensitive thin-film field-effect transistor sensor for ammonia with the DPP-bithiophene conjugated polymer entailing thermally cleavable tert-butoxy groups in the side chains. *ACS Appl. Mater. Interfaces* **2016**, *8*, 3635–3643.

9. Han, S.J.; Zhuang, X.M.; Shi, W.; Yang, X.; Li, L.; Yu, J.S. Poly(3-hexylthiophene)/polystyrene (P3HT/PS) blends based organic field-effect transistor ammonia gas sensor. *Sens. Actuator B Chem.* **2016**, *225*, 10–15.

10. Besar, K.; Yang, S.; Guo, X.G.; Huang, W.; Rule, A.M.; Breysse, P.N.; Kymissis, I.J.; Katz, H.E. Printable ammonia sensor based on organic field effect transistor. *Org. Electron.* **2014**, *15*, 3221–3230.

11. Chen, D.J.; Lei, S.; Chen, Y.Q. A single polyaniline nanofiber field effect transistor and its gas sensing mechanisms. *Sensors* **2011**, *11*, 6509–6516.

12. Yu, S.H.; Cho, J.; Sim, K.M.; Ha, J.U.; Chung, D.S. Morphology-driven high-performance polymer transistor-based ammonia gas sensor. *ACS Appl. Mater. Interfaces* **2016**, *8*, 6570–6576.

13. Wang, L.; Swensen, J.S. Dual-transduction-mode sensing approach for chemical detection. *Sens. Actuator B Chem.* **2012**, *174*, 366–372.

14. Liao, F.; Yin, S.; Toney, M.F.; Subramanian, V. Physical discrimination of amine vapor mixtures using polythiophene gas sensor arrays. *Sens. Actuator B Chem.* **2010**, *150*, 254–263.

15. Das, A.; Dost, R.; Richardson, T.; Grell, M.; Morrison, J.J.; Turner, M.L. A nitrogen dioxide sensor based on an organic transistor constructed from amorphous semiconducting polymers. *Adv. Mater.* **2007**, *19*, 4018–4023.

16. Cheon, K.H.; Cho, J.; Kim, Y.H.; Chung, D.S. Thin film transistor gas sensors incorporating high-mobility diketopyrrolopyrole-based polymeric semiconductor doped with graphene oxide. *ACS Appl. Mater. Interfaces* **2015**, *7*, 14004–14010.

17. Park, C.E. A composite of a graphene oxide derivative as a novel sensing layer in an organic field-effect transistor. *J. Mater. Chem. C* **2014**, *2*, 4539–4544.

18. Torsi, L.; Tanese, M.C.; Cioffi, N.; Gallazzi, M.C.; Sabbatini, L.; Zambonin, P.G.; Raos, G.; Meille, S.V.; Giangregorio, M.M. Side-chain role in chemically sensing conducting polymer field-effect transistors. *J. Phys. Chem. B* **2003**, *107*, 7589–7594.
19. Torsi, L.; Tafuri, A.; Cioffi, N.; Gallazzi, M.C.; Sassella, A.; Sabbatini, L.; Zambonin, P.G. Regioregular polythiophene field-effect transistors employed as chemical sensors. *Sens. Actuator B Chem.* **2003**, *93*, 257–262.
20. Lv, A.; Wang, M.; Wang, Y.; Bo, Z.; Chi, L. Investigation into the sensing process of high-performance H_2S sensors based on polymer transistors. *Chem. Eur. J.* **2016**, *22*, 3654–3659.

14 Power Efficiency Analysis of Low-Power Circuit Design Techniques in 90-nm CMOS Technology

Yelithoti Sravana Kumar, Tapaswini Samant, and Swati Swayamsiddha
Kalinga Institute of Industrial Technology, KIIT Deemed to be University

CONTENTS

14.1   Introduction 234
14.2   Existing Low-Power Techniques 235
    14.2.1   Conventional Complementary Metal Oxide Semiconductor........235
    14.2.2   Pass-Transistor Logic Style...236
    14.2.3   Differential Pass-Transistor Logic Style.......................237
    14.2.4   Transmission Gate Logic Style..237
    14.2.5   Gate Diffusion Input Logic Style 238
14.3   Proposed Low-Power Adiabatic Logic Techniques...................239
    14.3.1   Conventional Positive-Feedback Adiabatic Logic.......240
    14.3.2   Two-Phase Adiabatic Static Clocked Logic 241
14.4   Existing Design..241
    14.4.1   4×1 Multiplexer Using Conventional Complementary Metal
          Oxide Semiconductor 241
    14.4.2   4×1 Multiplexer Using Pass-Transistor Logic Style 241
    14.4.3   4×1 Multiplexer Using Differential Pass-Transistor Logic Style.... 242
    14.4.4   4×1 Multiplexer Using Transmission Gate Logic Style................242
    14.4.5   4×1 Multiplexer Using Gate Diffusion Input Logic Style 242
14.5   Proposed Design 243
    14.5.1   4×1 Multiplexer Using Conventional Positive-Feedback
          Adiabatic Logic 243
    14.5.2   4×1 Multiplexer Using Two-Phase Adiabatic Static
          Clocked Logic..245

14.6   Comparative Analysis..245
14.7   Conclusion 247
References..248

14.1   INTRODUCTION

The growing demand for low-power very large-scale integration (VLSI) can be addressed at various levels of design, including architecture, circuit, design, and process technology levels. At the circuit design level, there is a significant energy saving potential by properly selecting the logical style to implement the combinational circuit [1]. This is because all the important parameters that manage the switching capacity, the transition activity, the power consumption, and the short-circuit current are strongly influenced by the preferred logical style. Depending on the application, different types of circuits are implemented, and various design methods are used. As several aspects of performance are considered important, it is impossible to formulate universal optical rules of logical style.

This chapter analyzes 4-to-1 multiplexer using complementary metal oxide semiconductor (CMOS), transmission gate (TG), pass-transistor logic (PTL), dual pass-transistor logic (DPTL) styles, and gate diffusion input [2]. These implementations are compared based on the basis of transistor count, power dissipation, and delay [3].

A device that selects one of several analog or digital input signals and transfers the selected input to a single line is called as multiplexer. A multiplexer of inputs has n select lines, which are used to select which input line to send to the output; that is why, it is also called a data selector. Multiplexer can also be used to implement any combinational circuit [4]. So by simplifying the design of multiplexer, design of many combinational circuits can be simplified. Figure 14.1 and Table 14.1 show the block diagram and truth table for 4-to-1 multiplexer [5].

Power consumption is an important parameter in today's VLSI technology. In the adiabatic approach, the energy is stored in the charge capacitor instead of discharging to ground. The term "thermodynamics" means that there is no energy dissipation [6]. Decades of CMOS technology has created low-power apparatus, and the power consumption mainly occurs due to changing operation of the charge capacitor [7]. The two main types of dissipation are static and dynamic . Dynamic loss occurs due to the operation of the charge capacitor. Internal leakage occurs in the device that forms static electricity if circuit is not working, and dynamic power dissipation plays

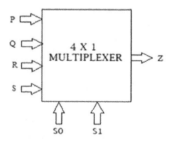

FIGURE 14.1   4×1 multiplexer block diagram.

TABLE 14.1
Truth Table of 4×1 Multiplexer

Selection Inputs		Inputs				Output
S1	S0	P	Q	R	S	Z
0	0	1	0	0	0	P
0	1	0	1	0	0	Q
1	0	0	0	1	0	R
1	1	0	0	0	1	S

a significant role in the circuit [8]. By decreasing the values of terminal capacitance and reducing the voltage supply in CMOS logic circuits, the power dissipation can be reduced [9]. Therefore, the effective model requires low power consumption. In this way, the energy is recovered by sending it back to the power supply, and this reduces overall power consumption. In this chapter, the performance of different adiabatic styles is evaluated, and also the simulation results show that it is the best approach in adiabatic logic [10].

14.2 EXISTING LOW-POWER TECHNIQUES

14.2.1 CONVENTIONAL COMPLEMENTARY METAL OXIDE SEMICONDUCTOR

The conventional or complementary CMOS logic gates consist of PMOS and NMOS as pull-up and pull-down logic networks. The CMOS logic style has the advantages of voltage scale and robustness over the size of the transistor, high noise margin, and reliable operation at low voltage. Connecting the input signal only to the transistor gate facilitates the use and characterization of the logic cells. The pairs of complementary transistors make the CMOS gate design efficient and easy. The main disadvantage of CMOS is that it has a significant number of large PMOS transistors and high input loads (Figure 14.2).

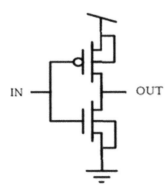

FIGURE 14.2 CMOS logic schematic diagram. CMOS, complementary metal oxide semiconductor.

CMOS circuits are constructed with the goal that each NMOS transistor requires contribution from a voltage source or other PMOS transistors. Likewise, all NMOS transistors require a contribution from the beginning additional NMOS transistor. When a low gate voltage is applied to the arrangement of PMOS transistor, it creates low resistance between source and drain, and when a high gate voltage is applied, it creates high resistance between source and drain. Similarly, the NMOS transistor arrangement makes high resistance between source and drain when a low gate voltage is applied, and when high gate voltage applied, it creates low resistance between source and drain [11]. However, during the switching time, when gate voltage passes from one state to another state, both MOSFETs (metal oxide semiconductor field-effect transistors) work for a short time.

This induces a short peak in energy consumption, which is a serious problem at high frequencies. CMOS complements all nMOSFETs with pMOSFET and connects both gates and both drains to achieve current reduction [12]. A high gate voltage causes the nMOSFET to drive and the pMOSFET to become nonconductive, but the opposite occurs when the gate voltage is low. This arrangement significantly reduces energy consumption and heat generation [13].

14.2.2 PASS-TRANSISTOR LOGIC STYLE

PTL is used to reduce the number of transistors to design different types of logics. It mainly depends on the inputs to control the gate and source to drain voltage. The advantage is that it is sufficient to perform different logical operations in a network of one-step transistors (NMOS or PMOS). Some styles of step transistor logic are considered to implement logics such as NMOS PTL, CMOS TG logic, and PLT [14].

It is the best of all these NMOS multiplexers. Two NMOS transistors are used, and that signal is selected which propagates through these two-pass transistors at the input. But the logic levels are degraded by pass transistors. Both the pass transistors have the same threshold voltage to perform proper operation (Figure 14.3).

The PTL requires a less number of transistors than the completely CMOS logic for the execution of a similar function, and it works faster and consumes less energy.

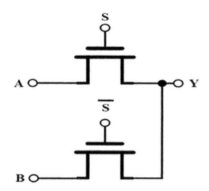

FIGURE 14.3   Schematic diagram for pass-transistor logic.

When XOR is implemented using simple logic gate, it requires more transistors in completely CMOS logic, but using PTL, it needs a less number of transistors.

14.2.3 DIFFERENTIAL PASS-TRANSISTOR LOGIC STYLE

The logic of the complementary pass transistor or the logic of the differential pass transistor refers to a logical family that is intended to a specific preferred position. In multiplexers and latches, it is the common logic family [15]. The complementary pass-transistor logic (CPL) utilizes a sequential transistor to select between the possible transforming values to estimations of the logics, and that output drives with the inverter. In the CMOS TG logic, NMOS and PMOS transistors are parallel in their connection. The CPL is utilized to show the execution style of the logic gate, where each gate is made out of a system of NMOS, only advance transistors followed by a CMOS output inverter.

CPL style implements logical gates using double-track coding. All CPL gates have two output cables for positive and complementary signals, eliminating the need for an inverter.

In the differential circuits, complementary inputs and outputs are available. The additional circuits require different signals for generation but can efficiently implement composite gates such as ADDERS, XOR, and MUX.

- The output is connected to V_{dd} or GND through a path of low resistance (high noise immunity), so CPL is a static gate.
- In the design technique, it accepted all the inputs, and it used same topology for all gates. This is a simple design of the gate library.

A twin PMOS transistor branch has been added to the N-tree DPL to avoid the problem of noise margin degradation in the CPL. This addition increases the input capacity.

In any case, the symmetric operation procedure and dual transmission characteristics compensate for the consequences of reduced speed and increased load. Full swing operation limits the threshold voltage scale and improves circuit performance at low supply voltages [16].

14.2.4 TRANSMISSION GATE LOGIC STYLE

A TG is a simple switch described as an electronic segment that clearly blocks or passes the signal level from input to output. This solid-state switch combines a PMOS transistor and an NMOS transistor [17]. The control gate is complementary to the one side with the goal that the two transistors are deactivated or activated. On the off chance that the voltage at node A is logic 1, the complementary logic 0 is applied to the dynamic low in node A, and when the two transistors are driving, the input signal is passed to the output. On the off chance that the voltage at node A is logic 0, the complementary logic 1 is applied to the dynamic high at node A, and if the two transistors are off, the high impedance state is limited at the input and output (Figure 14.4).

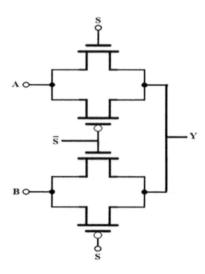

FIGURE 14.4   Transmission gate logic schematic diagram.

As a general rule, unlike conventional individual field effect transistors, TGs composed of two FETS do not have the substrate terminal (mass) inside associated with the source. In the TG logic, two transistors nMOSFET and pMOSFET are arranged in parallel, but source and drain terminals of the two transistors are connected to one another. Its gate terminals connected to one another through NOT gate (inverters) to frame control terminals.

Unlike discrete FET transistors, to represent the TG, two varieties of symbols are commonly used in this arrangement; the substrate terminal does not have the connection with the source [18]. Since the terminal of the substrate is connected to a specific supply voltage, the diode of the parasitic substrate (gate to the substrate) always reverses the polarization and does not affect the signal flow. Subsequently, the p-channel MOSFET substrate terminal is connected with the positive supply voltage potential, and the n-channel MOSFET substrate terminal is connected with the negative supply voltage potential.

14.2.5   GATE DIFFUSION INPUT LOGIC STYLE

The gate diffusion input (GDI) technique creates simple operations compared with other low-power techniques. It looks like a CMOS logic, and there are some important difference [19] (Figure 14.5).

1. The GDI cell contains three inputs and one output: G (common gate input for NMOS and PMOS), P-type source (source input/PMOS drain), N-type source (source input/NMOS drain), and D (common drain for NMOS and PMOS).
2. Both NMOS and PMOS transistors are connected to N or P (respectively), so they can be arbitrarily polarized instead of CMOS inverters.

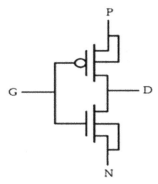

FIGURE 14.5   Gate diffusion input logic schematic diagram.

Note that not all functions are probable with standard P-well CMOS forms; however, they can be effectively executed with two-well CMOS or silicon-on-insulator (SOI) technology.

The different logical elements of the GDI cell for various input configurations significant to output are conventional polar, and there is a short circuit between N and P, resulting in static energy consumption, which is normal CMOS configuration.

There are inconveniences in implementation of OR, AND, and MUX. The impact can be reduced by executing the design in floating majority SOI technology that a full GDI library can implement.

14.3   PROPOSED LOW-POWER ADIABATIC LOGIC TECHNIQUES

The contemporary adaptable devices need a large live of energy backing. Energy ability is the primary concern for the present circuits. This can be accomplished by the low power utilization and low power scattering of the circuits. The charge regaining standard is employed within the adiabatic logic circuits. The energy is not isolated, but it is reused. The energy placed within the load capacitors is given by

$$E_{\text{stored}} = 1/2 \; C_L V_{dd}^2 \tag{14.1}$$

The amount of power worthless throughout the discharge of the load capacitor is given in Equation (14.1). CMOS assumes a significant job in decreasing control. However, later, a dynamic approach was intended to weary the weakness of CMOS logic. The adiabatic methodology used consumes less power and has low power dissipation. By using this adiabatic technique, the wasted power within the CMOS circuit is recovered [20]. There are two types of adiabatic logic square measure mentioned: partially adiabatic and fully adiabatic logic. The positive-feedback adiabatic logic (PFAL) and two-phase adiabatic static clocked logic (2PASCL) come under the partially adiabatic logic (Figure 14.6).

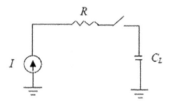

FIGURE 14.6   Adiabatic logic circuit logic schematic diagram.

14.3.1   CONVENTIONAL POSITIVE-FEEDBACK ADIABATIC LOGIC

PFAL is the new adiabatic technique that utilizes positive feedback. This level-regulated construction contains cross-coupled inverters with NMOS devices; the area unit is connected between the output and also the power clock. In PFAL, the pulse rail signal is used, referred to as the power clock that is apportioned into four stages:

1. Evaluation stage
2. Hold stage
3. Recovery stage
4. Wait stage

Throughout the evaluation stage, the outputs are measured from a steady information outstanding. Next during hold stage, the output is saved steady; in the recovery interval, the energy is recovered, and finally in the wait stage, outputs are inserted for the symmetry. PFAL is a dual-railing circuit that recognizes proportional commitments about one another and gives enhanced outputs with divided essential recovery [21].

The general schematic of the PFAL structure is shown in Figure 14.7. It consists of an adiabatic amplifier circuit and is created by the two PMOS and two NMOS

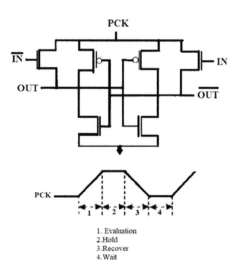

FIGURE 14.7   Conventional positive-feedback adiabatic logic schematic diagram.

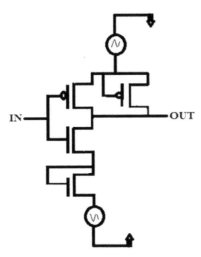

FIGURE 14.8 Two phase-adiabatic static clocked logic schematic diagram.

transistors; both output nodes are without any degradation in amplification level. The pMOSFETs of the adiabatic electronic equipment and frame a TG.

14.3.2 Two-Phase Adiabatic Static Clocked Logic

The 2PASCL comprises of two MOSFET diodes as shown in Figure 14.8. One is near the pull-up arrange and the other one is in the pull-down circuit. The 2PASCL works in two phases: one is evaluation stage and the other is hold stage. During the evaluation stage, when the data is at logic zero, the PMOS is ON and the NMOS is OFF. The capacitor charges and the output is zero, and it discharges through NMOS. The energy is sent back to the power supply.

During the hold stage, the NMOS is ON and the PMOS is OFF. The capacitor charges and the output is zero, and it releases through the PMOS. The vitality is sent back to the power supply at the draw-down system. The exchanging activity is diminished in view of the usage of sinusoidal air conditioning control supply in 2PASCL, subsequently diminishing force dissemination.

14.4 EXISTING DESIGN

14.4.1 4×1 Multiplexer Using Conventional Complementary Metal Oxide Semiconductor

See Figure 14.9.

14.4.2 4×1 Multiplexer Using Pass-Transistor Logic Style

See Figure 14.10.

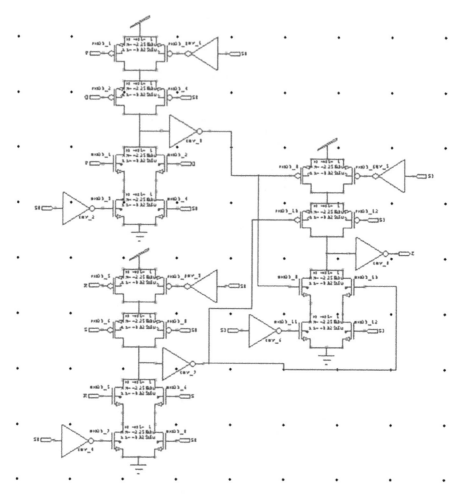

FIGURE 14.9 Schematic design diagram for 4×1 multiplexer using conventional CMOS logic. CMOS, complementary metal oxide semiconductor.

14.4.3 4×1 MULTIPLEXER USING DIFFERENTIAL PASS-TRANSISTOR LOGIC STYLE

See Figure 14.11.

14.4.4 4×1 MULTIPLEXER USING TRANSMISSION GATE LOGIC STYLE

See Figure 14.12.

14.4.5 4×1 MULTIPLEXER USING GATE DIFFUSION INPUT LOGIC STYLE

See Figure 14.13.

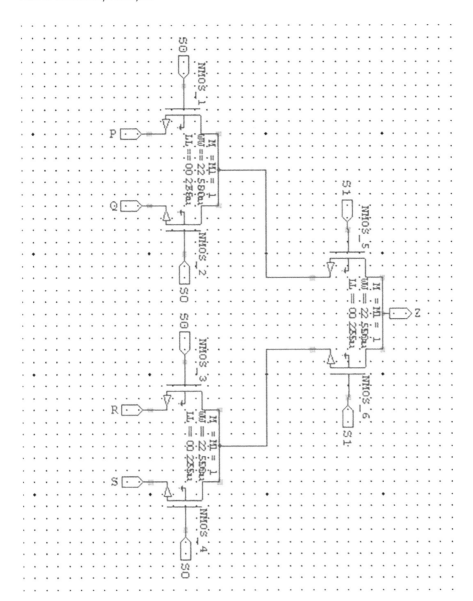

FIGURE 14.10 Schematic design diagram for 4×1 multiplexer using pass-transistor logic.

14.5 PROPOSED DESIGN

14.5.1 4×1 MULTIPLEXER USING CONVENTIONAL POSITIVE-FEEDBACK ADIABATIC LOGIC

See Figure 14.14.

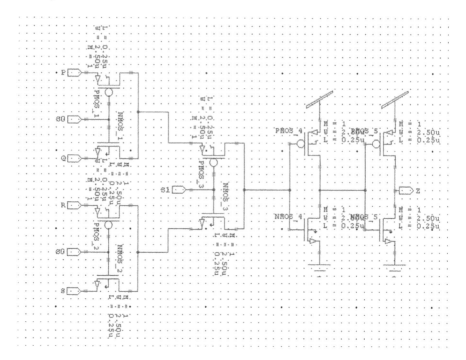

FIGURE 14.11 Schematic design diagram for 4×1 multiplexer using differential pass-transistor logic.

TABLE 14.2

Performance Comparison of Different Design Techniques with Respect to the Number of Transistors, Power Analysis, and Speed

Name of the Logic	Number of Transistors	Power Analysis		Speed	
		250 nm	90 nm	250 nm	90 nm
Conventional CMOS	26	10.511×10^{-8}	8.56×10^{-8}	21.892 ns	18.235 ns
Differential pass transistor	10	8.907×10^{-8}	5.26×10^{-8}	14.823 ns	10.549 ns
Pass transistor	6	8.657×10^{-8}	5.02×10^{-8}	16.548 ns	11.486 ns
Transmission gate	14	5.158×10^{-8}	3.56×10^{-8}	18.652 ns	13.586 ns
Gate diffusion input	6	4.163×10^{-8}	3.12×10^{-8}	19.548 ns	14.953 ns
Conventional PFAL	24	4.015×10^{-8}	3.09×10^{-8}	14.645 ns	10.561 ns
Proposed 2PASCL	26	3.374×10^{-8}	2.26×10^{-8}	12.562 ns	08.459 ns

CMOS, complementary metal oxide semiconductor; PFAL, positive-feedback adiabatic logic; 2PASCL, two-phase adiabatic static clocked logic.

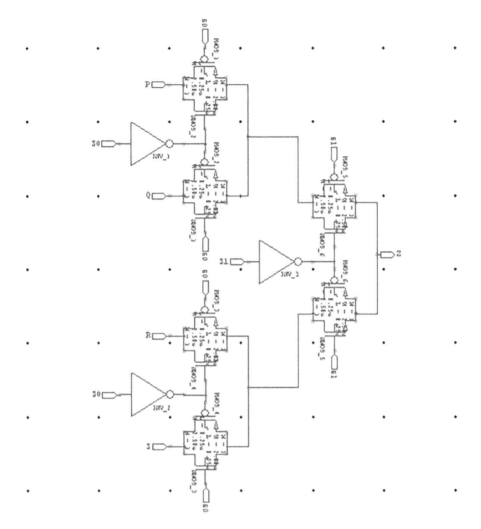

FIGURE 14.12 Schematic design diagram for 4×1 multiplexer using transmission gate logic.

14.5.2   4×1 Multiplexer Using Two-Phase Adiabatic Static Clocked Logic

See Figure 14.15.

14.6   COMPARATIVE ANALYSIS

In this section, performance analysis is done by comparing different aspects of parameters in different low-power techniques while they are compared with CMOS technology, which is used to simulate the multiplexer circuits under same position to comparison listed in the following:

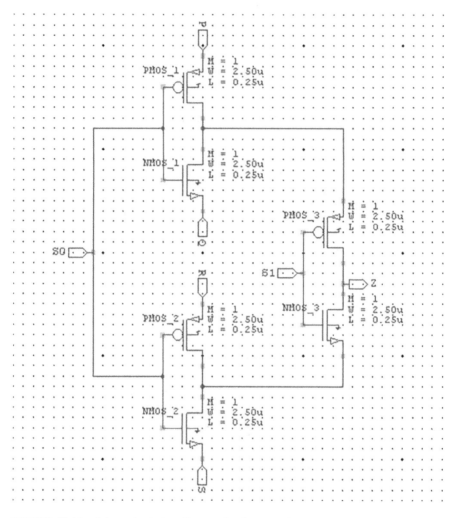

FIGURE 14.13 Schematic design diagram for 4×1 multiplexer using gate diffusion input logic.

The proposed changes are as per the following:

- The different multiplexer designs are enhanced on gate level through setting an increasingly accurate aspect relation to accomplish a lower power consumption.
- For the multiplexer, the power consumption is reduced by changing the supply voltages in the range of 0.1–5 V, so as to fit to the essential input dynamic range. In this way, the power consumption improvement rate is 12%.

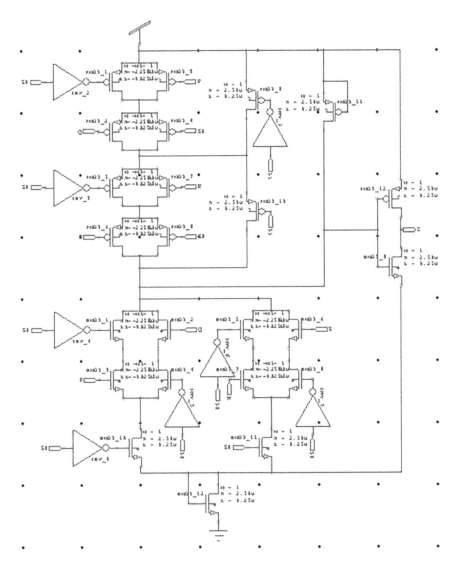

FIGURE 14.14 Schematic design diagram for 4×1 multiplexer using conventional positive-feedback adiabatic logic.

14.7 CONCLUSION

In this chapter, low-power multiplexers are implemented using different logic designs. From the simulation results, multiplexer design using 2PASCL gives low power consumption compared with PFAL, and various low-power and high-speed techniques, namely, TG logic, GDI logic, CMOS, DPTL, and PTL exist. The outcomes show that when compared with other proposed structures, CMOS has more power consumption

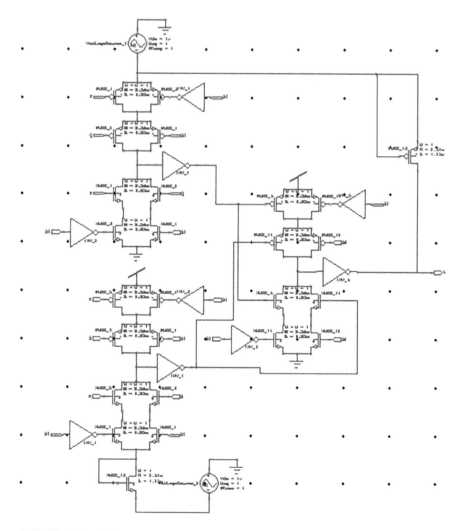

FIGURE 14.15 Schematic design diagram for 4×1 multiplexer using two-phase adiabatic static clocked logic.

and transistor count. The outcomes were simulated using Tanner EDA, and comparisons are carried out with respect to various parameters such as power dissipation, delay, and the total number of transistor. These points raise awareness about proposed techniques over CMOS, which make them effective and suitable to be utilized in digital circuits.

REFERENCES

1. E. Chitra, N. Hemavathi, V. Ganesan, "Energy Efficient design of logic circuits using adiabatic process", *International Journal of Engineering and Technology*, Vol. 9, no. 6, pp. 4504–4505, 2018.

2. D. Kamboj, A. Kumar, V. Kumar, "Design and implementation of optimized 4:1Mux using adiabatic technique", *International Journal of Advances in Engineering & Technology*, Vol. 3, no. 9, pp. 98–108, 2014.

3. D. Chaudhuri, A. Nag, S. Bose, S. Mitra, H. Ghosh, "Power and delay analysis of a 4 to 1 multiplexer implemented in different Logic Style", *International Journal of Innovative Research in Science, Engineering and Technology*, Vol. 4, no. 9, pp. 118–123, 2015.

4. I. Gupta, N. Arora, B.P Singh, "New design of high performance 2:1 multiplexer", *International Journal of Engineering Research and Applications (IJERA)*, Vol. 2, no. 2, pp. 1492–1496, 2012.

5. K. Ishii, H. Nosaka, M. Ida, K. Kurishima, S. Yamahata, T. Enoki, T. Shibata, S. Eiichi "4- bit multiplexer/demultiplexer chip set for 40- Gbit/s optical communication systems", *IEEE transactions on microwave theory and techniques*, Vol. 51, no. 11, pp. 2181–2187, 2003.

6. D. Patel, Dr. S.R.P Sinha, M. Shree, "Adiabatic logic circuits for low power VLSI applications", *International Journal of Scientific Research*, Vol. 5, no. 4, pp. 1585–1586, 2016.

7. A. Praveen, T. Tamil Selvi, Power efficient design of adiabatic approach for low power VLSI circuits. *ICEES 2019 Fifth International Conference on Electrical Energy Systems*, pp. 21–22, Chennai, India, 2019.

8. S. Machanooru, N. Bandi, A novel low power high speed positive feedback adiabatic logic. *International Conference on Innovations in Information, Embedded and Communication Systems (ICIIECS)*, 2017.

9. P. Saini, R. Mehra, "Leakage power reduction in CMOS VLSI circuits", *International Journal of Computer Applications*, Vol. 55, pp. 42–48, 2012.

10. B. Dilli Kumar, A. Chandra Babu, V. Prasad, "A comparative analysis of low power and area efficient digital circuit design", *International Journal of Computer Technology and Applications*, Vol. 4, no. 5, pp. 764–768, 2013.

11. P. Gupta, K. Mehrotra, K. Kashyap, H. Kaur, P. Dhall, "Area efficient, low power 4:1 multiplexer using NMOS 45nm technology", *International Journal on recent and innovation trends in computing and computation*, Vol. 4, no. 4, pp. 570–574, 2016.

12. V. Sikarwar, N. Yadav, S. Akashe, Design analysis of CMOS ring oscillator using 45 nm technology, *3rd International IEEE conference on Advance Computing (IACC)*, pp. 1491–1495, 2013.

13. R. Verma, R. Mehra, CMOS based design simulation Adder/Subtractor using different foundries, *National Conference on Recent Advances on Electronics and Communication Engineering*, pp. 1–7, 2014.

14. A. Sharma, R. Mehra, "Area and power efficient CMOS adder by hybridizing PTL and GDI Technique" *International Journal of Computer Applications*, Vol. 66, no. 4, pp. 15–22, 2013.

15. R. Singh, R. Mehra, "Power efficient design of multiplexer using adiabatic logic", *International Journal of Advances in Engineering and Technology*, Vol. 6, no. 1, pp. 246–254, 2013.

16. M.C. Wang, "Low power and area efficient finfet circuit design", *World Congress on Engineering and Computer Science (WCECS), San Francisco (USA)*, Vol. 1, pp. 20–22, 2009.

17. M. Mishra, S. Akashe, "High performance, low power 200 Gb/s 4:1 MUX with TGL in 45 nm technology", *Springer, Applied Nanoscience*, Vol. 4, no. 3, pp. 271–277, 2014.

18. A. Shrama, R. Singh, R. Mehra, Low power TG full adder design using CMOS nano technology, parallel distributed and grid computing, *IEEE 2nd International Conference*, pp. 310–213, 2012.

19. B. Mukherjee, A. Ghosal, Design & study of a low power high speed full adder using GDI multiplexer, *IEEE 2nd International Conference on Recent Trends in Information Systems*, pp. 465–470, 2015.

20. S. Sayedsalehi, K. Navi, "A novel architecture for quantum-dot cellular automata multiplexer", *International Journal of Computer Science*, Vol. 8, no. 6, pp. 55–60, 2011.
21. A. Agarwal, T. K. Gupta, A. K. Dadoria, "Ultra low power adiabatic logic using diode connected DC biased PFAL logic", *Advances in Electrical and Electronic Engineering*, Vol. 55, no. 1, pp. 46–54, 2017.

15 Macromodeling and Synthesis of Analog Circuits

B. S. Patro and Sushanta Kumar Mandal
Kalinga Institute of Industrial Technology,
KIIT Deemed to be University

CONTENTS

15.1 Introduction 251
15.2 Parametric-Based Macromodeling 252
    15.2.1 Symbolic Modeling 253
    15.2.2 Posynomial Templates/Geometric Programming 256
    15.2.3 Model Order Reduction 256
15.3 Nonparametric Macromodeling 259
    15.3.1 Artificial Neural Network 259
    15.3.2 Support Vector Machine 261
    15.3.3 Extreme Learning Machine 262
15.4 Conclusions 264
References 264

15.1 INTRODUCTION

The analog, mixed-signal, radio frequency (RF), and digital circuits are the major components of modern system-on-chips (SoCs). Demand for very high-end smart devices such as smart phones with multitasking features has been increased to improve the quality of life. Day by day the complexities of the SoCs are increasing, as more functions are included and as there is an increase in the number of transistors in a chip, which follows Moore's law (Schaller 1997; Patro and Vandana 2016). Due to a rapid growth in SoC functions and competition for time-to-market (TTM), understanding the features and designing an optimized analog circuit for a specific task have become challenging jobs for the circuit designers. Existing design and verification tools are not able to provide the feasible solution in desired limited time (Chen 2009). One of such powerful tools is simulation program with integrated circuit emphasis (SPICE)–based simulation modeling. It is a widely used simulation tool, able to handle many types of analysis with very high accuracy (Najm 2010). But this type of simulation method cannot handle the increasing complexities of the circuits and consumes lot of time, which increases TTM.

Recently, researchers have focused on various alternative modeling methods to improve the computational speed of the design tools (Zhang and Shi 2018; Kundu and Mandal 2018; Chen, Peng, and Yu 2018). Many existing modeling methods are available to formulate simple equation-based models (Ferent and Doboli 2013; Daems, Gielen, and Sansen 2003; Dong and Roychowdhury 2008). These macro-models are very effective for performance analysis of complex circuits. The main drawback of these modeling techniques is that these models can be developed by expert designers who have good core knowledge of the circuits. As an alternative, there are other machine learning (ML) models that do not require core knowledge competency. The behavioral macromodels of any device or circuit can be formulated by training the data sets generated from SPICE or similar simulators. These data sets contain the performance variation of the circuit output parameters with respect to the input parameters.

ML concept deals with the techniques involved to help the machine to learn the features from the data sets. One such ML modeling is support vector machine (SVM). The main advantage of SVM is that it can formulate an effective model even if the data sets are highly nonlinear. SVM has exhibited many better results in different fields of applications such as bioinformatics, medical diagnosis, biosurveillance, stock market analysis, spam filtering, and many more (Manavalan, Shin, and Lee 2018; Akhmadeev et al. 2018; McCowan, Moore, and Fry 2006; Wang, Li, and Bao 2018; Singh and Bhardwaj 2018). Recently, researchers have started using SVM in the field of very large-scale integration (VLSI) circuits as well. Similarly, another type of ML modeling technique is extreme learning machine (ELM). This is similar to neural network modeling. The main advantage of using ELM is that it has the feature of fast modeling and has the capability to handle big data. But most importantly, artificial neural network (ANN) has its own way of handling the fitting of the data, and in many research works, it has shown very good results. The detailed macromodeling flow and synthesis flow for analog circuits are shown in Figures 15.1 and 15.2.

Macromodeling of analog circuits is essential to develop the models that are required for the automatic and quick synthesis of analog circuits. These synthesized circuits are the building blocks for the integrated circuits (ICs). So it is highly required to generate the models as fast as possible, even for the complex circuits consisting of a larger number of transistors without affecting the accuracy compared with the original circuit. Many researchers are working on various methods to formulate such effective and robust models. Various types of macromodeling and synthesis techniques are discussed here.

Macromodeling can be classified into two categories: (i) knowledge based or parametric based and (ii) black box or nonparametric based. This detailed classification is shown in Figure 15.3. This chapter describes various types of macromodeling techniques reported in the literature.

15.2 PARAMETRIC-BASED MACROMODELING

This type of macromodeling is also known as knowledge-based macromodeling. As the name suggests, the modeling formulation requires detailed knowledge about

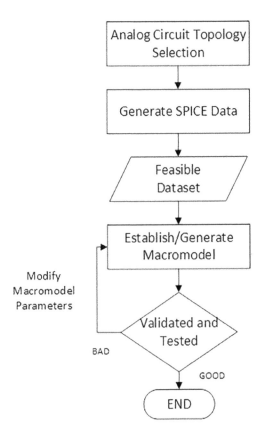

FIGURE 15.1   Macromodeling flow for the analog circuits.

the circuits and their mathematical analysis. So this modeling method is mostly preferred by the researchers who have sound notions about the circuits. In this technique, human interference plays a greater role in formulating the models. Various techniques are reported in literature. But in this section, three important methods are discussed which are on the basis of their approach to formulate the model. They are (i) symbolic modeling, (ii) posynomial templates, and (iii) model order reduction (MOR).

15.2.1   Symbolic Modeling

This is one of the oldest and most commonly used modeling methods used by the VLSI circuit designers (Gielen, Wambacq, and Sansen 1994; Shi and Tan 2000). Symbolic analysis is a systematic approach to obtaining the knowledge of analog building blocks in an analytic form. It is an essential complement to numerical simulation. This modeling approach provides robust and accurate models, which can be further used for the optimization or synthesis purpose. It helps to calculate the behavioral parameters of the circuits with respect to the independent quantities. In order to calculate performance parameters, a circuit net list is required.

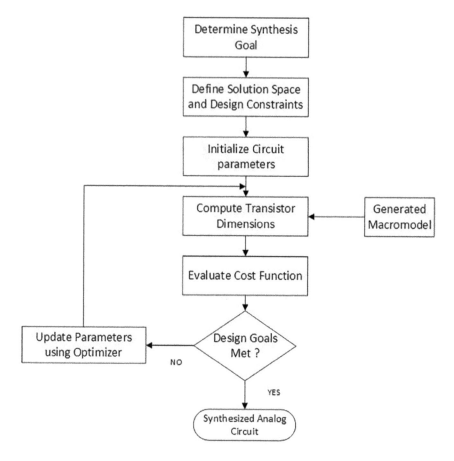

FIGURE 15.2   Analog synthesis flow.

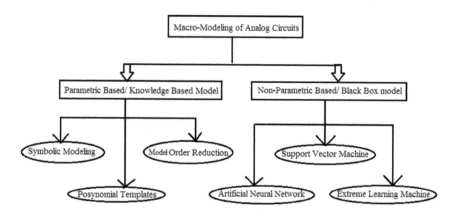

FIGURE 15.3   Classification of various types of macromodeling of analog circuits.

The end result is a set of algebraic equations for the performance parameters which describe the behavior of the circuits. So this technique can be related to numerical analysis.

Some good symbolic analyzers are reported in Kolka, Biolek, and Biolkova (2008); McConaghy and Gielen (2009); Gielen and Sansen (2012); Rutenbar, Gielen, and Roychowdhury (2007); and Shi and Tan (2000). In these analyzers, the modified nodal analysis (MNA) method was principally used to formulate the system of equations. In Vazzana, Grasso, and Pennisi (2017), the importance and usefulness of MNA can be easily observed from its implemented tool using matrix laboratory (MATLAB). Then, the recursive determinant expansion techniques provide the performance of these generated symbolic models. Other methods such as dead rows and V/I methods were also reported (Fakhfakh, Tlelo-Cuautle, and Fernández 2012). These methods remove the redundancy and hence complication of the circuits to some extent. Some other methods such as nodal admittance matrix (NAM) (Sánchez-López, Cante-Michcol, et al. 2013; Sánchez-López, Ochoa-Montiel, et al. 2013) are also used for such type of reduction mechanism for the circuits. Similarly, pathological equivalents have been used instead of active devices for synthesis purposes (Tlelo-Cuautle, Sánchez-López, and Moro-Frías 2010; Saad and Soliman 2010). Binary decision diagram (Shi 2013; Zhang and Shi 2011) was earlier used for logic synthesis and verification. But afterward, this method is used more frequently for symbolic model generation and synthesis of analog circuits. This method not only removes the data redundancy but also provides a mechanism for explicit enumeration, which decreases the complexity to a certain extent.

There are also some works (Ferent and Doboli 2013; Shokouhifar and Jalali 2014, 2015; Shi, Hu, and Deng 2017; Vazzana, Grasso, and Pennisi 2017) that concentrate mainly on automation of analog circuit synthesis. Ferent and Doboli (2013) provide a method for comparison of analog circuits on the basis of symbolic performance data generation. Shokouhifar and Jalali (2014, 2015) provide the automation methods for increasing the efficiency of symbolic models during the synthesis process by the introduction of evolutionary algorithms. Shi, Hu, and Deng (2017) discussed the advancement of automation for a symbolic macromodel generation for the ICs especially for the RLC type of elements. Similarly, a data structure–based determinant decision diagram has shown some fruitful results in developing automatic symbolic models [33].

Literature shows that this approach for generation of macromodels is most complicated and time taking. Also, automated tools are limited to simple circuits. For complex circuits, it uses various heuristic methods that somewhat simplifies, but still the models generated are highly intricate. This led to the generation of complex equations which further increases the memory storage and hence computing time. Again, these models show unexpected behavior in several regions of the design space. While reducing the redundancy using NAM, MNA, or other methods, some of the key parameters will not be taken care of for the removed values during synthesis time. These key features may be the vital parameters for the circuits during the final fabrication time. But still, it is not able to satisfy the present growing demand for fast design of circuits.

15.2.2 POSYNOMIAL TEMPLATES/GEOMETRIC PROGRAMMING

This is also a similar approach to symbolic modeling technique. It provides a set of posynomial equations representing the models for the circuits. The equations obtained that are the posynomial forms are a closed form of equations. With the help of these equations, the circuits are synthesized or optimized on the basis of specific application purposes. The most suitable optimization procedure used for posynomial templates is a geometric programming method.

A posynomial function is defined as

$$f(x_1,...,x_n) = \sum_{k=1}^{K} c_k x_1^{\alpha_{1k}} x_2^{\alpha_{2k}} ... x_n^{\alpha_{nk}}, \quad x_i > 0a \qquad (15.1)$$

where n is a positive real number, c_k is a nonnegative coefficient, and a_{ij} is a real coefficient.

Here also, the end result is a set of equations that are used to fit the geometry of the feature space, which represents the behavior of the circuit. The important thing that needs to be considered while formulating the posynomial model is the fitness of the feature space. The accuracy is affected to a large extent while formulating this feature space. Most of the time, it was found that the feature space obtained is not convex in nature. Hence, many times the posynomial templates are generated manually. This can be done by performing the piecewise linear fitting of the posynomial models to generate the overall model. The model fitting and identifying the correctness of the templates is very much essential to obtain an effective model. But still, it was found that there is a large difference in the accuracy of the model developed from the SPICE-simulated results. Some papers (Sáenz Noval et al. 2010; Mandal and Visvanathan 2001; DasGupta and Mandal 2009) have provided more accurate models that are easier to solve. The generation of posynomial models and the formulation of the optimization problem is difficult. Daems, Gielen, and Sansen (2003) provides a method of conversion of a posynomial model to a monomial model (the simplest form of posynomial). Rather, this method is not helpful always as it will become difficult for complicated circuits. Figure 15.4 shows the methodology for generation of posynomial models.

It is observed that there has been a trade-off between the speed of solutions and the scale of solutions. It has also been seen that the results are concentrated in local minima. Some efforts have been given for automatic generation of the template (Martins et al. 2017; Kundu and Mandal 2014). But considering the accuracy and time taken to generate the mathematical formulation, it also needs a lot of improvement to speed up the TTM of the ICs. This led to more alternatives to implement these posynomial templates for synthesis purposes.

15.2.3 MODEL ORDER REDUCTION

This technique has helped the designers a lot when the transistor count of the circuits is more than 10. It reduces the complexities of the circuits by generating a

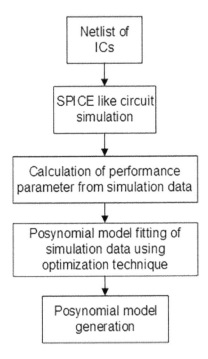

FIGURE 15.4 Flow diagram of generating posynomial models.

number of small-size models. When these models are again grouped together, then it will behave as similar to the original circuit. Like posynomial expressions, the final reduced equations provide the models that are represented in the closed form of mathematical equations. This technique has a very good advantage of modeling the circuits whose behavior is linear in nature. It provides robust models that are helpful for solving the mathematical equation in a faster manner. Pillage et al. (Shi, Hu, and Deng 2017) used the explicit moments matching via Pade approximations (Antoulas 2005), and based on that, they provided the asymptotic waveform evaluation (AWE) approach. But this approach becomes unstable and provides less accurate results for transfer functions of higher order. Gallivan et al. (1994) provided the Pade via Lanczos method that applies Krylov space projections (Freund 2000) for MOR in circuit simulation. This helps to overcome the drawbacks of the AWE approach. The Lanczos algorithm (Antoulas and Sorensen 2001) implicitly computes the first moment or a transfer function. The passive reduced-order interconnect macromodeling (PRIMA) (Odabasioglu, Celik, and Pileggi 1997) and the structure-preserving reduced-order interconnect macromodeling (SPRIM) (Freund 2004) MOR methods help in reducing large-scale RLC networks having multiple terminals without hampering the structure and passivity nature of the circuits. This is not observed in Feldmann's method (Feldmann 2004) who uses the SVDMOR technique. In this method, the correlation between circuit responses at different terminals is used to obtain the model, but the passiveness of the circuit is not preserved during the MOR process. This limitation is encountered by Liu et al. (2006) and resolved with the

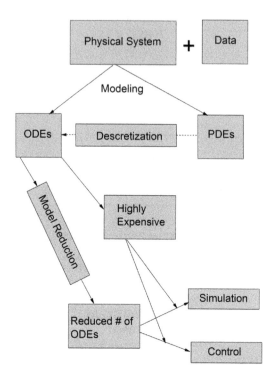

FIGURE 15.5   Generation of reduced-order equations using MOR technique. MOR, model order reduction. ODE, ordinary differential equation; PDE, partial differential equation.

help of extended-SVDMOR method. It reduces the number of input and output terminals separately. Figure 15.5 shows the procedure to generate reduced-order equations using the MOR technique.

MOR has the major limitation for nonlinear circuit. Chen (1999) proposed a method for the reduction of the nonlinear circuit using quadratic Taylor series expansion and Krylov space projections. The method is limited to weakly nonlinear circuits and is inaccurate for strongly nonlinear circuits. Gu (2011) proposed separate approaches for generating different reduced models for each circuit analysis domain called the ManiMOR and the QLMOR. These methods are difficult to use, scale poorly with the number of nonlinear terms, and still need some investigations to be put in practice for complex nonlinear circuit models. Philips (2000) presented a method for automatically extracting macromodels for weakly nonlinear circuits based on the Volterra series and variational analysis theory (Rugh 1981). Feng (2005) proposed a two-sided projection method to enhance this class of MOR methods based on the variational analysis. De Jonghe and Gielen (2012) presented a methodology to approximate nonlinear analog circuit models with a compact set of analytical behavioral models. The trajectory piece-wise linear (TPWL) (Rewieński and White 2006) method is a MOR method that consists of aggregating local linear approximations around expansion points from state trajectories driven by training inputs. An enhancement of the TPWL MOR method, which consists of an adaptive sampling of the linearization points across the model trajectory based on the error between the nonlinear model and its linearized form, is proposed in Nahvi, Nabi, and Janardhanan (2012).

Farooq et al. (2013) and Dong and Roychowdhury (2008) followed the main lines of the TPWL MOR method but replaced the local linear models with piecewise polynomial and Chebyshev interpolating polynomial models, respectively. This approach improved the accuracy of the local reduced models but increased their on-the-fly evaluation time. De Jonghe and Gielen (2012) presented a methodology to approximate nonlinear analog circuit models with a compact set of analytical behavioral models. This method is based on transient simulations and curve-fitting techniques and enhances the MOR method automation but does not overcome the input dependency problem. Although some recent works (Aridhi, Zaki, and Tahar 2015, 2016) have shown significant improvement in improving the modeling analysis of MOR using some heuristic approaches, still it needs a lot of development for solving different types of circuits.

15.3 NONPARAMETRIC MACROMODELING

This type of macromodeling is also known as black-box modeling. This method is mainly automatic in nature and needs less manual effort. It depends on the data sets generated from the SPICE simulation to develop models. Once the model is generated, it can replace the existing SPICE simulator to perform the optimization or synthesis of the circuit. So this type of modeling can be performed by anyone who has the basic idea of controlling the behavior of the model. It does not require an in-depth knowledge of the VLSI circuits. It can be further subdivided into various macromodeling methods. But this chapter tries to concentrate on the behavior of three important modeling methods that have very high accuracy and speed. They are (i) ANN, (ii) SVM, and (iii) ELM which are discussed in the following sections.

15.3.1 ARTIFICIAL NEURAL NETWORK

The growing demand for automation has led the researchers in the area of the circuit design to implement similar methods for developing the models for the analog circuits. ANN is one such advanced technique that can be used to generate the macromodels. Neural networks are referred to as the connection systems inspired by the biological neural networks, which are found in human and animal brains and bodies. Similar to the human brain, ANN learns progressively by providing some learning algorithms.

This method of computation was first proposed by Warren McCulloch and Walter Pitts (1943). Furthermore, it is improved for various applications with the introduction of perceptron by Rosenblatt (1958). In the mid-1980s, computational algorithm was widely used in many applications. This is due to Werbos' backpropagation algorithm (Werbos 1988). Later on, the SVM algorithm became popular in the 1990s and earlier part of the 21st century. Usually, SVM provides accurate models for complex data sets. After 2006, various algorithms have been formulated to improve the accuracy and modeling behavior of neural networks.

Several research works have been reported in literature, especially in generating the models for ICs (Creech et al. 1997; Mandal, Sural, and Patra 2008; Adhikari et al. 2015; Wolfe and Vemuri 2003). They generate the black-box models that do not require the inner qualitative properties of the circuits. They basically map between the input and output parameters. Earlier in the circuit modeling domain, they were mainly used in RF and microwave sectors (Fang et al. 2000; Devabhaktuni, Yagoub, and Zhang 2001; Zhang, Gupta, and Devabhaktuni 2003). Neural networks can model any type of functionalities of the circuits. Even if the circuit has hyperdimensionality or high nonlinear characteristics, they can easily model without compromising the speed and accuracy. They can interpolate functional values of the circuit parameters, which may not be present in lookup tables. Lookup tables are one of the methods to analyze the circuit behavior, which is enlisted in the table. This method is very much helpful when circuit behavior is nonlinear in nature.

In order to generate the macromodel, a training data set is required. This data set is generated from the SPICE simulation data set. Similarly, independent data set is also extracted from the same SPICE data, which may be called testing data, which are not present in the training data set to verify the accuracy and other working parameters of the model. One thing that must be taken care of while extracting the training or testing data is that the data sets must be feasible in the design space. After the model is properly validated and tested, then it can work as a replacement of the SPICE model, which requires a large amount of time to provide the results. This trained macromodel can be utilized for synthesis purpose to extract more designs and more input parameters by using different optimization methods. Using this process, macromodels are generated for OPAMP (Patro and Mandal 2017b).

As shown in Figure 15.6, it can be observed that ANN is basically composed of three main layers:

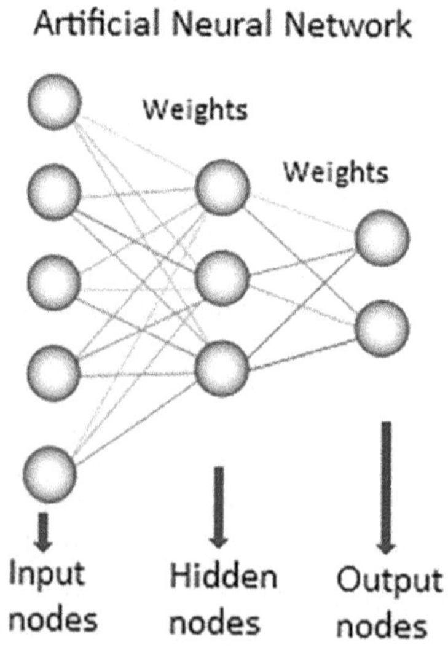

FIGURE 15.6   An artificial neural network structure.

- Input nodes
- Hidden nodes
- Output nodes

The input layer or nodes are composed of input variables. The hidden layer or nodes are composed of hidden neurons with calculated weights. They can be multilayers. Output layer is composed of output parameters of the model. There can be one node for a single output and many nodes for a multioutput model based on macromodeling technique.

A basic ANN composed of three types of parameters:

- Connection pattern between the layers of the neural network
- Weights of the connections
- Activation function

The connection pattern is referred by the parameter "bias." The weights of the connections are updated during the training of the neural network, which is also called as a learning process. The activation function converts the neuron's weighted input to its output activation.

15.3.2 Support Vector Machine

SVM is a type of ML technique used to generate a model based on statistical data analysis. It can be used for classification or regression type of problems. In the case of classification type of problems, it generates a hyperplane that can classify a data set into two distinctive sets. Many works have been carried out to make it multidimensional in nature (Komura et al. 2004; Cyganek, Krawczyk, and Woźniak 2015; Ji et al. 2016). In many other applications, it can be used to solve real-life problems such as data mining, pattern analysis, clustering, and so on (Borchani et al. 2015; Bishop 2006; Winters-Hilt and Merat 2007).

SVM was first proposed by Vapnik et al. (1997) and Cortes and Vapnik (1995). These works provide the implementation of SVM for classification problems. Furthermore, it is extended to the regression type of problems. In this type of problem, SVM finally provides a model that is used to estimate a performance parameter or function. Smola and Schölkopf (2004) and Scholkopf and Smola (2001) provide a detailed analysis of the implementation of SVM in the field of regression-based problems. Some research works (Suykens and Vandewalle 1999; Suykens, Van Gestel, and De Brabanter 2002) have been reported in the field of circuit synthesis with the help of performance models generated using least square SVM (LSSVM). Estimating the performance parameters is similar to the regression-based problems, and LSSVM has been fruitful for providing an accurate estimation of the parameters.

Earlier circuit synthesis was performed by parametric macromodeling. But some papers (De Bernardinis, Jordan, and Vincentelli 2003; Kiely and Gielen 2004) provided better perspectives for circuit modeling using black-box models. The generation of macromodels for the analog circuits has been reported in (Pandit et al. 2008;

Pandit, Mandal, and Patra 2009, 2010; Khandelwal, Garg, and Boolchandani 2015). In these papers, robust and accurate macromodels have been generated, which is observed from the optimized synthesis results (Boolchandani, Kumar, and Sahula 2009; Barros, Guilherme, and Horta 2007). Here, one vital thing is inferred that to maintain the robustness of the model generated by LSSVM, kernel functions play a major role.

Kernel functions are the type of functions used for pattern analysis and estimation. They establish a relationship for different data sets and provide effective results even if the data size and dimensions are very large. They provide a feature map that is explicitly obtained from the data sets. They help in providing a very high-quality performance along with SVM. They reduce the complex computation and hence SVM implementation for generating the macromodels with a cheap computational cost (Chang and Lin 2011; Basak, Pal, and Patranabis 2007; Patro and Mandal 2017a,c; Patro, Panigrahi, and Mandal 2012).

The main aspect of this SVM to solve the regression type of problems is ε sensitive loss. This means that the function estimated by the model must not deviate from the actual targets by the factor of ε. In other words, the errors less than ε can be ignored. The deviation of more than ε is the cost factor that needs to be optimized. So this optimization problem is solved by its dual formulation and with the help of Karush–Kuhn–Tucker condition. This is the key to the extension of this problem even toward the nonlinear type of regression problems.

15.3.3 EXTREME LEARNING MACHINE

ELM is another type of macromodeling method. It was first proposed by Huang, Zhu, and Siew in 2004.. Afterward, many works have been carried out in various fields. Theoretical studies have shown that even with randomly generated hidden nodes, ELM maintains the universal approximation capability of single-hidden layer feedforward neural networks (SLFNs) (Huang and Chen 2007, 2008; Huang, Zhu, and Siew 2006a). With commonly used activation functions, ELM can attain the almost optimal generalization bound of traditional false nearest neighbor (FNN) algorithm in which all the parameters are learned. Efficiency and generalization performance ELM over traditional FNN algorithms have been demonstrated on a wide range of problems from different fields (Huang et al. 2012; Huang, Zhu, and Siew 2006b). During the past decade, theories and applications of ELM have been extensively studied. From a learning efficiency point of view, the original design objects of ELM have threefolds: least human intervention, high learning accuracy, and fast learning speed. From a theoretical aspect, the universal approximation capability of ELM has been further studied in (Huang et al. 2012). The generalization ability of ELM has been investigated in the framework of statistical learning theory (Lin et al. 2015; Liu, Gao, and Li 2012) and the initial localized generalization error model (LGEM) (Xi-Zhao et al. 2013).

Very few works using ELM have been reported in the field of VLSI circuits (Xiong, Tian, and Yang 2016; Yu, Sui, and Wang 2016). These works are related to the fault diagnosis purpose. This is a separate domain of VLSI not related to the synthesis of analog circuits, which deals with performance-based modeling. So there

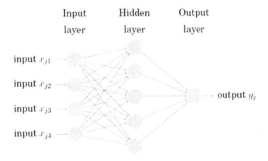

Input layer   Hidden layer   Output layer

input x_{j1}

input x_{j2}

input x_{j3} output y_j

input x_{j4}

FIGURE 15.7   A schematic overview of an ELM. ELM, extreme learning machine.

is a huge scope in the field of analog circuits macromodeling, and recently Patro and Mandal (2016) focused on developing macromodel using this ML method for operational transconductance amplifier (OTA).

An overview of the structure of an ELM is shown in Figure 15.7. This is similar to neural network, or in other words, it is the improvement of single-layer feedforward network. But the main difference is that the hidden layer is not tuned.

Table 15.1 summarizes different modeling techniques with their applications and complexities. ANN, SVM, and ELM have large area of applications. Due to low complexity and less time required for model generation, these ML techniques are used to generate macromodels for analog circuits.

TABLE 15.1
List of Macromodeling Techniques with Their Complexities

Sl. No.	Macromodeling Technique	Applications	Performance Analysis	Complexity
1.	Symbolic	Different types of electrical and electronics circuits and networks, mathematical models, etc.	Less time for known elements/circuits	Medium to large
2.	Posynomial	Circuits, aerospace and structural design, etc.	Less time for linear behavioral systems	Low to high
3.	MOR	Control system, MEMS, etc.	Less time for small and linear systems	Low to high
4.	ANN	Almost all fields where mathematical or data analysis is required	Less time for small systems	Low to high
5.	SVM	Almost all fields where mathematical or data analysis is required	Less time for small systems	Low to medium
6.	ELM	Almost all fields where mathematical or data analysis is required	Less time for small systems	Low

ANN, artificial neural network, ELM, extreme learning machine; MEMS, microelectromechanical system; MOR, model order reduction; SVM, support vector machine.

15.4 CONCLUSIONS

This chapter provides a detailed analysis of several existing techniques for macromodeling and generation of macromodels of analog circuits. These generated macromodels can be used in the synthesis flow of analog circuits using various optimization techniques such as genetic algorithm, particle swarm optimization, and so on. At the end, comparison of different macromodeling techniques has been made based on performance and complexity. It is found that ANN, SVM, and ELM show better performance due to less complexity.

REFERENCES

Adhikari, Shyam Prasad, Kim Hyongsuk, Budhathoki Ram Kaji, Yang Changju, and O Chua Leon. 2015. "A Circuit-Based Learning Architecture for Multilayer Neural Networks with Memristor Bridge Synapses." *IEEE Transactions on Circuits and Systems I: Regular Papers* 62 (1). IEEE: 215–223.

Akhmadeev, Konstantin, Aya Houssein, Said Moussaoui, Einar A Høgestøl, Steffan D Bos-Haugen, Hanne F Harbo, David Laplaud, Jennifer Graves, and Pierre-Antoine Gourraud. 2018, July. "SVM-based Tool to Detect Patients with Multiple Sclerosis Using a Commercial EMG Sensor." In *2018 IEEE 10th Sensor Array and Multichannel Signal Processing Workshop (SAM)* (pp. 376–379). IEEE..

Antoulas, Athanasios C. 2005. *Approximation of Large-Scale Dynamical Systems.* SIAM.

Antoulas, Athanasios C, and Dan C Sorensen. 2001. "Approximation of Large-Scale Dynamical Systems: An Overview." *International Journal of Applied Mathematics and Computer Science* 11: 1093–1121.

Aridhi, Henda, Mohamed H Zaki, and Sofiene Tahar. 2016. "Enhancing Model Order Reduction for Nonlinear Analog Circuit Simulation." *IEEE Transactions on Very Large Scale Integration (VLSI) Systems* 24 (3). IEEE: 1036–1049.

Aridhi, Henda, Mohamed H Zaki, and Sofiène Tahar. 2015. "Fast Statistical Analysis of Nonlinear Analog Circuits Using Model Order Reduction." *Analog Integrated Circuits and Signal Processing* 85 (3). Springer: 379–394.

Barros, Manuel, Guilherme Jorge, and Horta Nuno. 2007. GA-SVM feasibility model and optimization kernel applied to analog IC Design automation. In *Proceedings of the 17th ACM Great Lakes Symposium on VLSI,* 469–472.

Basak, Debasish, Pal Srimanta, and Patranabis Dipak Chandra. 2007. "Support Vector Regression." *Neural Information Processing-Letters and Reviews* 11(10): 203–224.

Bishop, Christopher M. 2006. *Pattern Recognition and Machine Learning.* Springer.

Boolchandani, D, Kumar Anupam, and Sahula Vineet. 2009. Multi-Objective Genetic Approach for Analog Circuit Sizing Using SVM Macro-Model. In *Proceedings of the TENCON 2009-2009 IEEE Region 10 Conference,* 1–6.

Borchani, Hanen, Varando Gherardo, Bielza Concha, and Larrañaga Pedro. 2015. "A Survey on Multi-Output Regression." *Wiley Interdisciplinary Reviews: Data Mining and Knowledge Discovery* 5 (5). Wiley Online Library: 216–233.

Chang, Chih-Chung, and Lin Chih-Jen. 2011. "LIBSVM: A Library for Support Vector Machines." *ACM Transactions on Intelligent Systems and Technology (TIST)* 2 (3). ACM: 27.

Chen, Wai-Kai. 2009. *Analog and VLSI Circuits.* CRC Press.

Chen, Yong. 1999. *Model Order Reduction for Nonlinear Systems.* Massachusetts Institute of Technology.

Chen, Pai-Yu, Peng Xiaochen, and Yu Shimeng. 2018. "NeuroSim: A Circuit-Level Macro Model for Benchmarking Neuro-Inspired Architectures in Online Learning." *IEEE*

Transactions on Computer-Aided Design of Integrated Circuits and Systems 37 (12), pp. 3067–3080.

Cortes, Corinna, and Vapnik Vladimir. 1995. "Support Vector Machine." *Machine Learning* 20 (3): 273–297.

Creech, Gregory L, Bradley J Paul, Christopher D Lesniak, Thomas J Jenkins, and Mark C Calcatera. 1997. "Artificial Neural Networks for Fast and Accurate EM-CAD of Microwave Circuits." *IEEE Transactions on Microwave Theory and Techniques* 45 (5). IEEE: 794–802.

Cyganek, Bogusław, Krawczyk Bartosz, and Woźniak Michał. 2015. "Multidimensional Data Classification with Chordal Distance Based Kernel and Support Vector Machines." *Engineering Applications of Artificial Intelligence* 46. Elsevier: 10–22.

Daems, Walter, Gielen Georges, and Sansen Willy. 2003. "Simulation-Based Generation of Posynomial Performance Models for the Sizing of Analog Integrated Circuits." *IEEE Transactions on Computer-Aided Design of Integrated Circuits and Systems* 22 (5). IEEE: 517–534.

Daems, Walter, Gielen Georges, and Sansen Willy. 2007. "Posynomial modeling, sizing, optimization and control of physical and non-physical systems." Google Patents.

DasGupta, Samiran, and Mandal Pradip. 2009. An automated design approach for CMOS LDO regulators. In *Design Automation Conference, 2009. ASP-DAC 2009. Asia and South Pacific*, 510–515.

De Bernardinis, Fernando, Jordan Michael I, and Vincentelli A Sangiovanni. 2003. Support vector machines for analog circuit performance representation. In *Proceedings of the 40th Annual Design Automation Conference*, 964–969.

De Jonghe, Dimitri, and Gielen Georges. 2012. "Characterization of Analog Circuits Using Transfer Function Trajectories." *IEEE Transactions on Circuits and Systems I: Regular Papers* 59 (8). IEEE: 1796–1804.

Devabhaktuni, Vijay K, Yagoub Mustapha C E, and Zhang Qi-Jun. 2001. "A Robust Algorithm for Automatic Development of Neural-Network Models for Microwave Applications." *IEEE Transactions on Microwave Theory and Techniques* 49 (12). IEEE: 2282–2291.

Dong, Ning, and Roychowdhury Jaijeet. 2008. "General-Purpose Nonlinear Model-Order Reduction Using Piecewise-Polynomial Representations." *IEEE Transactions on Computer-Aided Design of Integrated Circuits and Systems* 27 (2). IEEE: 249–264.

Fakhfakh, Mourad, Tlelo-Cuautle Esteban, and Fernández Francisco V. 2012. *Design of Analog Circuits through Symbolic Analysis.* Sharjah: Bentham Science Publishers.

Fang, Yonghua, Yagoub Mustapha C E, Wang Fang, and Zhang Qi-Jun. 2000. "A New Macromodeling Approach for Nonlinear Microwave Circuits Based on Recurrent Neural Networks." *IEEE Transactions on Microwave Theory and Techniques* 48 (12). IEEE: 2335–2344.

Farooq, Muhammad Umer, Xia Likun, Hussin Fawnizu Azmadi Bin, and Malik Aamir Saeed. 2013. Automated model generation of analog circuits through modified trajectory piecewise linear approach with Cheby Shev Newton interpolating polynomials. In *2013 4th International Conference on Intelligent Systems Modelling & Simulation (ISMS)*, 605–609.

Feldmann, Peter. 2004. Model order reduction techniques for linear systems with large numbers of terminals. In *Design, Automation and Test in Europe Conference and Exhibition, 2004. Proceedings*, 2: 944–947.

Feng, Lihong. 2005. "Review of Model Order Reduction Methods for Numerical Simulation of Nonlinear Circuits." *Applied Mathematics and Computation* 167 (1). Elsevier: 576–591.

Ferent, Cristian, and Alex Doboli. 2013. "Symbolic Matching and Constraint Generation for Systematic Comparison of Analog Circuits." *IEEE Transactions on Computer-Aided Design of Integrated Circuits and Systems* 32 (4). IEEE: 616–629.

Freund, Roland W. 2000. "Krylov-Subspace Methods for Reduced-Order Modeling in Circuit Simulation." *Journal of Computational and Applied Mathematics* 123 (1). Elsevier: 395–421.

Freund, Roland W. 2004. SPRIM: Structure-preserving reduced-order interconnect macromodeling. In *Proceedings of the 2004 IEEE/ACM International Conference on Computer-Aided Design*, 80–87.

Gallivan, Kyle, Grimme Eric, and Van Dooren Paul. 1994. "Asymptotic Waveform Evaluation via a Lanczos Method." *Applied Mathematics Letters* 7 (5). Elsevier: 75–80.

Gielen, Georges, and Sansen Willy. 2012. *Symbolic Analysis for Automated Design of Analog Integrated Circuits*. Vol. 137. Springer Science & Business Media.

Gielen, Georges, Piet Wambacq, and Willy M Sansen. 1994. "Symbolic Analysis Methods and Applications for Analog Circuits: A Tutorial Overview." *Proceedings of the IEEE* 82 (2). IEEE: 287–304.

Gu, Chenjie. 2011. *Model Order Reduction of Nonlinear Dynamical Systems*. University of California.

Huang, Guang-Bin, and Chen Lei. 2007. "Convex Incremental Extreme Learning Machine." *Neurocomputing* 70 (16–18). Elsevier: 3056–3062.

Huang, Guang-Bin, and Chen Lei. 2008. "Enhanced Random Search Based Incremental Extreme Learning Machine." *Neurocomputing* 71 (16–18). Elsevier: 3460–3468.

Huang, Guang-Bin, Zhou Hongming, Ding Xiaojian, and Zhang Rui. 2012. "Extreme Learning Machine for Regression and Multiclass Classification." *IEEE Transactions on Systems, Man, and Cybernetics, Part B (Cybernetics)* 42 (2). IEEE: 513–529.

Huang, Guang-Bin, Zhu Qin-Yu, and Siew Chee-Kheong. 2004. Extreme learning machine: A new learning scheme of feedforward neural networks. In *2004 IEEE International Joint Conference on Neural Networks, 2004. Proceedings*, 2: 985–90.

Huang, Guang-Bin, Zhu Qin-Yu, and Siew Chee-Kheong. 2006a. "Extreme Learning Machine: Theory and Applications." *Neurocomputing* 70 (1). Elsevier: 489–501.

Huang, Guang-Bin, Zhu Qin-Yu, and Siew Chee-Kheong. 2006b. "Real-Time Learning Capability of Neural Networks." *IEEE Trans. Neural Networks* 17 (4): 863–878.

Ji, Guoli, Lin Yang, Lin Qianmin, Huang Guangzao, Zhu Wenbing, and You Wenjie. 2016. Predicting DNA-binding proteins using feature fusion and MSVM-RFE. In *2016 10th IEEE International Conference on Anti-Counterfeiting, Security, and Identification (ASID)*, 109–112.

Khandelwal, Sapna, Garg Lokesh, and Boolchandani Dharmendar. 2015. "Reliability-Aware Support Vector Machine-Based High-Level Surrogate Model for Analog Circuits." *IEEE Transactions on Device and Materials Reliability* 15 (3). IEEE: 461–463.

Kiely, Tholom, and Gielen Georges. 2004. Performance modeling of analog integrated circuits using least-squares support vector machines. In *Proceedings of the Conference on Design, Automation and Test in Europe* Vol. 1, 10448.

Kolka, Zdenek, Biolek Dalibor, and Biolkova Viera. 2008. "Symbolic Analysis of Linear Circuits with Modern Active Elements." *WSEAS Transactions on Electronics* 5(6): 88–96.

Komura, Daisuke, Nakamura Hiroshi, Tsutsumi Shuichi, Aburatani Hiroyuki, and Ihara Sigeo. 2004. "Multidimensional Support Vector Machines for Visualization of Gene Expression Data." *Bioinformatics* 21 (4). Oxford University Press: 439–444.

Kundu, Sudip, and Mandal Pradip. 2014. "ISGP: Iterative Sequential Geometric Programming for Precise and Robust CMOS Analog Circuit Sizing." *Integration, the VLSI Journal* 47 (4). Elsevier: 510–531.

Kundu, Sudip, and Mandal Pradip. 2018. "An Efficient Method of Pareto-Optimal Front Generation for Analog Circuits." *Analog Integrated Circuits and Signal Processing* 94 (2). Springer: 289–316.

Lin, Shaobo, Liu Xia, Fang Jian, and Xu Zongben. 2015. "Is Extreme Learning Machine Feasible? A Theoretical Assessment (Part II)." *IEEE Transactions on Neural Networks and Learning Systems* 26 (1). IEEE: 21–34.

Liu, Xueyi, Gao Chuanhou, and Li Ping. 2012. "A Comparative Analysis of Support Vector Machines and Extreme Learning Machines." *Neural Networks* 33. Elsevier: 58–66.

Liu, Pu, Tan Sheldon X-d, Yan Boyuan, and McGaughy Bruce. 2006. An extended SVD-based terminal and model order reduction algorithm." In *Behavioral Modeling and Simulation Workshop, Proceedings of the 2006 IEEE International*, 44–49.

Manavalan, Balachandran, Shin Tae Hwan, and Lee Gwang. 2018. "DHSpred: Support-Vector-Machine-Based Human DNase I Hypersensitive Sites Prediction Using the Optimal Features Selected by Random Forest." *Oncotarget* 9 (2). Impact Journals, LLC: 1944.

Mandal, Sushanta K, Sural Shamik, and Patra Amit. 2008. "ANN-and PSO-Based Synthesis of on-Chip Spiral Inductors for RF ICs." *IEEE Transactions on Computer-Aided Design of Integrated Circuits and Systems* 27 (1). IEEE: 188–192.

Mandal, Pradip, and Visvanathan V. 2001. "CMOS Op-Amp Sizing Using a Geometric Programming Formulation." *IEEE Transactions on Computer-Aided Design of Integrated Circuits and Systems* 20 (1). IEEE: 22–38.

Martins, R, Lourenço N, Póvoa R, Canelas A, Horta N, Passos F, Castro-López R, Roca E, and Fernández F. 2017. Layout-aware challenges and a solution for the automatic synthesis of radio-frequency IC blocks. In *2017 14th International Conference on Synthesis, Modeling, Analysis and Simulation Methods and Applications to Circuit Design (SMACD)*, 1–4.

McConaghy, Trent, and Gielen Georges G E. 2009. "Template-Free Symbolic Performance Modeling of Analog Circuits via Canonical-Form Functions and Genetic Programming." *IEEE Transactions on Computer-Aided Design of Integrated Circuits and Systems* 28 (8). IEEE: 1162–1175.

McCowan, Iain, Moore Darren, and Fry Mary-Jane. 2006. Classification of cancer stage from free-text histology reports. In *Engineering in Medicine and Biology Society, 2006. EMBS'06. 28th Annual International Conference of the IEEE*, 5153–5156.

McCulloch, Warren S, and Pitts Walter. 1943. "A Logical Calculus of the Ideas Immanent in Nervous Activity." *The Bulletin of Mathematical Biophysics* 5 (4). Springer: 115–133.

Nahvi, S A, M Nabi, and Janardhanan S. 2012. Adaptive sampling of nonlinear system trajectory for model order reduction. In *2012 Proceedings of International Conference on Modelling, Identification & Control (ICMIC)*, 1249–1255.

Najm, Farid N. 2010. *Circuit Simulation*. John Wiley & Sons.

Odabasioglu, Altan, Celik Mustafa, and Pileggi Lawrence T. 1997. PRIMA: Passive reduced-order interconnect macromodeling algorithm. In *Proceedings of the 1997 IEEE/ACM International Conference on Computer-Aided Design*, 58–65.

Pandit, Soumya, Bhattacharya Sumit K, Mandal Chittaranjan, and Patra Amit. 2008. "A Fast Exploration Procedure for Analog High-Level Specification Translation." *IEEE Transactions on Computer-Aided Design of Integrated Circuits and Systems* 27 (8). IEEE: 1493–1497.

Pandit, Soumya, Mandal Chittaranjan, and Patra Amit. 2009. Systematic methodology for high-level performance modeling of analog systems. In *2009 22nd International Conference on VLSI Design*, 361–366.

Pandit, Soumya, Mandal Chittaranjan, and Patra. 2010. "An Automated High-Level Topology Generation Procedure for Continuous-Time $\Sigma\Delta$ Modulator." *Integration, the VLSI Journal* 43 (3). Elsevier: 289–304.

Patro, B Shivalal, and Mandal Sushanta K. 2017b. "Macro-Modeling of OTA Using ANN for Fast Synthesis." *International Journal of Engineering Sciences and Management*.

Patro, Shivalal, and Mandal Sushanta Kumar. 2017c. "A Multi Output Formulation for Analog Circuits Using MOM-SVM." *Indonesian Journal of Electrical Engineering and Computer Science* 7(1): 90–96.

Patro, B Shivalal, and Mandal Sushanta K. 2016. "A Novel Modeling Technique for Operational Amplifier Using RBF-ELM." *Journal of Engineering Science and Technology Review* 9(4): 74–76.

Patro, B Shivalal, and Mandal Sushanta K. 2017a. "Support Vector Machine Based Macro-Modeling of Voltage Controlled Oscillator for Fast Synthesis Purpose." *Journal of Advanced Research in Dynamical and Control Systems*, no. Special: 655–663.

Patro, B S, Panigrahi J K, and Mandal Sushanta K. 2012. A 6--17 GHz linear wide tuning range and low power ring oscillator in 45nm CMOS process for electronic warfare. In *2012 International Conference on Communication, Information & Computing Technology (ICCICT)*, 1–4.

Patro, B Shivalal, and Vandana B. 2016. "Low Power Strategies for beyond Moore's Law Era: Low Power Device Technologies." *Design and Modeling of Low Power VLSI Systems.* IGI Global, 27.

Phillips, Joel R. 2000. Automated extraction of nonlinear circuit macromodels. In *Custom Integrated Circuits Conference, 2000. CICC. Proceedings of the IEEE 2000*, 451–454.

Rewieński, Michał, and White Jacob. 2006. "Model Order Reduction for Nonlinear Dynamical Systems Based on Trajectory Piecewise-Linear Approximations." *Linear Algebra and Its Applications* 415 (2–3). Elsevier: 426–454.

Rosenblatt, Frank. 1958. "The Perceptron: A Probabilistic Model for Information Storage and Organization in the Brain." *Psychological Review* 65 (6). American Psychological Association: 386.

Rugh, Wilson John. 1981. *Nonlinear System Theory.* Johns Hopkins University Press.

Rutenbar, Rob A, Gielen Georges G E, and Roychowdhury Jaijeet. 2007. "Hierarchical Modeling, Optimization, and Synthesis for System-Level Analog and RF Designs." *Proceedings of the IEEE* 95 (3). IEEE: 640–669.

Saad, Ramy A, and Soliman Ahmed M. 2010. "A New Approach for Using the Pathological Mirror Elements in the Ideal Representation of Active Devices." *International Journal of Circuit Theory and Applications* 38 (2). Wiley Online Library: 148–178.

Sáenz Noval, Jorge Johanny, Roa Fuentes Elkim Felipe, Ayala Pabón Armando, and Van Noije Wilhelmus. 2010. "A methodology to improve yield in analog circuits by using geometric programming." In *Proceedings of the 23rd Symposium on Integrated Circuits and System Design*, 140–145.

Sánchez-López, C, Cante-Michcol B, Morales-López F E, and Carrasco-Aguilar M A. 2013. "Pathological Equivalents of CMs and VMs with Multi-Outputs." *Analog Integrated Circuits and Signal Processing* 75 (1). Springer: 75–83.

Sánchez-López, C, Ochoa-Montiel R, Ruiz-Pastor A, and González-Contreras B M. 2013. Symbolic nodal analysis of fully-differential analog circuits. In *2013 IEEE Fourth Latin American Symposium on Circuits and Systems (LASCAS)*, 1–4.

Schaller, Robert R. 1997. "Moore's Law: Past, Present and Future." *IEEE Spectrum* 34 (6). IEEE: 52–59.

Scholkopf, Bernhard, and Smola Alexander J. 2001. *Learning with Kernels: Support Vector Machines, Regularization, Optimization, and Beyond.* MIT Press.

Shi, Guoyong. 2013. "A Survey on Binary Decision Diagram Approaches to Symbolic Analysis of Analog Integrated Circuits." *Analog Integrated Circuits and Signal Processing* 74 (2). Springer: 331–343.

Shi, C-JR, and Tan Xiang-Dong. 2000. "Canonical Symbolic Analysis of Large Analog Circuits with Determinant Decision Diagrams." *IEEE Transactions on Computer-Aided Design of Integrated Circuits and Systems* 19 (1). IEEE: 1–18.

Shi, Guoyong, Hu Hanbin, and Deng Shuwen. 2017. "Topological Approach to Automatic Symbolic Macromodel Generation for Analog Integrated Circuits." *ACM Transactions on Design Automation of Electronic Systems* 22 (3). ACM: 1–25. doi:10.1145/3015782.

Shokouhifar, Mohammad, and Jalali Ali. 2014. Automatic symbolic simplification of analog circuits in MATLAB using ant colony optimization. In *2014 22nd Iranian Conference on Electrical Engineering (ICEE)*, 407–412.

Shokouhifar, Mohammad, and Jalali Ali. 2015. "An Evolutionary-Based Methodology for Symbolic Simplification of Analog Circuits Using Genetic Algorithm and Simulated Annealing." *Expert Systems with Applications* 42 (3). Elsevier: 1189–1201.

Singh, Vishal Kumar, and Bhardwaj Shweta. 2018. "Spam Mail Detection Using Classification Techniques and Global Training Set." *Intelligent Computing and Information and Communication*, 623–632. Springer.

Smola, Alex J, and Bernhard Schölkopf. 2004. "A Tutorial on Support Vector Regression." *Statistics and Computing* 14 (3). Springer: 199–222.

Suykens, Johan A K, and Vandewalle Joos. 1999. "Least Squares Support Vector Machine Classifiers." *Neural Processing Letters* 9 (3). Springer: 293–300.

Suykens, Johan A K, Van Gestel Tony, and De Brabanter Jos. 2002. *Least Squares Support Vector Machines*. World Scientific.

Tlelo-Cuautle, E, Sánchez-López C, and Moro-Frías D. 2010. "Symbolic Analysis of (MO) (I) CCI (II)(III)-Based Analog Circuits." *International Journal of Circuit Theory and Applications* 38 (6). Wiley Online Library: 649–659.

Vapnik, Vladimir, Golowich Steven E, Smola Alex, and others. 1997. "Support Vector Method for Function Approximation, Regression Estimation, and Signal Processing." *Advances in Neural Information Processing Systems*. Morgan Kaufmann Publishers, 281–287.

Vazzana, Giorgio Antonino, Grasso Alfio Dario, and Pennisi Salvatore. 2017. A toolbox for the symbolic analysis and simulation of linear analog circuits. In *2017 14th International Conference on Synthesis, Modeling, Analysis and Simulation Methods and Applications to Circuit Design (SMACD)*, 1–4.

Wang, Shuheng, Li Guohao, and Bao Yifan. 2018. "A Novel Improved Fuzzy Support Vector Machine Based Stock Price Trend Forecast Model." *ArXiv Preprint ArXiv*: 1801.00681.

Werbos, Paul J. 1988. "Generalization of Backpropagation with Application to a Recurrent Gas Market Model." *Neural Networks* 1 (4). Elsevier: 339–356.

Winters-Hilt, Stephen, and Merat Sam. 2007. "SVM Clustering." *BMC Bioinformatics*, 8: S18.

Wolfe, Glenn, and Vemuri Ranga. 2003. "Extraction and Use of Neural Network Models in Automated Synthesis of Operational Amplifiers." *IEEE Transactions on Computer-Aided Design of Integrated Circuits and Systems* 22 (2). IEEE: 198–212.

Xi-Zhao, Wang, Qing-Yan Shao, Qing Miao, and Jun-Hai Zhai. 2013. "Architecture Selection for Networks Trained with Extreme Learning Machine Using Localized Generalization Error Model." *Neurocomputing* 102. Elsevier: 3–9.

Xiong, Jian, Tian Shulin, and Yang Chenglin. 2016. "Fault Diagnosis for Analog Circuits by Using EEMD, Relative Entropy, and ELM." *Computational Intelligence and Neuroscience* 2016. Hindawi.

Yu, Wen Xin, Sui Yongbo, and Wang Junnian. 2016. "The Faults Diagnostic Analysis for Analog Circuit Based on FA-TM-ELM." *Journal of Electronic Testing* 32 (4). Springer: 459–465.

Zhang, He, and Shi Guoyong. 2011. Symbolic behavioral modeling for slew and settling analysis of operational amplifiers. In *2011 IEEE 54th International Midwest Symposium on Circuits and Systems (MWSCAS)*, 1–4.

Zhang, Ailin, and Shi Guoyong. 2018. "A Fast Symbolic SNR Computation Method and Its Verilog-A Implementation for Sigma-Delta Modulator Design Optimization." *Integration, the VLSI Journal* 60. Elsevier: 190–203.

Zhang, Qi-Jun, Gupta Kuldip C, and Devabhaktuni Vijay K. 2003. "Artificial Neural Networks for RF and Microwave Design-from Theory to Practice." *IEEE Transactions on Microwave Theory and Techniques* 51 (4). IEEE: 1339–1350.

16 Performance-Linked Phase-Locked Loop Architectures: Recent Developments

Umakanta Nanda
VIT-AP University

Debiprasad Priyabrata Acharya
NIT Rourkela

Prakash Kumar Rout and Debasish Nayak
Silicon Institute of Technology

Biswajit Jena
Koneru Lakshmaiah Education Foundation

CONTENTS

16.1 Introduction 271
16.2 Performance-Linked Phase-Locked Loop Components 272
16.3 Recent Performance-Linked Architectures of Phase-Locked Loop
       Blocks 274
16.4 Design Challenges 284
16.5 Conclusion 285
References 285

16.1 INTRODUCTION

Phase-locked loops (PLLs) [1–5] are used in almost all communication systems as frequency synthesizers and clock generators. Sometimes the PLL itself is deliberated as a unique system, and its design is one of the most challenging tasks, requiring dedicated efforts. Design of high-performance PLLs contains combination of several components of different features, covering the entire field of circuit designs that are analog, digital, and mixed signal. To minimize the trade-off between power consumption and phase noise [6,7] is the major challenge. The techniques and strategies essential for implementing each building block of the PLL shown in Figure 16.1

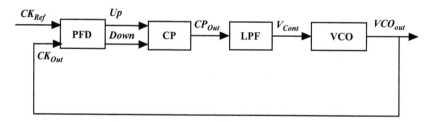

FIGURE 16.1   Basic CP-PLL. CP-PLL, charge pump phase-locked loop; LPF, low-pass filter; PFD, phase frequency detector; VCO, voltage-controlled oscillator.

differ. Each circuit needs a specific methodology, and it has to be driven by an appropriate design flow that certifies that the final design meets the traditional requirements. The charge pump PLL (CP-PLL) is the most promising one for realization in CMOS technology.

A basic CP-PLL contains four major components: a phase frequency detector (PFD), a CP, a loop filter, and a voltage-controlled oscillator (VCO). The PFD compares the input reference signal (CK_{Ref}) and the feedback signal (CK_{Out}) with respect to their phase and frequency and generates two signals, namely, *Up* and *Down* in response [8]. The CP gets these two signals as its input and generates a constant current into the loop filter (second order) having a capacitor CP and two resistors R_1 and R_2 that filter the error voltage produced due to the phase difference between CK_{Ref} and CK_{Out}. The filtered output voltage is termed as control voltage of the VCO that changes the VCO frequency in the direction that reduces the phase difference between the CK_{Ref} and CK_{Out} signals. When the control voltage matches the average frequencies of the CK_{Ref} and CK_{Out}, the loop is said to be locked, and there exists a cycle of CK_{Out} for each cycle of CK_{Ref}. However, they may not exactly match with respect to their phase and have a constant or fluctuating phase difference; however, excessive phase difference can cause loss of lock.

16.2 PERFORMANCE-LINKED PHASE-LOCKED LOOP COMPONENTS

The scopes for improvement in the performance parameters by linking them to the PLL components (Figure 16.2) are discussed in this section.

A. Phase Frequency Detector:

Analyzing the generic PLL [3], a simple PFD produces two signals (*Up* and *Down*) [3], responding to the CK_{Ref} and CK_{Out}. The noise added from the PFD modulates these pulse widths [8], which generate an arbitrary element in the CP output current. This difference increases the phase noise of the PFD, which is also referred to timing jitter [9]. Recently, a number of tristate PFDs have been designed in literature, which includes precharge-type PFD [9,10] and latch-based PFD [11,12]. Out of these PFD topologies, the latch-based PFD is mostly employed for its high operating speed, wide input range, and low power consumption. However, by employing delay

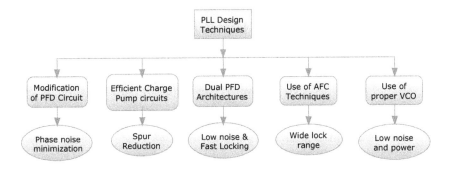

FIGURE 16.2 Classification of PLL design techniques and their achievement. AFC, adaptive frequency calibration; PFD, phase frequency detector; PLL, phase-locked loop; VCO, voltage-controlled oscillator.

element in the PFD, a dead zone–free PLL is reported in Refs. [13–19] with low phase noise and fast locking capability.

For high-speed microprocessors and high-performance digital communication systems with clock generators, a fast locking PLL [20–22] is generally adopted. These kinds of application can sustain a certain amount of phase noise but need to be fast. On the other hand, frequency synthesizers for RF applications need to have ultralow phase noise. Hence, in modern ICs, fast frequency acquisition, low phase noise, and wide lock range are required in all most all PLL applications.

To reduce noise and jitter, designers focused on individual PLL component's optimization [9,23, and 24]. They also minimized the loop bandwidth [25,26] to acquire improved noise performance, whereas bandwidth should be enhanced to minimize the lock time. Hence, there is a tight trade-off between the phase noise and frequency acquisition time. Hence, to achieve simultaneously fast and low noise capability, dual/multi-PFD architectures [29–31] are designed and reported in the past decade.

B. Charge Pump:

CP-PLLs are widely deployed in modern communication systems for their larger gain, wide frequency acquisition range, and fast locking capability. However, nonideal effects such as current mismatch between the *Up* and *Down* and increased glitches at the CP output motivated the designers [32–38] to reduce current mismatch between *Up* and *Down* networks of the CP and supply a signal with absolute no glitch in the control voltage to the VCO. For this, they adopted only one current source that supplies both the *Up* and *Down* currents. Transmission gates (TGs) were also used to diminish the nonideal effects such as clock feedthrough and charge injection to overcome the glitches in VCO control voltage. Current mismatch can also occur due to charge sharing [37–42] effect.

C. Voltage-Controlled Oscillator:

Due to the increasing demand for high-frequency multiband and multistandard transceivers in modern wireline, wireless communication, and broadcasting systems, a wide-band fast locking PLL achieving low phase noise is also highly imperative [43, 44]. Therefore, requirement of a VCO of high-tuning range, low phase noise is of paramount need [45,46]. The leakage current due to the device mismatch gives rise to reference spur in a CP-PLL [23]. It also affects the common mode voltage of a current-starved VCO (CS-VCO) [47] over a large frequency range. However, if we implement a high-tuning range VCO having large gain, the phase noise will be degraded, which threatens the stability of the PLL further. In this case, we should come up with a solution to design a wide-range PLL without sacrificing the phase noise. In the past decade or so, various adaptive frequency calibration (AFC) techniques [48–54] are realized to design wide-range PLLs.

As VCOs are also used for frequency translation, it uses inductor and capacitor (LC) tank VCO topology [55–57] for its relatively high oscillation frequency and low phase noise. Numerous researchers have come up with diverse LC VCO designs [58–61] for achieving low phase noise, but the power consumption and silicon area in these cases are very high.

Another type of VCO is in the form of differential designs. These VCOs can have a faster oscillation frequency due to their current mode logic. It contains a source-coupled differential pair and symmetric loads that provide good control over delay and high dynamic supply noise rejection. The major problem with this design is that the tuning range is very small.

D. Loop Filter:

The loop filter or low-pass filter is a vital component as it determines the stability of a CP-PLL. It is used to suppress the noise and high-frequency signal components from the PFD and provide a dc-controlled signal for the VCO. As active loop filters are more complex and generate more noise, usually a passive filter is preferred for PLL design.

Calculating proper loop filter component values depends on filter topology, order, bandwidth, phase margin, and pole ratios. The stability is related to the phase margin of the PLL. Higher phase margin may decrease peaking response of the loop filter at the expense of degraded lock time. Further choosing a very small loop bandwidth improves the reference spur, but the lock time also gets increased, whereas choosing a loop bandwidth of too wide, the lock time is decreased, but the reference spur gets worsen.

16.3   RECENT PERFORMANCE-LINKED ARCHITECTURES OF PHASE-LOCKED LOOP BLOCKS

A. PFD Architectures in PLL to Reduce Dead Zone:

A simple PFD architecture can be referred from Ref. [62]. Here two resettable D flip-flops as depicted in Figure 16.3 are used to detect the phase difference between the reference signal and the feedback signal. Depending on the position of the CK_{Ref} and CK_{Out}, this PFD produces *Up* and *Down* that switches the CP current.

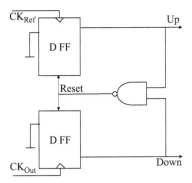

FIGURE 16.3   Conventional PFD. PFD, phase frequency detector.

The transfer function of the above PFD is illustrated in Figure 16.4 that shows the dead zone (φ_{dz}) [13] area in phase difference axis. It is the minimum pulse width of the PFD output that is needed to turn on the CP completely. Here, the PFD is unable to detect the phase difference smaller than dead zone. So the output of the PLL fluctuates within this range, triggering jitter and phase noise further [63].

To prevent dead zone, a delay element block having delay of T_D is usually inserted in the reset path of the PFD. A new technique is deployed in Ref. [13] to minimize the reference spur by maintaining the dead zone. To begin with, they experimented with fixed delay elements. Then, controlling the delay length by the feedback from the CP, a variable delay element is used. The reference spurs stated here are reduced by around 20 and 24 dB at 50 and 100 MHz frequencies, respectively, (compare to Ref. [38]).

Dead zone, alternatively blind zone [14], appears when the input phase difference goes near 2π. Here, the next rising edge of one of the inputs following that of the other input at the time of reset cannot be detected by the PFD. This topology uses only 16 transistors that eliminates blind zone

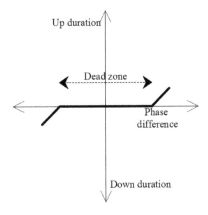

FIGURE 16.4   Dead zone (transfer function of PFD). PFD, phase frequency detector.

and achieves high acquisition speed. This topology avoids the reset process when the phase difference ranges between π and 2π. In 0.5 µm CMOS process, this PLL can operate at 800 MHz by using this PFD.

In Ref. [15], the precharging time for the internal parasitic capacitances is responsible for the blind zone. Hence, using two extra transistors, they have proposed a PFD that minimizes the precharging problem. A high-speed PFD [12] with a delay cell and two additional transistors is designed here. After simulation, the blind zone is compared and listed among the PFD of this work [11,12]. In this work, the blind zone is stated to be 61 ps, whereas in Refs. [11] and [12], it is reported as 156 and 221 ps, respectively. In Ref. [16], the authors have adopted a modified dynamic logic style PFD for a delay-locked loop (DLL) design. Here, they make one output of the PFD *high* at a time even for a small phase difference of the inputs to the PFD. Hence, the dead zone of the designed PLL at 800 and 84 MHz frequencies is limited to 10.25 and 127.3 ps, respectively. As the jitter is directly translated from the dead zone, it is also limited in this design. In Ref. [17], the two types of PFDs have been developed where the operation of the PFDs does not depend on any reset process, thereby completely eliminating the dead zone concept. The designers of Ref. [18] have considered double-gate MOSFET (DG-MOSFET) for the NOR gate of the CP PFD, making it more area efficient than its conventional counterpart. The reduced transistor count also helps in lowering the parasitic capacitance that increases the speed of the PFD. More importantly, for a phase difference of 60 and 80 ps, this PFD generates the output to initiate the CP, which reaches a required threshold level of logic high in 32- and 45-nm technologies, respectively. Hence, this faster rise time at a very small phase error makes PFD to avoid dead zone in a better manner than the PFD having conventional NOR gate. A new technique is adopted by deploying a modified Dickson CP design with charge transfer switches (CTSs) in Ref. [19] to shrink the blind zone of a latch-based PFD. Here the MOS transistor switches with proper on/off cycles are employed as an alternative to the diodes to have the flow of charges in pumping process, mentioned to as CTSs. The CTSs have been deployed to realize the CPs and display better voltage pumping gain than the diodes.

The dead zone in Ref. [64] is reduced by making the pulse widths of *Up* and *Down* half of their actual size. A predelay element is injected in Ref. [65] for the reference clock and the local signals, respectively, before they are alternatively provided to the reset inputs of the S–R latches. In this work, the dead zone is removed and merged into the intervals of π or $-\pi$. By inserting a delay of 1.2 ns, a reduced dead zone of 0.42 ns is achieved. A new difference PFD [66] that can operate at high frequency uses an edge detector circuit to avoid dead zone between two input signals. A complementary PFD is used in Ref. [67], where a phase detector and a frequency detector operate in parallel. When the phase difference reaches below 180°, the frequency detector is turned off. To make it dead zone free even at 4 GHz operating frequency, the AND gates between the inputs of the phase detector are replaced by TGs. In Ref. [68], the reset part and the delay part are separated in the PFD block ensuring no erroneous condition at the rising edge of the input when the delay is active. In a very simple design, two inverters acting as a buffer are used in Ref. [69] to get rid of the dead zone. A double-edge-checking PFD is implemented in Ref. [70] that avoids *Up* and *Down* signals to rise simultaneously reducing the dead zone. A delayed version of the two inputs is generated in Ref. [73] where only the rising edge of the inputs is delayed to minimize the blind zone of the PLL.

To get rid of the dead zone in Refs. [74,75] the PFD performs a fast reset operation from modified tristate inverters acting as a D flip-flop where the design can detect input phase differences as small as 750 fs at input operating frequencies of 38 kHz–2.5 GHz in 1.2 V and 90 nm CMOS technology. Some recent PFD performance parameters are summarized in Table 16.1.

B. Charge Pump Circuits in PLL for Reference Spur Reduction:

Usually CP circuits paired with PFD as shown in Figure 16.5, ideally generate a constant voltage commonly called as control voltage for the VCO [76]. The DC level of this voltage is controlled by the *Up* and *Down* signals of the PFD [62]. Hence, it depends on the phase difference between the two inputs of the PFD. In the previous section, we discussed about the modified PFD circuits that reduce the dead zone. In most cases, delay elements are introduced. However, due to this delay in PFD architectures, reference spurs may seriously increase [30]. Reference spurs are nothing but the highly enhanced phase noise at multiple reference frequency offsets resulting from the periodic ripples in the control voltage. The reference spur [20] is expressed as,

$$S_r = 20\log\left|\frac{\sqrt{2}\frac{I_{CP}R}{2\pi}\varphi_e K_{VCO}}{2 f_{CK_{Ref}}}\right| - 20\log\frac{f_{CK_{Ref}}}{f_{P_{LPF}}} \ (dBc) \tag{16.1}$$

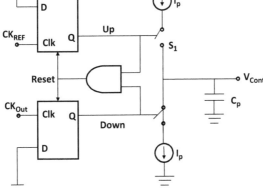

FIGURE 16.5 Conventional PFD and CP. CP, charge pump; PFD, phase frequency detector.

TABLE 16.1

Summary of the Literature Reporting Dead Zone Minimization in PFD

Performance Parameters	[14] 2007	[71] 2009	[15] 2010	[72] 2010	[66] 2010	[16] 2012	[17] 2013	[19] 2014	[74] 2015
Simulation type	Post layout	Schematic	Experimental	Schematic	Schematic	Schematic	Post layout		Experimental
Technology (nm)	500	180	130	130	130	180	180	180	90
V_{DD} (V)	5	1.8	1.2	1.2	1.2	1.8	1.8	1.8	1.2
Lock-in time (ns)	150	–	–	2800	90	265	4500	–	–
Maximum frequency (GHz)	0.8	2.5	2.9	1.25	1.5	0.8	4	1	–
Dead zone/blind zone (ps)	Nil	Nil	52	Nil	Nil	10.25	Nil	Nil	0.7
Power consumption (mW)	–	0.006 @ 50 MHz	0.496 @ 128 MHz	0.062	0.01	5	12.1	0.381	–

PFD, phase frequency detector.

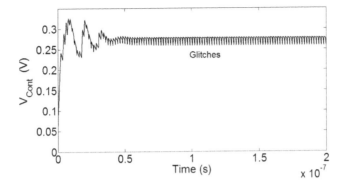

FIGURE 16.6   Conventional charge pump nonideal effects.

where R is the resistor value of the loop filter, I_{CP} is the CP output current, φ_e is the phase offset, f_{CKRef} is the frequency of the reference signal, and f_{PLPF} is the frequency of the pole of the loop filter.

Nonidealities (*leakage current, current mismatch*, and *timing mismatch*) of the CP give rise to the glitches indicated in Figure 16.6. Even after 80 ns, although the PLL is locked, the glitches are yet present, which can bring a challenging amount of reference spur.

Architectures proposed in literature to improve the performance of CP are broadly categorized into two types: "differential CP" and "single-ended CP." Differential CPs are more advantageous than single-ended [20]. For example, the overall performance is not affected by the switch mismatch between the NMOS and PMOS transistors. Moreover, the switches are made up of only NMOS transistors, and due to its fully symmetric operation, the inverter delays between *Up* and *Down* signals do not generate any offset. Compared with the single-ended, this CP doubles the range of output compliance. Since the leakage current is a result of common mode offset, this configuration is not affected by it. Better immunity is provided by two on-chip loop filters. However, differential CPs also suffer from some critical drawbacks. To control the current mismatch, gain boosting circuits [21,22] are used. When the *Down* is active, MN_2 and MN_3 work in concurrence with MN_1 to deliver a gain boosting circuit. It rises the output resistance of the CP and improves the current matching features; however, to improve reference spur, the additional switching errors were not analyzed.

To minimize the reference spur level, the phase offset can be reduced, as these are directly proportional to each other (Equation 16.1). To lessen the phase offset, the turn-on time of the PFD can be reduced. Perfect current matching characteristics are achieved in Ref. [24] by using an error amplifier and reference current sources. To reduce the current mismatch, a second compensation circuit is deployed in a CP, making it to have two push–pull CPs and two replica-feedback biasing circuits working as compensators [26]. In Ref. [27], the CP having a rail-to-rail error Op Amp (operational amplifier) with reference circuit and cascade current mirror matches the current over a wide range of output voltage. In Ref. [28], an enhanced CP design with reduced charge injection and current mismatch is presented. By changing the position of the switching transistor and adding extra compensation transistors, the charge injection is improved. A transconductance amplifier is added to overcome the current mismatch of the basic CP. To enhance the output resistance to match the output current, two high-swing cascade current mirrors, each for *Up* and *Down* networks, are used in Ref. [29]. A ratioed current CP is designed [30] to suppress the magnitude of reference spur. This CP is implemented by properly sizing the source and drain networks of the CP. The sizing can be done by using the relation (16.2) between the ratioed current and the size of the transistors.

Considering the aspect ratios of P_1, P_2 and N_1, N_2 to be same, the ratio of the *Down* and *Up* currents can be written as,

$$\frac{I_{Down}}{I_{Up}} = \frac{\left(\left|V_{GSP3}\right| - \left|V_{Tp}\right|\right)^2 \left(1 + \lambda_P \left|V_{DSP3}\right|\right)}{\left(\left|V_{GSP1}\right| - \left|V_{Tp}\right|\right)^2 \left(1 + \lambda_P \left|V_{DSP1}\right|\right)} \cdot \frac{W_{N1} \left(V_{GSN1} V_{TN}\right)^2 \left(1 + \lambda_N V_{DSN1}\right)}{W_{N3} \left(V_{GSN3} V_{TN}\right)^2 \left(1 + \lambda_N V_{DSN3}\right)} \quad (16.2)$$

where I_{Down} and I_{Up} are the currents due to *Down* and *Up* networks, respectively, V_{GSP3}, V_{GSP1}, V_{DSP3}, and V_{DSP1} are the gate-to-source voltages and drain-to-source voltage of PMOS 3 and 1, respectively, λ_P and λ_N are the channel length modulation coefficients of PMOS and NMOS, respectively, V_{TP} and V_{TN} are the threshold voltages of PMOS and NMOS, respectively, and W_{N1} and W_{N3} are the widths of NMOS 1 and 3, respectively.

As the glitches in the CP output current are mostly responsible in the reference spur, in Ref. [77], an additional buffer stage having a delay element, a delay capacitor, and a MOSFET is introduced so that the glitches will appear with some time delay. Moreover, the capacitance offered by the delay capacitor and the transistor (parasitic) attenuates the magnitude of the glitches. In Ref. [78], a differential CP design technique is adopted to accomplish low charge sharing and charge injection. Incorporating a sampled data common-mode feedback circuit, the output voltage range of the CP is increased while lowering the mismatch between charging and discharging current is realized. A dynamic current follow technology is implemented in Ref. [79] to match the *Up* and *Down* currents, and two differential pair inverters are incorporated to make the CP faster. The authors in Ref. [80] claim that as the *Up* and *Down* currents are derived from a common reference current source, as long as the aspect ratios of the transistors are same, there will not be any mismatch between *Up* and *Down* currents. Introducing a new CP [81] to reduce charge sharing, current mismatch, and charge injection problems, using differential current-steering switches with DC reference voltage biasing at one side, feedthrough of the input pulses can be eliminated. In Ref. [82], current mismatch compensator circuits are introduced so that the CP will not be influenced by process variation and power supply noise, which are also responsible for current mismatch and consequently phase offset. Similarly, in Ref. [83], a β-multiplier circuit with a differential amplifier is employed to produce the reference current independent of the supply voltage. In Ref. [84], the feedback stability issue associated with Ref. [81] is resolved, deploying differential current-steering switches with one side connected to a DC reference voltage. This circuit also minimizes the charge sharing, timing mismatch, and feed-through of the input pulses. Presenting a replica biasing using feedback, it reduces the current mismatch between the *Up* and *Down* modules of the CP. In Ref. [85], a digital calibration technique was reported. Here the static phase difference of the PFD input is measured. This amount of phase difference is associated to current mismatch and used for calibration. Self-calibrated CPs with a voltage scaler are deployed in Ref. [86] to reduce the static phase error, thereby minimizing the reference spur and jitter of a low-temperature polysilicon thin-film transistor (LTPS-TFT) PLL. However, in Refs. [85,86] are difficult to be applied where the phase error is not static. In Ref. [87], a dynamic auxiliary path in the CP is adopted. In each reference cycle interval, this path is utilized to detect and correct the current mismatch between *Up* and *Down*. By getting a mirrored controlled voltage, a timing control unit determines when to activate the current mismatch detection module, where a comparator is used to detect the current difference. Getting the difference from the comparator, a finite state machine is used to update the *Up* to match with *Down* current. Unlike most existing calibration techniques where the calibration is performed only once at the beginning of the locking process of the PLL, this technique dynamically adjusts the *Up* current to achieve a better matching performance. In Ref. [88], a wide input ranged rail-to-rail Op Amp and self-biasing cascode current mirror is used to reduce the CP current mismatch in a large output voltage range. Additionally, a precharging current source is deployed to enhance the initial charge current that will speed up the settling time of the PLL. Included with a lock detector, a high-resolution phase detector, and a 5-bit successive approximation register-controlled controller, CP current mismatch calibration scheme is reported in Ref. [89] to reduce reference spurs. Out of the aforementioned, some recent CP performance parameters are summarized in Table 16.2.

C. Low-Noise Fast PLLs Using Multi-PFD Architectures:

Design of a PLL system that simultaneously improves more than one performance parameter has also been an important area of research. In Ref. [90], both a phase detector and a frequency detector are incorporated, because the appropriate type of detector is pretty application dependent. A multiplying phase detector is suitable for locking of a data pulse stream, whereas a PFD is suitable for frequency synthesis since the input signal does not have missing transitions. Second, frequency detection is required to improve the locking time. Another example of multi-PFD is Ref. [41], where, to improve the acquisition time and lock range, a sequential PFD and a precharged PFD are used, respectively. A lock detector circuit is used to enable the respective CP of the PFD to be operated for a particular

TABLE 16.2

Summary of the Literature Reporting Spur Reduction Using New Charge Pump Circuits

Performance Parameters	[21] 2006	[22] 2007	[23] 2008	[24] 2000	[26] 2009	[27] 2009	[28] 2010	[80] 2005	[84] 2007	[87] 2009	[88] 2011
Simulation type	Schematic	Post layout	Post layout	Post layout	Schematic	Post layout	Experimental	Schematic	Schematic	Schematic	Experimental
Technology (nm)	350	90	180	250	130	180	130	180	180	180	180
V_{DD} (V)	3.3	2.5	1.8	2.5	1.2	1.8	1.5	1	1.2	1.8	1.8
Reference spur (dBc/Hz)	–	–	–	–75	–	–	–	–	–	–55	–
Current mismatch (%)	0.1	0.6	0.01	1	3.2	0.1	1	Nil	0.5	–	0.4
Area (mm²)	–	0.289	15	40	–	0.036	3.96	–	–	–	0.036
Power consumption (mW)	–	–	–	–	0.132	3	1	0.028	0.85	1	0.9

span of time. A PLL with phase error detector (PED) circuit is reported in Ref. [42] to reduce both the power consumption and acquisition time. The PED circuit delivers a dual-slope PFD and CP characteristic that effectively shrinks the power consumption. A coarse-tuning current is activated to track the large phase difference for fast locking, and to complete the fine adjustment near small phase difference, a fine-tuning current is initiated.

Similarly, in another work [91], a dual-slope PFD and CP architecture to accomplish fast locking of PLL is analyzed. The periodic ripples on the control voltage of the VCO are randomized by a novel random clock generator [40] where random selection of the PFDs is performed to reduce the reference spur at the output of the PLL in locked state. In Ref. [92], the PLL uses two PFDs (fast locking and low noise) for adaptively to cater the need of locking performance and noise reduction purposes. The performance parameters of the reported multi-PFD architectures are summarized in Table 16.3.

D. Wide-Lock Range PLLs Using AFC Techniques:

Due to the increasing demand for high-frequency multiband and multi-standard transceivers in modern communication systems, a wide-band fast locking PLL achieving low phase noise is also highly essential. To improve the locking process, lock range, and lock time and to reduce the total number of comparisons, a code optimization along with a binary search algorithm is used in Ref. [48]. The frequency range is found here to be more than 400 MHz, and the lock time is less than 65 μs when measured in 180-nm technology. To enhance the frequency range, both the discrete and continuous tuning mechanisms are adopted in Ref. [49]. An AFC technique is used where an auxiliary digital loop to select a particular band of VCO is incorporated. This PLL was simulated and implemented on FPGA using Xilinx

TABLE 16.3

Summary of the Literature Reporting Performance Improvement Using Multi-PFD Architectures

	PLL with Multi PFD Architecture			
Parameters	**[40]**	**[41]**	**[42]**	**[91]**
Simulation type	Measured	Measured	Schematic	Measured
Technology (nm)	180	1500	350	350
Frequency (GHz)	2.5	0.16	2.4	0.8
Lock range (GHz)	2.5–2.7	0.12–0.25	1.8–2.5	0.358–1.44
Lock-in time (ns)	N/A	4500	150	3000
Phase noise (dBc/Hz at 1 MHz offset)	−105	N/A	N/A	N/A
Power consumption (mw)	20	18.68	18.5	23.1
Layout area (mm^2)	1.56	845	N/A	87.1

PFD, phase frequency detector; PLL, phase-locked loop.

system generator. The lock-in time achieved here is 1.7 μs. In Ref. [50], the VCO incorporates a 5-bit differential switched capacitor array to build a tuning range from 2 to 3.2 GHz. The calibration time achieved here is less than 6 μs. Another AFC technique implemented in Ref. [51] works in two different modes, namely, frequency calibration mode and store/load mode. In the first mode, a new frequency detector is employed to reduce the lock time to 16 μs, whereas in second mode of operation by loading the calibration results stored after frequency calibration, the AFC makes the VCO come back to the calibrated frequency in about 1 μs. In a 900-MHz PLL [52], an automatic switched-capacitor discrete-tuning loop is deployed to have 20% more tuning range than the conventional one with a calibration time of 2 ms when measured in 0.6-μm technology. However, this method lacks in the calibration time. A better solution is presented in Ref. [53] where both the switched-capacitor bank LC VCO and the AFC technique are used to get a tuning range of 600 MHz, which is as wide as 40% of the highest frequency. However, this technique also suffers from high calibration time of the order of tens of μs.

16.4   DESIGN CHALLENGES

A number of PLL architectures along with their merits and drawbacks by linking them with the vital performance indices are presented in the previous sections. Synchronizing the phase and frequency of a signal has become a tedious job, because carrying out algebraic operations on frequencies is more challenging than on other electrical parameters such as voltages or currents. This challenge has taken different directions throughout the years, encouraging the invention of numerous architectures and circuit techniques. This section presents some of the major challenges in the design of PLLs at both architecture and circuit levels.

A. Charge Pump Current Mismatch:
Charge pump circuits in the PLL frequently use current mirror to match the output currents due to the variation of *Up* and *Down* signals generated for the PFD. But the current mirrors also have threshold voltage mismatch in their MOSFETs and finite output impedance due to which the charge pump current mismatch cannot be reduced to null though it is in control in several architectures discussed in the previous section.

B. Loop Filter Leakage Current:
The high impedance nodes of the loop filter are responsible for the leakage current that discharges the integrator capacitor. It occurs due to the reverse current of drain/source diffusion diodes and gate leakage in deep submicron CMOS FETs.

C. Dead Zone:
As discussed in the previous section, though dead zone is minimized by adopting various circuit techniques, it cannot be made zero. The phase

offset will always remain between the reference and feedback signals, which increases the jitter.

D. Delay in Feedback Loop:

In mixed-signal PLL circuits, the delay introduced due to the digital blocks in the feedback loop such as buffer, divider, and PFD produces a linear fluctuating phase shift as a function of frequency where the phase shift is directly proportional to the operating frequency. Due to this, the phase margin is also degraded that puts a question mark on the stability of the PLL. Further for large-bandwidth PLL, the feedback loop is always troublesome.

E. PLL Sampling Effect:

PFD does not compare the phases of the reference and feedback signal continuously. Practically, it is done only based on their edge position which makes the PFD a sampled data system. A hold process is thought of during the inactive time of the PFD when there is no signal to compare. The process of phase sample and hold in a PLL leads a phase lag in feedback loop. This degrades the PLL phase margin and again puts the doubt on stability.

16.5   CONCLUSION

PLL is a very widely used circuit technique in almost all modern electronic systems. In this chapter, recent PLL architectures are discussed in relation with their performance. The performance parameters are compared for each category of designing. This chapter presents a comprehensive view of the PLL architectures.

REFERENCES

1. J. P. Frazier, J. Page, "Phase-lock loop frequency acquisition study," *IRE Transactions on Space Electronics and Telemetry*, vol. 8, no. 3, pp. 210–227, Sept. 1962.
2. F. M. Gardner, "Charge-pump phase-lock loops," *IEEE Transactions on Communications*, vol. 28, no. 11, pp. 1849–1858, Nov. 1980.
3. R. E. Best, *Phase Locked Loops Design, Simulation and Applications*, 5th ed. New York: McGraw-Hill Publication, 2003.
4. U. Nanda, D. P. Acharya, S. K. Patra, Design of a low noise PLL for GSM application. *International Conference on Circuits, Controls and Communications (CCUBE), Bengaluru*, pp. 1–4, 2013.
5. F. M. Gardner, *Phaselock Techniques*, 3rd ed. New York: Wiley-Interscience, 2005.
6. C. Toumazou, G. S. Moschytz, B. Gilbert, *Trade-Offs in Analog Circuit Design: The Designer's Companion*. New York: Springer-Verlag, 2004.
7. P. K. Rout, D. P. Acharya, U. Nanda, Advances in analog integrated circuit optimization: a survey, Applied optimization methodologies in manufacturing systems. IGI Global, USA, pp. 309–333, 2018.
8. I. Thompson, P. V. Brennan, "Phase noise contribution of the phase/frequency detector in a digital PLL frequency synthesiser," *IEE Proceedings of Circuits, Devices and Systems*, vol. 150, no. 1, pp. 1–5, Feb. 2003.
9. H. Johansson, "A simple precharged CMOS phase frequency detector," *IEEE Journal of Solid-State Circuits*, vol. 33, no. 2, pp. 295–299, Feb. 1998.

10. H. Kondoh, H. Notani, T. Yoshimura, H. Shibata, Y. Matsuda, "A 1.5-V 250-MHz to 3.0-V 622-MHz operation CMOS phase-locked loop with precharge type phase-frequency detector," *IEICE Transaction on Electronics*, vol. E78-C, no. 4, pp. 381–388, Apr. 1995.

11. G.-Y. Tak, S.-B. Hyun, T. Y. Kang, B. G. Choi, S. S. Park, "A 6.3–9-GHz CMOS fast settling PLL for MB-OFDM UWB applications," *IEEE Journal of Solid-State Circuits*, vol. 40, no. 8, pp. 1671–1679, Aug. 2005.

12. W.-H. Lee, J.-D. Cho, S.-D. Lee, "A high speed and low power phase frequency detector and charge-pump," Proceeding of *Asia South Pacific Design Automation Conference*, vol. 1, pp. 269–272, Jan. 1999.

13. C. T. Charles, D. J. Allstot, "A calibrated phase/frequency detector for reference spur reduction in charge-pump PLLs," *IEEE Transactions on Circuits and Systems II: Express Briefs*, vol. 53, no. 9, pp. 822–826, Sept. 2006.

14. W. Hu, L. Chunglen, X. Wang, "Fast frequency acquisition phase-frequency detector with zero blind zone in PLL," *Electronics Letters*, vol. 43, no. 19, pp. 1018–1020, 13 Sept. 2007.

15. W.-H. Chen, M. E. Inerowicz, B. Jung, "Phase frequency detector with minimal blind zone for fast frequency acquisition," *IEEE Transactions on Circuits and Systems II: Express Briefs*, vol. 57, no. 12, pp. 936–940, Dec. 2010.

16. H. S. Raghav, S. Maheshwari, M. Srinivasarao, B. P. Singh, Design of low power, low jitter DLL tested at all five corners to avoid false locking. *10th IEEE International Conference on Semiconductor Electronics (ICSE), 2012*, pp. 526–531, 19–21 Sept. 2012.

17. K. K. A. Majeed, B. J. Kailath, "Low power, high frequency, free dead zone PFD for a PLL design," *IEEE Faible Tension Faible Consommation (FTFC), 2013*, pp. 1–4, 20–21 Jun. 2013.

18. S. Laha, S. Kaya, Dead zone free area efficient charge pump phase frequency detector in nanoscale DG-MOSFET. *IEEE 56th International Midwest Symposium on Circuits and Systems (MWSCAS), 2013*, pp. 920–923, 4–7 Aug. 2013.

19. A. C. Kailuke, P. Agrawal, R. V. Kshirsagar, Design of phase frequency detector and charge pump for low voltage high frequency PLL. *International Conference on Electronic Systems, Signal Processing and Computing Technologies (ICESC)*, pp. 74–78, 9–11 Jan. 2014.

20. B. Razavi, *Phase-Locking in High-Performance Systems: From Devices to Architectures*. New York: Wiley-IEEE Press, Feb. 2003.

21. C. Quemada, G. Bistue, I. Adin, *Design Methodology for RF CMOS Phase Locked Loops*. Boston, MA: Artech Publications, 2009.

22. M. Mansuri, D. Liu, C.-K.K. Yang, "Fast frequency acquisition phase-frequency detectors for Gsamples/s phase-locked loops," *IEEE Journal of Solid-State Circuits*, vol. 37, no. 10, pp. 1331–1334, Oct. 2002.

23. U. Nanda, D. P. Acharya, S. K. Patra, "A new transmission gate cascode current mirror charge pumpfor fast locking low noise PLL," *Circuits, Systems, and Signal Processing*, vol. 33, no. 9, pp. 2709–2718, Sept. 2014.

24. P. K. Rout, D. P. Acharya, G. Panda, "A multi objective optimization based fast and robust design methodology for low power and low phase noise current starved VCO," *IEEE Transaction on Semiconductor Manufacturing*, vol. 27, no. 1, pp. 43–50, Feb. 2014.

25. J. Lee, B. Kim, "A low-noise fast-lock phase-locked loop with adaptive bandwidth control," *IEEE Journal of Solid-State Circuits*, vol. 35, pp. 1137–1145, 2002.

26. Crowley, Phase locked loop with variable gain and bandwidth, *U.S. Patent 4,156,855*, 29 May 1979.

27. J. G. Maneatis, "Low-jitter process-independent DLL and PLL based on self-biased techniques," *IEEE Journal of Solid State Circuits*, vol. 31, no. 11, pp. 1723–1732, Nov. 1996.

28. C. S. Vaucher, "An adaptive PLL Tunning System architecture Combining High Spectral purity and Fast settling time," *IEEE Journal of Solid State Circuits*, vol. 35, no. 4, pp. 490–502, Apr. 2000.

29. T.-W. Liao, J.-R. Su, C.-C. Hung, "Spur-reduction frequency synthesizer exploiting randomly selected PFD," *IEEE Transaction on Very Large Scale Integration (VLSI) Systems*, vol. 21, no. 3, pp. 589–592, Mar. 2013.

30. Y. Woo, Y. M. Jang, M. Y. Sung, "Phase-locked loop with dual phase frequency detectors for high-frequency operation and fast acquisition," *Microelectronics Journal*, Elsevier, vol. 33, pp. 245–252, 2002.

31. Y.-F. Kuo, R.-M. Weng, C.-Y. Liu, A fast locking PLL with phase error detector. *IEEE Conference on Electron Devices and Solid-State Circuits, 2005*, pp. 423–426, 19–21 Dec. 2005.

32. W. Rhee, Design of high-performance CMOS charge pumps in phase-locked loops *Proceedings of the IEEE International Symposium on Circuits and Systems, ISCAS'99. 1999*, vol. 2, pp. 545–548, Jul. 1999.

33. Y.-S. Choi, D.-H. Han, "Gain-boosting charge pump for current matching in phase-locked loop," *IEEE Transaction on Circuits and Systems II: Express Briefs*, vol. 53, no. 10, pp. 1022–1025, Oct. 2006.

34. R. H. Mekky, M. Dessouky, Design of a low-mismatch gain-boosting charge pump for phase-locked loops, *International Conference on Microelectronics, ICM 2007.* pp. 321–324, 29–31 Dec. 2007.

35. J. Zhou, Z. Wang, A high-performance CMOS charge-pump for phase-locked loops *International Conference on Microwave and Millimeter Wave Technology, 2008*, vol. 2, pp. 839–842, 21–24 Apr. 2008.

36. J.-S. Lee, M.-S. Keel, S.-Il Lim, S. Kim, "Charge pump with perfect current matching characteristics in phase-locked loops," *Electronics Letters*, vol. 36, no. 23, pp. 1907–1908, 9 Nov. 2000.

37. M. El-Hage, Y. Fei, Architectures and design considerations of CMOS charge pumps for phase-locked loops. *Canadian Conference on Electrical and Computer Engineering, 2003.IEEE CCECE 2003.* vol. 1, pp. 223–226, 4–7 May 2003.

38. M.-S. Hwang, J. Kim, D.-K. Jeong, "Reduction of pump current mismatch in charge-pump PLL," *Electronics Letters*, vol. 45, no. 3, pp. 135–136, 29 Jan. 2009.

39. N. Hou, Z. Li, Design of high performance CMOS charge pump for phase-locked loops synthesizer. *15th Asia-Pacific Conference on Communications*, pp. 209–212, 8–10 Oct. 2009.

40. M. Jung, A. Ferizi, R. Weigel, A charge pump with enhanced current matching and reduced clock-feedthrough in wireless sensor nodes *Asia-Pacific Microwave Conference Proceedings (APMC)*, pp. 2291–2294, 7–10 Dec. 2010.

41. C. Zhang, T. Au, M. Syrzycki, A high performance NMOS-switch high swing cascode charge pump for phase-locked loops. *IEEE 55th International Midwest Symposium on Circuits and Systems (MWSCAS)*, pp. 554–557, 5–8 Aug. 2012.

42. N. Kamal, S. F. Al-Sarawi, D. Abbott, "Reference spur suppression technique using ratioed current charge pump," *Electronics Letters*, vol. 49, no. 12, pp. 746–747, 6 Jun. 2013.

43. M. Brandolini, P. Rossi, D. Manstretta, F. Svelto, "Toward multistandard mobile terminals -fully integrated receivers requirements and architectures," *IEEE Transaction on Microwave Theory and Techniques*, vol. 53, no. 3, pp. 1026–1038, 2005.

44. U. Nanda, D. P. Acharya, S. K. Patra, "Low noise and fast locking phase locked loop using a variable delay element in the phase frequency detector," *Journal of Low Power Electronics, American Scientific Publishers*, vol. 10, no. 1, pp. 53–57, 2014.

45. J. W. M. Rogers, F. F. Dai, M. S. Cavin, D. G. Rahn, "A multiband ΔΣ fractional-N frequency synthesizer for a MIMO WLAN transceiver RFIC," *IEEE Journal of Solid-State Circuits,* vol. 40, no. 3, pp. 678–689, 2005.

46. N. H. W. Fong, J. O. Plouchart, N. Zamdmer, L. Duixian, L. F. Wagner, C. Plett, N. G. Tarr, "Design of wide-band CMOS VCO for multiband wireless LAN applications," *IEEE Journal of Solid-State Circuits,* vol. 38, no. 8, pp. 1333–1342, 2003.

47. R. Holzer, A 1 V CMOS PLL designed in high-leakage CMOS process operating at 10–700 MHz. *Proceedings of IEEE International Solid-State Circuits Conference,* pp. 272–273, 2002.

48. K.-S. Lee, E.-Y. Sung, I.-C. Hwang, B.-H. Park, Fast AFC technique using a code estimation and binary search algorithm for wideband frequency synthesis. *Proceedings of the 31st European Solid-State Circuits Conference, 2005, ESSCIRC 2005,* pp. 181–184, 12–16 Sept. 2005.

49. M. R. Saadat, M. Momtazpour, B. Alizadeh, Simulation and improvement of two digital adaptive frequency calibration techniques for fast locking wide-band frequency synthesizers. *International Conference on Design & Technology of Integrated Systems in Nanoscale Era,* pp. 136–141, 2–5 Sept. 2007.

50. H. Zhang, "A broadband VCO with an adaptive frequency calibration circuit," *International Journal of Digital Content Technology and its Applications(JDCTA),* vol. 6, no. 17, pp. 47–56, Sept. 2012.

51. Y. Yadong, Y. Yuepeng, L. Weiwei, D. Zhankun, "A fast lock frequency synthesizer using an improved adaptive frequency calibration," *Journal of Semiconductors,* vol. 31, no. 6, p. 065011, Jun. 2010.

52. T.-H. Lin and W. J. Kaiser, "A 900-MHz 2.5-mA CMOS frequency synthesizer with an automatic SC tuning loop," *IEEE Journal of Solid-State Circuits,* vol. 36, pp. 424–431, Mar. 2001.

53. H.-I. Lee, J.-K. Cho, K.-S. Lee, I.-C. Hwang, T.-W. Ahn, K.-S.Nah, B.-H. Park, "A Σ-Δ fractional-N frequency synthesizer using a wideband integrated VCO and a fast AFC technique for GSM/GPRS/WCDMA applications," *IEEE Journal of Solid-State Circuits,* vol. 39, pp. 1164–1169, Jul. 2004.

54. J. Shin, H. Shin, "A fast and high-precision VCO frequency calibration technique for wideband Δ-Σ fractional-N frequency synthesizers," *IEEE Transaction on Circuits and Systems I: Regular Papers,* vol. 57, no. 7, pp. 1573–1582, Jul. 2010.

55. D. Baek, T. Song, E. Yoon, S. Hong, "8-GHz CMOS quadrature VCO using transformer-based LC tank," *IEEE Microwave and Wireless Components Letters,* vol. 13, pp. 446–448, Oct. 2003.

56. M. Tsai, Y. Cho, H. Wang, "A 5-GHz low phase noise differential Colpitts CMOS VCO," *IEEE Microwave and Wireless Components Letters,* vol. 15, pp. 327–329, May 2005.

57. Y. Eo, K. Kim, B. Oh, "Low noise 5 GHz differential VCO using InGaP/GaAs HBT Technology," *IEEE Microwave and Wireless Components Letters,* vol. 13, pp. 259–261, Jul. 2003.

58. J.-H. Duan, J.-P. Li, C. Qin, A 2.5 GHz low phase noise LC VCO for WLAN applications in 0.18 um CMOS technology. *3rd IEEE International Symposium on Microwave, Antenna, Propagation and EMC Technologies for Wireless Communications,* 27–29 Oct. 2009.

59. J. Jung, P. Upadyaya, P. Liu, D. Heo, Compact sub-1mW low phase noise CMOS LC-VCO based on power reduction technique. *IEEE MTT-S International Microwave Symposium Digest (MTT)* 2011.

60. Y. Xu, Z. Li A CMOS LC VCO in 0.5μm process. *IEEE International Conference on Industrial Technology,* 2008.

61. B. Han, J. Wu, C. Hu, A 2.5 GHz low phase noise LC VCO in 0.35 um SiGeBiCMOS technology. *7th International Conference on ASICON'07 by IEEE*, 22–25 Oct. 2007.

62. R. J. Baker, H. W. Li, D. E. Boyce, CMOS circuit design, layout, and simulation. *IEEE Press Series on Microelectronic Systems*, 2002.

63. G. S. Singh, D. Singh, S. Moorthi, Low power low jitter phase locked loop for high speed clock generation. *Asia Pacific Conference on Postgraduate Research in Microelectronics and Electronics (PrimeAsia), 2012*, pp. 192, 196, 5–7 Dec. 2012.

64. G. B. Lee, P. K. Chan, L. Siek, "A CMOS phase frequency detector for charge pump phase-locked loop," *42nd Midwest Symposium on Circuits and Systems*, vol. 2, pp. 601–604, 1999.

65. D.-C. Juang, D.-S. Chen, J.-M. Shyu, C.-Y. Wu, A low-power 1.2 GHz 0.35 μm CMOS PLL. *Proceedings of the Second IEEE Asia Pacific Conference on ASICs*, pp. 99–102, 2000.

66. K.-H. Cheng, T.-H. Yao, S.-Y. Jiang, W.-B. Yang, A difference detector PFD for low jitter PLL. *The 8th IEEE International Conference on Electronics, Circuits and Systems*, pp. 43–46, 2001.

67. M. Renaud, Y. Savaria, A CMOS three-state frequency detector complementary to an enhanced linear phase detector for PLL, DLL or high frequency clock skew measurement. *Proceedings of the 2003 International Symposium on Circuits and Systems, ISCAS '03.*, vol. 3, pp. III-148–III-151, 25–28 May 2003.

68. K.-S. Lee, B.-H. Park, H.-I. Lee, M. J. Yoh, Phase frequency detectors for fast frequency acquisition in zero-dead-zone CPPLLs for mobile communication systems. *Proceedings of the 29th European Solid-State Circuits Conference, ESSCIRC '03.*, pp. 525–528, 16–18 Sept. 2003.

69. M. M. Kamal, E. W. El-Shewekh, M. H. El-Saba, Design and implementation of a low-phase-noise integrated CMOS frequency synthesizer for high-sensitivity narrow-band FM transceivers. *Proceedings of the 15th International Conference on Microelectronics, 2003*, pp. 167–175, 9–11 Dec. 2003.

70. C.-P. Chou, Z.-M. Lin, J.-D. Chen, A 3-ps dead-zone double-edge-checking phase-frequency-detector with 4.78 GHz operating frequencies. *Proceedings of the IEEE Asia-Pacific Conference on Circuits and Systems*, vol. 2, pp. 937–940, 6–9 Dec. 2004.

71. N. M. H. Ismail, M. Othman, CMOS phase frequency detector for high speed applications, *International Conference on Microelectronics*, pp. 201–204, 19–22, Dec. 2009.

72. J. Lan, Y. Wang, L. Liu, R. Li, A nonlinear phase frequency detector with zero blind zone for fast-locking phase-locked loops. *International Conference on Anti-Counterfeiting Security and Identification in Communication*, pp. 41–44, 18–20 Jul. 2010.

73. C. Zhang, M. Syrzycki, Modifications of a dynamic-logic phase frequency detector for extended detection range. *53rd IEEE International Midwest Symposium on Circuits and Systems (MWSCAS)*, pp. 105–108, 1–4 Aug. 2010.

74. J. Strzelecki, S. Ren, "Near-zero dead zone phase frequency detector with wide input frequency difference," *Electronics Letters*, vol. 51, no. 14, pp. 1059–1061, 2015.

75. M. K. Hati, T. K. Bhattacharyya, A PFD and charge pump switching circuit to optimize the output phase noise of the PLL in 0.13-μm CMOS. *International Conference on VLSI Systems, Architecture, Technology and Applications*, pp. 1–6, 8–10 Jan. 2015.

76. B. Razavi, *Design of Analog CMOS Integrated Circuits*. New York: McGraw-Hill, 2000.

77. R. A. Baki, M. N. El-Gamal, A new CMOS charge pump for low-voltage (1V) high-speed PLL applications. *Proceedings of the International Symposium on Circuits and Systems, ISCAS '03.*, vol. 1, pp. I-657–I-660, 25–28 May 2003.

78. B. Terlemez, J. Uyemura, The design of a differential CMOS charge pump for high performance phase-locked loops. *Proceedings of the 2004 International Symposium on Circuits and Systems, ISCAS '04.*, vol. 4, pp. IV, 56 1–4, 23–26 May 2004.

79. Z. Tao, Z. Xuecheng, S. Xubang, A CMOS charge pump with a novel structure in PLL. *Proceedings of 7th International Conference on Solid-State and Integrated Circuits Technology*, vol. 2, pp. 1555–1558, 18–21 Oct. 2004.

80. H. Yu, Y. Inoue, Y. Han, A new high-speed low-voltage charge pump for PLL applications. *6th International Conference on ASIC*, vol. 1, pp. 387–390, Oct. 2005.

81. S. J. Byun, B. Kim, C.-H. Park, Charge pump circuit for a PLL. *US Patent 6,952,126 B2*, 4 Oct. 2005.

82. K. S. Ha, L. S. Kim, Charge pump reducing current mismatch in DLLs and PLLs. *Proceedings of ISCAS, Kos, Greece*, pp. 2221–2224, May 2006.

83. N. T. Hieu, T. W. Lee, H. H. Part, A perfectly current matched charge pump of CP-PLL for chip-to-chip optical link. *Pacific Rim Conference on Lasers and Electro-Optics, Seoul, Korea*, pp. 1–2, Aug 2007.

84. Y. Sun, L. Siek, P. Song, Design of a high performance charge pump circuit for low voltage phase-locked loops. *International Symposium on Integrated Circuits, ISIC '07*, pp. 271–274, 26–28 Sept. 2007.

85. C.-F. Liang, S.-H. Chen, S.-I. Liu, "A digital calibration technique for charge pumps in phase-locked systems," *IEEE Journal of Solid-State Circuits*, vol. 43, no. 2, pp. 390–398, Feb. 2008.

86. W.-M. Lin, S.-I. Liu, C.-H. Kuo, C.-H. Li, Y.-J. Hsieh, C.-T. Liu, "A phase-locked loop with self-calibrated charge pumps in 3 um LTPS-TFT technology," *IEEE Transactions on Circuits and Systems II: Express Briefs*, vol. 56, no. 2, pp. 142–146, Feb. 2009.

87. W.-H. Chiu, T.-S. Chang, T.-H. Lin, A charge pump current miss-match calibration technique for $\Delta\Sigma$ fractional-N PLLs in 0.18-μm CMOS. *IEEE Asian Solid-State Circuits Conference, A-SSCC 2009*, pp. 73–76, 16–18 Nov. 2009.

88. L. Zhiqun, Z. Shuangshuang, H. Ningbing, "Design of a high performance CMOS charge pump for phase-locked loop synthesizers," *Journal of Semiconductors*, vol. 32, no. 7, p. 075007, Jul. 2011.

89. Y.-W. Chen, Y.-H. Yu, Y.-J. E. Chen, "A 0.18 μm CMOS dual-band frequency synthesizer with spur reduction calibration," *IEEE Microwave and Wireless Components Letters*, vol. 23, no. 10, pp. 551–553, Oct. 2013.

90. K. M. Ware, H.-S. Lee, C.G. Sodini, "A 200-MHz CMOS phase-locked loop with dual phase detectors," *IEEE Journal of Solid-State Circuits*, vol. 24, no. 6, pp. 1560–1568, Dec. 1989.

91. K.-H. Cheng, W.-B. Yang, "A dual-slope phase frequency detector and charge pump architecture to achieve fast locking of phase-locked loop," *IEEE Transactions on Circuits and Systems II: Analog and Digital Signal Processing*, vol. 50, no. 11, pp. 892–896, Nov. 2003.

92. U Nanda, D P Acharya, Adaptive PFD "Selection Technique for Low Noise and Fast PLL in Multi- standard Radios," *Microelectronics Journal*, Elsevier, vol. 64, pp. 92–98, 2017.

17 Review of Analog-to-Digital and Digital-to-Analog Converters for A Smart Antenna Application

B. S. Patro, A. Senapati, and T. Pradhan
Kalinga Institute of Industrial Technology,
KIIT Deemed to be University

CONTENTS

17.1 Introduction 291
    17.1.1 Butler Matrix 294
    17.1.2 Architecture of Smart Antenna System............................ 294
        17.1.2.1 Receiver.. 294
        17.1.2.2 Transmitter.. 296
17.2 Analog-to-Digital Converters and Digital-to-Analog Converters 297
    17.2.1 Basics of Analog-to-Digital Conversion 297
    17.2.2 Sampling 298
    17.2.3 Quantization 299
    17.2.4 Successive Approximation Register Analog-to-Digital Converter 300
        17.2.4.1 Analog-to-Digital Converter........................ 301
    17.2.5 Flash Analog-to-Digital Converter.............................. 303
    17.2.6 Pipelined Analog-to-Digital Converter 304
    17.2.7 Delta–Sigma Analog-to-Digital Converter.................... 305
17.3 Digital-to-Analog Converter... 305
    17.3.1 Feedback Digital-to-Analog Converter 305
    17.3.2 Decimator 305
17.4 Conclusion 306
References... 307

17.1 INTRODUCTION

In recent years, there is an enormous rise in traffic for mobile and personal communication systems. The rise in traffic put a demand on both manufacturers and operators to increase capacity (Murch and Letaief 2002; Fletcher and Darwood 1998; Tsai and Woerner 2001; Poormohammad and Farzaneh 2018). Capacity enhancement is a major challenge for service providers as the radiation environment is affected by negative factors to limit the capacity (Consortium and others 2005; Shivapanchakshari and Aravinda 2019).

The spectral efficiency of the networks is maximized by the wireless carriers by exploring novel ways in order to improve their return on investment (Godara 2004). Lots of efforts have been put by the researchers to improve the wireless systems performance. Smart antennas (SAs) have shown improved quality and coverage efficiently in the wireless network systems (Chryssomallis 2000). SA systems have received much attention in the past few years (Chryssomallis 2000; Andersson et al. 1991; Tsoulos et al. 1998; Kohno 1998; Rappaport 1998; Tsoulos 2000) because they can increase system capacity (very important in urban and densely populated areas) by dynamically tuning out interference while focusing on the intended user (Boukalov and Haggman 2000; Liberti and Rappaport 1999) along with impressive advances in the field of digital signal processing.

Base station antennas in a cellular system are either omnidirectional or sectorized. There is a waste of resources because the majority of transmitted signal power radiates in directions other than desired user directions. Also, the signal power radiated over the coverage area of the network will experience interference by other undesired users. An SA system solves this problem by transmitting power toward the desired directions only. Adaptive beamforming capability in SA systems enables flexible synthesis and steering of antenna beams for optimized signal-to-noise and signal-to-interference ratio (SIR) performances.

SAs, also known as adaptive antennas and digital antenna arrays, are antenna arrays with smart signal-processing algorithms. Reconfigurable antennas have a similar capability as SAs, but they are single-antenna elements and not antenna arrays. SA techniques are used in acoustic signal processing, track and scan radar, radio astronomy, and radio telescope and cellular systems such as W-CDMA, UMTS, and LTE.

The deployment of SAs for wireless communications has emerged as one of the leading technologies for achieving high-efficiency networks that maximize capacity and improve quality and coverage. SA systems have received much attention in the past few years because they can increase system capacity (very important in urban and densely populated areas) by dynamically tuning out interference while focusing on the intended user along with impressive advances in the field of digital signal processing. The International Telecommunication Union (ITU) suggested an SA system as one of the key technologies for the fourth generation (4G) and beyond to improve wireless network capacity and mobility. Selected control algorithms, with predefined criteria, provide adaptive arrays (AAs) the unique ability to alter the radiation pattern characteristics (nulls, side-lobe level, main beam direction, and beamwidth). These control algorithms originate from several disciplines and target-specific applications (e.g., in the field of seismic, underwater, aerospace, and more recently cellular communications). The commercial introduction of SAs is a great promise for a big increase in system performance in terms of capacity, coverage, and signal quality, all of which will ultimately lead to increased spectral efficiency.

The main functions of SAs are direction-of-arrival (DOA) estimation and beam-forming. DOA estimation: In order to maximize the performance of the adaptive SA, the accurate estimation of the DOA of all signals transmitted to the AA antenna is required. The DOA estimation methods is categorized as a nonsubspace method and the subspace method. Nonsubspace methods are simpler, but performance is not good in terms of resolution. It depends on the spatial spectrum, and DOAs are obtained as locations of peaks in the spectrum. The main advantage of these techniques is that it can be used in situations where there is a lack of information about the properties of the signal. Various DOA estimation algorithms available in the literature are MUSIC, Improved MUSIC, and ESPIRIT (Gross 2005; Balanis and Ioannides 2007; Basha, Sridevi, and Prasad 2013).

Beamforming: After the DOA estimation, the function of the beamforming block is to generate the main beam toward the user and nulls toward interferers. Various beamforming algorithms available in literature are least mean square (LMS), sample matrix inversion (SMI), recursive least square (RLS), constant modulus algorithm (CMA), and conjugate gradient method (CGM) (Gross 2005; Patel, Makwana, and Parmar 2016; Awan et al. 2017; Lakshmi, Sivvam, and Rajyalakshmi 2018). A functional block diagram of the digital signal–processing part of an AA antenna system is shown in Figure 17.1. The SA system downconverts the received signals to baseband and digitizes them. Then using DOA algorithm, the signal-of-interest and signal-not-of-interest is located by dynamically changing the complex weights [w_0, w_1, ..., w_{M-1}]. Then the adaptive algorithm such as LMS, RLS, or SMI computes the appropriate weights to produce an optimum radiation pattern, i.e., main beam toward user and null toward interferer.

Smart antenna configurations:

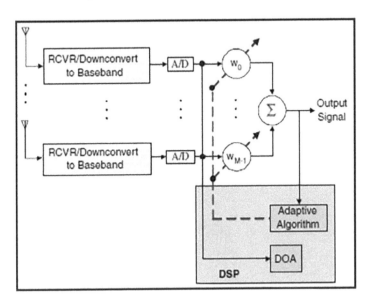

FIGURE 17.1 Smart antenna block diagram. DOA, direction of arrival; DSP, digital signal processor.

Two types of SAs are available:

1. **Switched Beam Antenna**: In some SAs, the beam patterns are fixed. This kind of SAs are known as switched beam (SB) antennas. Depending upon the requirements of the system, which beam has to be selected has to be decided accordingly. The benefits of SB antennas are that these can be easily fed into the present cell structures and are of low cost.
2. **Adaptive Antenna Array**: Adaptive SAs are the ones that let the beam drive itself in any direction and at the same time eliminate the interfering signals. The beam direction can be evaluated by the DOA estimation methods. It is a highly rated SA and consists of a complex transceiver.

In SB antenna, fixed beam patterns are available; depending on the requirement, beam has been selected. But in adaptive antenna array, beam has been generated as desired. The difference between SB and adaptive antenna array has been described. These two types of SAs are available in this literature.

17.1.1 BUTLER MATRIX

Fixed beams can be easily created through Butler matrices. Details of the derivation can be found in Butler (1961) and Shelton and Kelleher (1961). Phase shifters are used for producing several simultaneous fixed beams, and multiple beams can be achieved by exciting two or more beam ports with radio frequency (RF) signals at the same time.

17.1.2 ARCHITECTURE OF SMART ANTENNA SYSTEM

17.1.2.1 Receiver

SA receiver with M antenna elements, a radio unit, a beamforming unit, and a signal-processing unit is shown in Figure 17.2 (Lehne 1999).

In the radio unit, there are *M* downconversion chains, complex analog-to-digital converters (ADCs), amplifier, and channel decoder. The received signal is combined

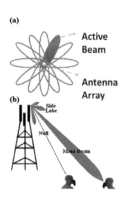

FIGURE 17.2   Smart antennas: (a) switched beam antenna and (b) adaptive antenna array.

and multiplied with complex weights $[w_1, w_2, w_3 ..., w_M]$, which are calculated by the signal-processing unit. The weight updating is done using various adaptive signal processing algorithms such as LMS and its variants, SMI, RLS, CMA, and CGM. Linearly constrained minimum variance (LCMV) is used for multiple-input multiple-output (MIMO) antenna system (Wang, Chen, and Jiang 2018). SA with electrically steerable parasitic array radiator (ESPAR) is implemented using an adaptive beamforming algorithm based on simultaneous perturbation stochastic approximation (Ganguly, Ghosh, and Kumar 2019). Particle swarm optimization is used for power and phase of excitation delivered to array elements using field-programmable gate array (FPGA) (Greda et al. 2019). Weights are estimated to maximize the power of the received signal from the desired user (SB or phased array [PA]) and to maximize the SIR by suppressing the received signal from interfering sources (AA). FPGA is used for real-time implementation of SA arrays with ADC and DAC.

The method for calculating the weights differs depending on the type of optimization criterion. When the SB is used, the receiver will test all the predefined weight vectors (corresponding to the beam set) and choose the best one giving the strongest received signal level. If the PA approach is used, which consists of directing a maximum gain beam toward the strongest signal component, the weights are calculated after the DOA is first estimated. A number of well-documented methods exist for estimating the DOA. In the AA approach, where maximization of SIR is needed, the optimum

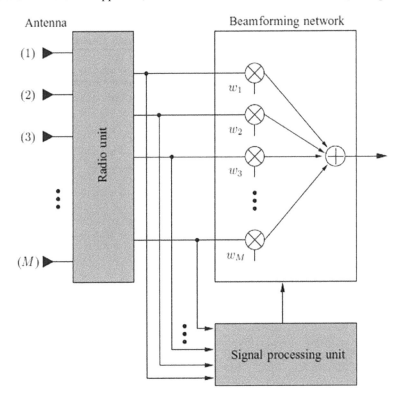

FIGURE 17.3 Smart antenna receiver.

weight vector (of dimension M) w_{opt} can be computed using a number of algorithms such as optimum combining and others that will follow (Balanis and Ioannides 2007).

When the beamforming is done digitally (after ADC), the beamforming and signal-processing units can normally be integrated with the same unit (digital signal processor). The separation in Figure 17.3 is done to clarify the functionality. The beamforming can be performed at either RF or intermediate frequency.

17.1.2.2 Transmitter

The transmission part of an SA system is shown in Figure 17.4. It is schematically similar to its reception part. The signal is split into N branches, which are weighted by the complex weights $[w_1, w_2, ..., w_N]$ in the beamforming unit. The weights are calculated by the signal-processing unit. The radio unit consists of D/A converters and the upconverter chains. The principal difference between uplink and downlink is that since there are no SAs applied to the user terminals (mobile stations), there is only limited knowledge of the channel state information available. Therefore, the optimum beamforming in downlink is difficult, and the same performance as the uplink cannot be achieved (Balanis and Ioannides 2007).

The main blocks for any SA systems are ADC and DAC. Without this improved and faster antenna system, to provide and process the signal is a tedious task. The components that are essential for the receiver part as well as transmitter part are an ADC and a DAC. ADC is used at the receiver side and DAC is used at the transmitter

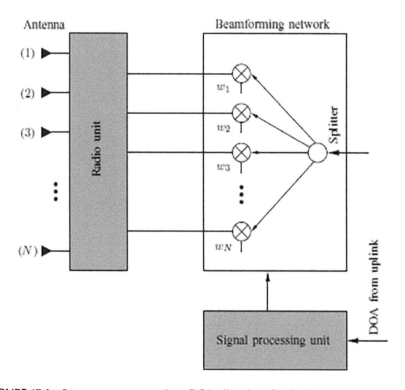

FIGURE 17.4   Smart antenna transmitter. DOA, direction of arrival.

side of the SA array. The receiver will receive the RF signal, but it is needed to be translated to an accurate digital signal at each antenna element. The job of translating an analog RF signal to an accurate digital signal is carried out by an appropriate ADC. Similar is the case with transmitter of the antenna where the reverse of ADC happens. At the transmitter end, the input signal is mostly digital, whereas the output is of real-time analog signal, and this conversion of digital to analog is done by a DAC. The next section will cover about this essential part of an antenna with various types of ADCs and DACs.

17.2 ANALOG-TO-DIGITAL CONVERTERS AND DIGITAL-TO-ANALOG CONVERTERS

The growth of digital computing and signal processing in electronics usually narrated as "the world is becoming more digital every day." One factor that has given an advantage to digital circuits is that, compared with their analog counterparts, they are less sensitive to disturbances and more robust in supply and process variations, allowing easier design and offering more extensive programmability. But the primary factor that has made digital circuits ubiquitous in all aspects of our lives is the boost in their performance, as a result of advances in integrated circuit technologies.

Nevertheless, the world intended as the sum of all-natural occurring signals is analog (human beings, included). Thus, a logical consequence, since the digital devices have to interact with the analog world, is that the more "the world is becoming digital," the more the devices are required that interface the analog world with the world of the digital processors. These devices are the ADCs. ADCs are essential part of many electronic devices, e.g., in mobile phones to encode the voice, in digital cameras to encode the signals generated by the image sensor, and in telephone modems to encode the incoming electrical signals.

17.2.1 BASICS OF ANALOG-TO-DIGITAL CONVERSION

ADCs are electronic systems that perform the transformation of analog signals that are continuous in time and in amplitude into digital signals that are discrete in time and amplitude as well. Figure 17.5 illustrates the general block diagram of an ADC intended for the conversion of low-pass signals, which essentially consists of an antialiasing filter (AAF), sampler, and quantizer. First, the analog input signal $x_a(t)$ passes through the AAF, a low-pass analog filter that prevents out-of-band components from folding over the signal bandwidth B_w during the subsequent sampling, which would corrupt the signal information. The resulting band-limited signal $x(t)$ is sampled at a rate f_s by the sample and hold (S/H) circuit, thus yielding a discrete time signal $x_s(n)=x(nT_s)$, where $T_s=1/f_s$. Finally, the values of $x_s(n)$ are quantized using N bits, so that each continuous-valued input sample is mapped onto the closer discrete-valued level out of the 2^N that covers the input range, yielding the converter digital output $y_d(n)$.

As shown in Figure 17.6, the fundamental processes involved in the analog-to-digital conversion are sampling and quantization.

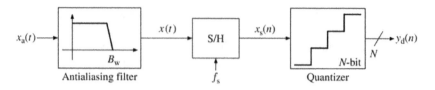

FIGURE 17.5 General block diagram of an ADC. ADC, analog-to-digital converter; S/H, sample and hold.

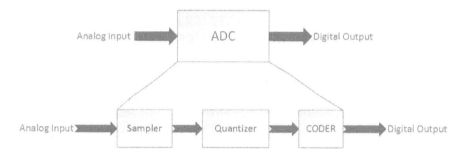

FIGURE 17.6 Basic ADC. ADC, analog-to-digital converter.

If we consider any basic ADC, there are three basic blocks as sampler, quantizer, and coder. The sampler determines the output data rate and is responsible for issues such as antialiasing in Nyquist ADC. Quantizer takes the analog-sampled signal as input and maps it to the discrete quantized value. Quantization noise is added to the ADC at the quantizer block not at the sampler. Coder is basically a digital filter or digital coder that collects the data from the quantizer. So the most important components of an ADC are sampler and quantizer.

17.2.2 SAMPLING

It is a process that converts the continuous input to a discrete signal in time and imposes a limit on the bandwidth of the analog input signal. According to the Nyquist theorem, to prevent information loss, $x(t)$ must be sampled at a minimum rate of $f_N=2B_W$, often referred to as the Nyquist frequency, and thus the ADCs in which analog input signal is sampled at the minimum rate $(f_s=f_N)$ are called Nyquist ADCs. Conversely, ADCs in which $f_s>f_N$ are called oversampling ADCs. The speed prior to the Nyquist rate for an ADC is expressed in terms of the oversampling ratio, defined as

$$OSR = \frac{f_s}{2B_w}$$

Whether oversampling is used or not in an ADC has a noticeable influence on the requirements of its AAF. Since in Nyquist ADCs the input signal bandwidth

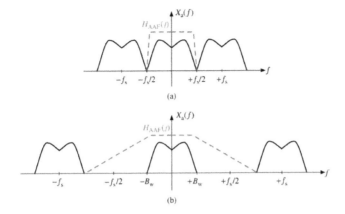

FIGURE 17.7 Antialiasing filter for (a) Nyquist-rate ADC and (b) oversampling ADC. ADC, analog-to-digital converter.

B_w aligns with $f_s/2$, then aliasing will occur if $x_a(t)$ contains frequency components above $f_s/2$. High-order analog AAFs are thus required to implement sharp transition bands capable of removing out-of-band components with no significant attenuation of the signal band, as illustrated in Figure 17.7a for the low-pass case. Conversely, as $f_s/2 > B_w$ in oversampling ADCs, the replicas of the input signal spectrum that are created by the sampling process are farther apart than in Nyquist ADCs. As illustrated in Figure 17.7b, frequency components of the input signal in the range $[B_w, f_s - B_w]$ do not alias within the signal band, so that the filter transition band can be smoother, which greatly reduces the order required for the AAF and simplifies its design.

17.2.3 QUANTIZATION

The quantization process also introduces a limitation on the performance of an ideal ADC, because an error is generated while performing the continuous-to-discrete transformation of the input signal in amplitude, commonly referred to as quantization error. The operation of quantizers is illustrated in Figure 17.8. As a matter of example, Figure 17.8c depicts the I/O characteristic of a quantizer with $N=2$, although results apply to a generic N-bit quantizer. Input amplitudes within the full-scale input range $[-X_{FS}/2, +X_{FS}/2]$ are rounded to 1 out of the 2^N different output levels, which are usually encoded into a binary digital representation. If these levels are equally spaced, the quantizer is said to be uniform, and the separation between adjacent output levels is defined as the quantization step:

$$\Delta = \frac{Y_{FS}}{2^N - 1}$$

where Y_{FS} stands for the full-scale output range. As X_{FS} and Y_{FS} are not necessarily equal, the quantizer may exhibit a gain different from unity, as indicated in

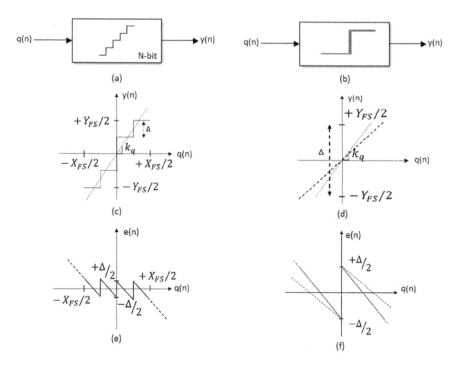

FIGURE 17.8 Illustration of the quantization process: (a) multibit quantizer block, (b) single-bit quantizer block, (c) I/O characteristic of a multibit quantizer, (d) I/O characteristic of a single-bit quantizer, (e) multibit quantization error, and (f) single-bit quantization error.

Figure 17.8c by the slope k_q. As shown in Figure 17.8e, the quantizer operation thus inherently generates a rounding error that is a nonlinear function of the input. Note that, if $q(n)$ is kept within the range $[-X_{FS}/2, +X_{FS}/2]$, the quantization error $e(n)$ is bounded within $[-\Delta/2, +\Delta/2]$. The former input range is known as the nonoverload region of the quantizer, as opposed to ranges with $|q(n)| > \Delta/2$, for which the magnitude of $e(n)$ grows monotonously.

Figure 17.8 also shows the operation of a single-bit quantizer ($N=1$). From Figure 17.8d, it can be noted that, compared with the multibit case, the output of a single-bit quantizer is determined by the input sign only, regardless of its magnitude. Therefore, the gain k_q is undefined and can be arbitrarily chosen.

The following section will discuss about various ADCs and DACs with their origins and various characteristics. Figure 17.9 shows the classification of various types of ADCs.

17.2.4 SUCCESSIVE APPROXIMATION REGISTER ANALOG-TO-DIGITAL CONVERTER

Successive approximation register (SAR) ADC is one of the most popular ADCs for its accuracy. The main components for this type of ADCs are (i) SAR, (ii) (DACs, (iii) S/H, and (iv) comparator. Figure 17.10 explains the basic structure of SAR-ADC where it is clearly shown that with each successive approximation, the digital value

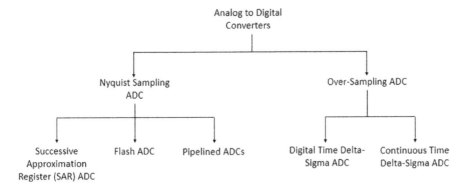

FIGURE 17.9 Various types of ADCs. ADC, analog-to-digital converter.

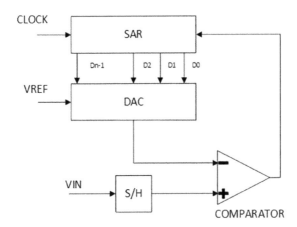

FIGURE 17.10 Structure of SAR-ADC. SAR-ADC, successive approximation register analog-to-digital converter.

is again converted to respective analog value and compared with the real-time input value. The compared value is again fed to SAR logic where it again approximated the appropriate digital value, and this process is repeated until a minimum accuracy is reached. The main disadvantage of this ADC is low speed; it is slower than other ADCs due to the approximation procedure.

17.2.4.1 Analog-to-Digital Converter

The story of different types of ADCs started away back in the 1970s when the first SAR-ADC) has been introduced by McCreary JL. Precisely the first SAR-ADC has been introduced in *IEEE Journal of Solid-State Circuits* in 1975 with the title as "All-MOS Charge Redistribution Analog-to-Digital Conversion Techniques—Part I" by McCreary and Gray (n.d.). Here the authors discussed the use of binary-weighted capacitor array with inherent sample–hold function for the first time to design an SAR-ADC. In that era, the research about ADCs had just begun.

The research on SAR-ADC became stagnant with the invention of the pipelined ADC. The first pipelined ADC was invented in the late 1980s (Lewis and Gray 1987; Lewis et al. 1992). The later attracted more researchers due to its speed and resolution bit production at that time. Thus, SAR-ADC was out of the research scope for almost more than a decade.

In the year 2004, Dieter Draxelmayr presented a paper on time-interleaved SAR-ADC, which was a low-power high-speed SAR-ADC at that time (Draxelmayr 2003). Dieter Draxelmayr presented the paper in ISSCC (International Solid-State Circuits Conference). This was the time when flash ADCs were among the fastest ADCs but with a disadvantage of high-power consumption. Flash ADCs use a 2N number of comparators to compare VIN with their reference voltages simultaneously, thus making itself a power hunger design. SAR-ADC on the other hands uses only a single comparator with some switches, which made SAR-ADC compatible for CMOS process miniaturization and future proof. With the presentation of the paper by Dieter Draxelma at ISSCC, researchers were again inclined toward the SAR-ADC as the concern about the speed had been eliminated. This has been again proved from the increased number of papers published on SAR-ADCs after 2004 (Matsuura 2016).

Craninckx and Van der Plas in 2007 at IEEE ISSCC presented a paper on a dynamic low-power SAR-ADC with a passive charge-sharing technique and asynchronous controller. The prototype was designed without using any active circuits, which are in general used for increasing the speed. Thus, they cut off any chances of static power loss in their design. Although the concerns remained with the resolution that was just below 10 ENOB and speed of the design. Their prototype achieved a power consumption of 290 mu W in 90-nm digital CMOS technology at that time (Craninckx and Van Der Plas 2007).

Pieter Harpe, Eugenio Cantatore, and Arthur van Roermund published a journal with a theory of the Data-Driven Noise-Reduction Method (DDNR) at *IEEE Journals of Solid-State Circuits* in 2013 (Harpe, Cantatore, and Van Roermund 2013). In this paper, they introduced DDNR and also claimed that it has enhanced the noise performance of the comparator, selectively. They have sacrificed the power consumption factor to achieve a higher resolution in their design. With a self-oscillating comparator, they had managed to generate an internal clock for bit cycling, which limited the use of an external clock to generate only sample rate frequency. The design used a segmented capacitive DAC to reduce differential nonlinearity (DNL) error as well as to save power. The prototype gained a resolution up to 12 bit and also reduced the leakage power to ensure a maintained efficiency even at low sample rates.

In 2012, Akira Shikata, Ryota Sekimoto, Tadahiro Kuroda, and Hiroki Ishikuro proposed the use of a trilevel comparator for SAR-ADC with their published journal at *IEEE Journal of Solid-State Circuits*, where they used the trilevel comparator for reducing the load of the comparator with reference to the speed requirement. With that, they reduced the resolution of internal DAC by 1 bit. The paper implies that the authors had managed to operate their proposed design at a single low voltage of 0.5 V, and therefore, their design achieved a low power of 1.2 μW with 1.1 MHz sampling frequency. In this paper, the capacitor mismatch problem at the level of DAC was addressed by the capacitor array with the ability to be reconfigured and calibration process (Shikata et al. 2012). They have shown an improvement of 3 dB in resolution of ADC with the help of trilevel comparator, and they have also reduced the DNL and integral nonlinearity error by a factor of 10 with DAC calibration.

Marcus Yip and Anantha P. Chandrakasan presented a paper regarding single reconfigurable SAR-ADC in the IEEE International Solid-State Circuits Conference in 2011. Their proposed design had the ability of scalable power that depends on the change in resolution and sample rate of ADC. They suggested the use of a resolution-reconfigurable DAC, a fully dynamic architecture with boosted sampling and no bias currents and a low-leakage 65 nm process that enables the power scaling feature in their design (Yip and Chandrakasan 2011).

17.2.5 FLASH ANALOG-TO-DIGITAL CONVERTER

Flash ADCs are faster, but they lack of accuracy. Figure 17.11 shows the basic structure of flash ADC. Flash ADC was first invented in 1979 and published in *IEEE Journal of Solid-State Circuits* (Dingwall 1979). Here the author Andrew G. F. Dingwall was able to introduce a parallel ADC that was capable of operating at 40 MHz. This proposal was a combination of two parallel ADCs with each having the capability to work at 20 MHz. Andrew G. F. Dingwall named it as flash ADC. He has used the then commercial R-2R ladder network ADC for his prototype.

Flash ADC after its invention was quite popular for its speed of operation. The main concern was accuracy and power consumption. As time progresses, it remains the fastest among the ADCs and has attracted many researchers over time.

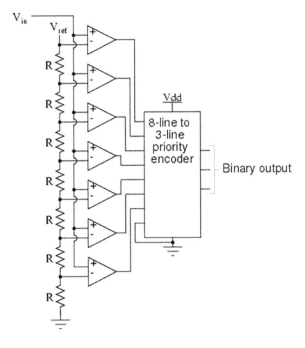

FIGURE 17.11   Structure of flash ADC. ADC, analog-to-digital converter.

In 2012, at International Conference on Green Technologies, a paper has been presented by George Tom Varghese and K.K. Mahapatra (Varghese and Mahapatra 2012), where they have addressed the issue of conversion from thermometer code to binary code for a flash ADC by introducing an encoder circuit in their proposal for translating the thermometer code to intermediary gray code and thus reducing the bubble error that was produced by test and hold circuit. They have used pseudo-NMOS logic to make the design faster with low power dissipation. The experiment was carried out in 90-nm technology with a sampling frequency of 5 GHz and a power dissipation of 0.3149 mW.

Pradeep Kumar and Amit Kolhe in 2011 in their published paper at *International Journal of Soft Computing and Engineering (IJSCE)* had proposed a 3-bit flash ADC architecture where they claimed their design to be low in hardware complexity and latency (Kumar and Kolhe 2011). They have used seven comparators with a response time of 6.82 ns for each comparator, which in turn gave a response time of 18.77 ns for the whole ADC structure. The range of the input voltage for their structure was from 0 to 3 V, and the output voltage had a deviation of 0.05 V with respect to the comparator voltage. The claimed power consumed by the structure was 36.273 mw.

Bui Van Hieu and Seunghwar Choi have proposed a new approach to deal with complex bubble errors that occurs during the conversion of thermometer code to binary code in flash ADCs, at International Midwest Symposium on Circuits and Systems in 2011 (Hieu et al. 2011). They had used a bubble error detection circuit rather than the bubble error correction circuit in thermometer code to the binary code encoder for their flash ADC, which happened to be helpful for less power consumption. The paper claimed that the proposed circuit can be used for the detection of a higher level of the bubble error and also has the ability to detect the previous version of bubble errors.

17.2.6 PIPELINED ANALOG-TO-DIGITAL CONVERTER

The ADC that has almost made SAR-ADC obsolete for more than a decade was pipelined ADC, which was first invented by Stephen H. Lewis and Paul R. Gray. They had published their paper in the *IEEE Journal of Solid-State Circuits* in which they proposed 5 M samples with 9-bit resolution pipelined ADC in the year 1987 (Lewis and Gray 1987). The invention was intended to minimize the silicon area that was used back then by the flash ADC without hampering the speed that was produced by the flash ADC. Stephen H. Lewis followed up his paper with another paper in 1992 (Lewis et al. 1992). The authors Stephen H. Lewis and R. Ramachandran published the paper in the *IEEE Journal of Solid-State Circuits*. This paper was about a 10-bit resolution and 20 M samples ADC where the input signal frequency used was 5 MHz. The revolutionary pipelined ADC became the research point for the researcher from thereon.

In 2010, Chun C. Lee and Michael P. Flynn presented a paper at Symposium on VLSI Circuits on SAR-Assisted 2-Stage Pipeline ADC (Lee and Flynn 2010). The paper proposed a 6-bit multiplying digital-to-analog converter (MDAC) for the first stage, which in fact enables large resolution and a 7-bit SAR-ADC for the second stage in the pipelined ADC. The use of SAR-ADC in place of flash ADC for sub-ADC in MDAC eliminated the power consumption factor and increased the open-loop gain. The second stage of this prototype used SAR-ADC to reduce power and increase resolution, which eliminated the use of more pipelines.

In 2019, Kyoung-Jun Moon, Dong-Shin Jo, Wan Kim, Michael Choi, Hyung-Jong Ko, and Seung-Tak Ryu at the *IEEE Journal of Solid-State Circuits* published their work regarding a pipeline SAR-ADC where their prototype has used current-mode residue processing (Moon et al. 2019). This proposed state of the art used different building blocks such as degenerated gm-cell as an open-loop residual amplifier; the S/H circuit has been replaced by a switched-current mirror and use of a split-current DAC for current-domain SAR conversion. There fabricated 28-nm prototype achieved a signal-to-noise dynamic range ratio (SNDR) of 56.6 dB with a figure of merit of a 21.7-fJ/conversion step. The residue processing used in this prototype enhanced the operation speed.

17.2.7 Delta–Sigma Analog-to-Digital Converter

There are two types of ADCs: first type of ADC follows the Nyquist sampling theorem, whereas other type of ADC follows oversampling theorem for operation. Delta–sigma ADC comes under the later type of ADCs. The demand for delta–sigma ADCs increased with a due period because of its low power consumption. The delta–sigma ADCs can be divided into two parts such as discrete-time delta–sigma ADC and continuous-time delta–sigma ADC. The demand for continuous-time delta–sigma ADC increased due to the capability of producing high resolution in low power (Pradhan 2017).

17.3 DIGITAL-TO-ANALOG CONVERTER

17.3.1 Feedback Digital-to-Analog Converter

The feedback DAC is employed to convert the digital modulator output to analog and subtract it from the modulator input. It is normally implemented with a charge-redistribution DAC in DT converters, whereas current-steering DACs are used in CT converters.

Switched-capacitor return-to-zero DAC is a combination of the low jitter sensitivity and low distortion. Due to low distortion of return-to-zero DAC, the output will be almost rail to rail, thus achieving a higher SNDR and signal-to-noise ratio. Furthermore, this DAC improves the performance at odd harmonics of the input signal for the continuous-time delta–sigma modulator (CTDSM).

17.3.2 Decimator

A decimator is a digital filter with more complexity which is responsible for downsampling the oversampled output of the CT$\Delta\Sigma$M to the Nyquist rate. The purposes of decimator are as follows:

1. It downsamples the oversampled modulated output.
2. It removes undesired band of frequencies with a filter.

Thus, increasing the resolution of complete CTΔΣ ADC, Amrith Sukumaran and Shanthi Pavan in their published journal at *IEEE Journal of Solid-State Circuits* in 2014 studied the use of finite impulse response (FIR) feedback DAC in single-bit continuous-time delta–sigma ADCs. They proposed a method to stabilize a CTDSM using an FIR feedback DAC (Sukumaran and Pavan 2014).

In 2010, in *IEEE Journal of Solid-State Circuits*, Gerry Taylor, Member, and Ian Galton published a paper where they have proposed a CTDSM with mostly having digital circuitry. The design was based on a voltage-controlled ring oscillator. It was claimed to be free from an analog integrator, feedback DACs, comparators, reference voltages, and requirement of a low-jitter clock. Because most of the parts are digital in this proposed prototype, the area of the design was claimed to be less than that of a conventional CTDSM (Taylor and Galton 2010).

In October 2019, Antonios Nikas, Sreenivas Jambunathan, Leonhard Klein, Matthias Voelker, and Maurits Ortmanns published a paper in *IEEE Journal of Solid-State Circuits* where they proposed a CTDSM using a current-reuse DAC and modified instrumentation amplifier (IA). The prototype used the modified IA as the integrator for the delta–sigma modulator, whereas the current-reuse DAC was used as the feedback circuit to that integrator. The authors claimed that the design consumes 22% less power in 180-nm CMOS technology with a dynamic range of 90 dB (Nikas et al. 2019).

In September 2019, a letter was published in *IEEE Solid-State Circuits Letters* by John Bell and Michael P. Flynn proposing a multiband continuous-time delta–sigma ADC. In the proposed prototype, a new class of continuous-time delta–sigma ADC demonstrated two simultaneous bands such as baseband and bandpass. The total added bandwidth of the ADC was 90 MHz while the bands were shown to be separated by 500 MHz. The authors claimed that this prototype was first to be presented by them (Bell, Member, and Flynn 2019).

In 2014, a paper by Astria Nur Irfansyah, Long Pham, Andrew Nicholson, Torsten Lehmann, Julian Jenkins, and Tara Julia Hamilton entitled "Nauta OTA in a Second-Order Continuous-Time Delta–Sigma Modulator" proposed the use of Nauta OTA as an integrator in a second-order CTDSM. They studied the structure for high-bandwidth operation and simple inverter-based structure (Irfansyah et al. 2014).

In 1992, a journal published in *IEEE Journal of Solid-State Circuits* by Bram Nauta entitled "A CMOS Transconductance-C Filter Technique for Very High Frequencies" first proposed the use of inverter-based tranconductors. He first studied the work of differential amplifier–based integrator and then proposed how can an improved result be achieved by inverter-based operational transconductance amplifier (Nauta 1992).

17.4 CONCLUSION

ADC and DAC are part of SA transmitter and receiver. This chapter elaborates different types of ADCs and DACs, which can be used for SA applications. SA is very much required to meet the growing demand of the users for faster communication and networking. So choosing the appropriate ADC and DAC is also an important task for the researchers considering and fulfilling all the design requirements of the users and service providers. The future scope includes performance comparison; further analysis can be done with different types of antenna arrays such as linear, planar, circular, and so on and their performance comparison.

REFERENCES

Andersson, Sören, Mille Millnert, Mats Viberg, and Bo Wahlberg. 1991. A study of adaptive arrays for mobile communication systems. In *Proceedings ICASSP 91: 1991 International Conference on Acoustics, Speech, and Signal Processing*, 3289–92.

Awan, Adnan Anwar, Shahid Khattak, Aqdas Naveed Malik, and others. 2017. Performance comparisons of fixed and adaptive beamforming techniques for 4G smart antennas. In *2017 International Conference on Communication, Computing and Digital Systems (C-CODE)*, 17–20.

Balanis, Constantine A, and Panayiotis I Ioannides. 2007. "Introduction to Smart Antennas." *Synthesis Lectures on Antennas* 2 (1). Morgan & Claypool Publishers: 1–175.

Basha, T S Ghouse, P V Sridevi, and M N Giri Prasad. 2013. "Beam Forming in Smart Antenna with Precise Direction of Arrival Estimation Using Improved MUSIC." *Wireless Personal Communications* 71 (2). Springer: 1353–64.

Bell, John, Student Member, and Michael P Flynn. 2019. "A Simultaneous Multiband Continuous-Time ADC With 90-MHz Aggregate Bandwidth in 40-Nm CMOS" 2 (9): 91–94. doi:10.1109/LSSC.2019.2933159.

Boukalov, Adrian O, and S-G Haggman. 2000. "System Aspects of Smart-Antenna Technology in Cellular Wireless Communications-an Overview." *IEEE Transactions on Microwave Theory and Techniques* 48 (6). IEEE: 919–29.

Butler, Jesse. 1961. "Beam-Forming Matrix Simplifies Design of Electronically Scanned Antenna." *Electron. Design* 9: 170–73.

Chryssomallis, Michael. 2000. "Smart Antennas." *IEEE Antennas and Propagation Magazine* 42 (3). IEEE: 129–36.

Consortium, International Engineering, and others. 2005. "Smart Antenna Systems." http://read.pudn.com/downloads157/doc/comm/699605/smart_antenna%20systems.pdf.

Craninckx, Jan, and Geert Van Der Plas. 2007. A 65fJ/conversion-step 0-to-50MS/s 0-to-0.7mW 9b charge-sharing SAR ADC in 90nm digital CMOS. In *Digest of Technical Papers - IEEE International Solid-State Circuits Conference*. doi:10.1109/ISSCC.2007.373386.

Dingwall, Andrew G.F. 1979. "Monolithic Expandable 6 Bit 20MHz CMOS/SOS A/D Converter." *IEEE Journal of Solid-State Circuits* 14 (6): 926–32. doi:10.1109/JSSC.1979.1051299.

Draxelmayr, Dieter. 2003. "A 6b 600MHz 10mW ADC Array in Digital 90nm CMOS." In *Digest of Technical Papers - IEEE International Solid-State Circuits Conference*, 47:212–213+536. doi:10.1109/isscc.2004.1332695.

Fletcher, P N, and P Darwood. 1998. "Beamforming for Circular and Semicircular Array Antennas for Low-Cost Wireless Lan Data Communications Systems." *IEE Proceedings-Microwaves, Antennas and Propagation* 145 (2). IET: 153–157.

Ganguli

Ganguly, Saurav, Jayanta Ghosh, and Puli Kishore Kumar. 2019. "Performance Analysis of Array Signal Processing Algorithms for Adaptive Beamforming." *Simulation* 2: 15.

Godara, Lal Chand. 2004. *Smart Antennas*. Boca Raton, FL: CRC press.

Greda, Lukasz A, Andreas Winterstein, Daniel L Lemes, and Marcos V T Heckler. 2019. "Beamsteering and Beamshaping Using a Linear Antenna Array Based on Particle Swarm Optimization." *IEEE Access*, 7: 141562–73.

Gross, F B. 2005. *Smart Antenna for Wireless Communication.* New York: Graw-Hill. Inc.

Harpe, Pieter, Eugenio Cantatore, and Arthur Van Roermund. 2013. "A 10b/12b 40 KS/s SAR ADC with Data-Driven Noise Reduction Achieving up to 10.1b ENOB at 2.2 FJ/Conversion-Step." *IEEE Journal of Solid-State Circuits* 48 (12). IEEE: 3011–18. doi:10.1109/JSSC.2013.2278471.

Hieu, Bui Van, Seunghwan Choi, Jongkug Seon, Youngcheol Oh, Chongdae Park, Jaehyoun Park, Hyunwook Kim, and Taikyeong Jeong. 2011. A new approach to thermometer-to-binary encoder of flash ADCs- bubble error detection circuit. In *2011 IEEE 54th International Midwest Symposium on Circuits and Systems (MWSCAS)*, 1–4. IEEE. doi:10.1109/MWSCAS.2011.6026403.

Irfansyah, Astria Nur, Long Pham, Andrew Nicholson, Torsten Lehmann, Julian Jenkins, and Tara Julia Hamilton. 2014. Nauta OTA in a second-order continuous-time delta-sigma modulator. In *Midwest Symposium on Circuits and Systems*, 849–52. Institute of Electrical and Electronics Engineers Inc. doi:10.1109/MWSCAS.2014.6908548.

Kohno, Ryuji. 1998. "Spatial and Temporal Communication Theory Using Adaptive Antenna Array." *IEEE Personal Communications* 5 (1). IEEE: 28–35.

Kumar, Pradeep, and Amit Kolhe. 2011. "Design & Implementation of Low Power 3-Bit Flash ADC in 0. 18μm CMOS." *International Journal of Soft Computing and Engineering* 1 (5): 71–74.

Lakshmi, T S Jyothi, Sandeep Sivvam, and V Rajyalakshmi. 2018. Performance evaluation of smart antennas using non blind adaptive algorithms. In *2018 Conference on Signal Processing and Communication Engineering Systems (SPACES)*, 66–71.

Lee, Chun C., and Michael P. Flynn. 2010. "A 12b 50MS/s 3.5mW SAR Assisted 2-Stage Pipeline ADC." *IEEE Symposium on VLSI Circuits, Digest of Technical Papers.* IEEE, 239–40. doi:10.1109/VLSIC.2010.5560243.

Lehne, Per H. 1999. "An Overview of Smart Antenna Technology for Mobile Communications Systems." *IEEE Communications Surveys, Fourth Quarter* 2 (4).

Lewis, Stephen H., and Paul R. Gray. 1987. "A Pipelined 5-Msample/s 9-Bit Analog-to-Digital Converter." *IEEE Journal of Solid-State Circuits* 22 (6): 954–61. doi:10.1109/JSSC.1987.1052843.

Lewis, Stephen H, H Scott Fetterman, George F Gross, R Ramachandran, T R Viswanathan, and Senior Member. 1992. "A 10-b 20-Msample / s Analog-to-Digital Converter." *IEEE Journal of Solid-State Circuits* 27 (3): 351–58.

Liberti, Joseph C, and Theodore S Rappaport. 1999. *Smart Antennas for Wireless Communications: IS-95 and Third Generation CDMA Applications.* Upper Saddle River, NJ: Prentice Hall PTR.

Matsuura, Tatsuji. 2016. "Recent Progress on CMOS Successive Approximation ADCs." *IEEJ Transactions on Electrical and Electronic Engineering* 11 (5): 535–48. doi:10.1002/tee.22290.

Mccreary, James L, and P R Gray. n.d. [1975] "All-MOS Charge Redistribution Analog-to-Digital Conversion Techniques." *IEEE Journal of Solid-State Circuits*, SC-10 (12): 371–79.

Moon, Kyoung-Jun, Dong-Shin Jo, Wan Kim, Michael Choi, Hyung-Jong Ko, and Seung-Tak Ryu. 2019. "A 9.1-ENOB 6-MW 10-Bit 500-MS/s Pipelined-SAR ADC with Current-Mode Residue Processing in 28-Nm CMOS." *IEEE Journal of Solid-State Circuits* 54 (9). IEEE: 2532–42. doi:10.1109/jssc.2019.2926648.

Murch, Ross D, and K Ben Letaief. 2002. "Antenna Systems for Broadband Wireless Access." *IEEE Communications Magazine* 40 (4). IEEE: 76–83.

Nauta Bram. 1992. "A CMOS Transconductance-C Filter Technique for Very High Frequencies." *IEEE Journal of Solid-State Circuits* 27 (2): 142–53.

Nikas, Antonios, Sreenivas Jambunathan, Leonhard Klein, Matthias Voelker, and Maurits Ortmanns. 2019. "A Continuous-Time Delta-Sigma Modulator Using a Modified Instrumentation Amplifier and Current Reuse DAC for Neural Recording." *IEEE Journal of Solid-State Circuits* 54 (10). IEEE: 2879–91. doi:10.1109/jssc.2019.2931811.

Patel, Dhaval N, B J Makwana, and P B Parmar. 2016. Comparative analysis of adaptive beam-forming algorithm LMS, SMI and RLS for ULA smart antenna. In *2016 International Conference on Communication and Signal Processing (ICCSP)*, 1029–33.

Poormohammad, Sarah, and Forouhar Farzaneh. 2018. "Proposed 2D and 3D Geometries Intended for Smart Antenna Applications, Including Direction Finding and Beamforming Implementation." *IET Radar, Sonar & Navigation* 13 (5). IET: 673–81.

Pradhan, Tuhinansu. 2017. "Design of a low-voltage low power dynamic latch comparator for a 1.2-V 0.4-mW CT delta sigma modulator with 41-dBm SNDR". In 2017 International Conference on Trends in Electronics and Informatics (ICEI) (pp. 815–20). IEEE.

Pradhan, T. 2017. Implementation of a low power second order continuous time delta sigma modulator in 0.09µm technology using Cadence. *M. Tech Thesis*, KIIT Deemed to be University.

Rappaport, Theodore S. 1998. *Smart Antennas: Adaptive Arrays, Algorithms, & Wireless Position Location*. Piscataway, NJ: Institute of Electrical and Electronics Engineers Inc.

Shelton, J, and K Kelleher. 1961. "Multiple Beams from Linear Arrays." *IRE Transactions on Antennas and Propagation* 9 (2). IEEE: 154–61.

Shikata, Akira, Ryota Sekimoto, Tadahiro Kuroda, and Hiroki Ishikuro. 2012. "A 0.5 v 1.1 MS/Sec 6.3 FJ/Conversion-Step SAR-ADC with Tri-Level Comparator in 40 Nm CMOS." *IEEE Journal of Solid-State Circuits* 47 (4). IEEE: 1022–30. doi:10.1109/JSSC.2012.2185352.

Shivapanchakshari, T G, and H S Aravinda. 2019. An efficient mechanism to improve the complexity and system performance in OFDM using switched beam smart antenna (SSA). *Computer Science On-Line Conference*, 60–66.

Sukumaran, Amrith, and Shanthi Pavan. 2014. "Low Power Design Techniques for Single-Bit Audio Continuous-Time Delta Sigma ADCs Using FIR Feedback." *IEEE Journal of Solid-State Circuits* 49 (11). IEEE: 2515–25. doi:10.1109/JSSC.2014.2332885.

Taylor, Gerry, and Ian Galton. 2010. "A Mostly-Digital Variable-Rate Continuous-Time Delta-Sigma Modulator ADC." *IEEE Journal of Solid-State Circuits* 45 (12). IEEE: 2634–46. doi:10.1109/JSSC.2010.2073193.

Tsai, Jiann-An, and Brian D Woerner. 2001. Adaptive beamforming of uniform circular arrays (UCA) for wireless CDMA system. In *Conference Record of Thirty-Fifth Asilomar Conference on Signals, Systems and Computers (Cat. No. 01CH37256)*, 1: 399–403.

Tsoulos, George V. 2000. *Adaptive Antennas for Wireless Communications*. Wiley-IEEE Press.

Tsoulos, George V, Georgia E Athanasiadou, Mark A Beach, and Simon C Swales. 1998. "Adaptive Antennas for Microcellular and Mixed Cell Environments with DS-CDMA." *Wireless Personal Communications* 7 (2–3). Springer: 147–69.

Varghese, George Tom, and K. K. Mahapatra. 2012. A high speed low power encoder for a 5 bit flash ADC." In *2012 International Conference on Green Technologies, ICGT 2012*, 41–5. doi:10.1109/ICGT.2012.6477945.

Wang, Xiaodong, Changlin Chen, and Weidong Jiang. 2018. Implementation of real-time LCMV adaptive digital beamforming technology. In *2018 International Conference on Electronics Technology (ICET)*, 134–7.

Yip, Marcus, and Anantha P. Chandrakasan. 2011. A resolution-reconfigurable 5-to-10b 0.4-to-1V power scalable SAR ADC. *Digest of Technical Papers - IEEE International Solid-State Circuits Conference*. IEEE, 190–1. doi:10.1109/ISSCC.2011.5746277.

18 Active Inductor–Based VCO for Wireless Communication

Aditya Kumar Hota
Veer Surendra Sai University of Technology

Shasanka Sekhar Rout
GIET University

Kabiraj Sethi
Veer Surendra Sai University of Technology

CONTENTS

18.1   Introduction 311
      18.1.1   Inductor–Capacitor Voltage-Controlled Oscillator 312
            18.1.1.1   Linear Feedback Approach.. 312
            18.1.1.2   Cross-Coupled Approach... 313
18.2   Ring Voltage-Controlled Oscillator.. 314
18.3   Active Inductor 314
18.4   Voltage-Controlled Oscillator Using Active Inductor 315
18.5   Discrete Fourier Transform for Voltage-Controlled Oscillator................. 319
18.6   Summary 320
References.. 322

18.1   INTRODUCTION

The recent wireless communication systems require very wide-band voltage-controlled oscillators (VCOs) to be integrated with other multimode, multiband systems, which is very difficult to achieve with a passive, untuned inductor-based VCO. The gigabit-per-second (Gbps) serial links demand small-width clock pulses, which are fulfilled by the ring VCOs. In fiber optic communication, maintaining the phase noise of the oscillator is a very challenging task. So the requirement is a very-high-Q-value VCO to keep the phase noise low. This can be done by designing a high-Q tank circuit having high-Q inductor or capacitor. In millimeter wave application, it is not possible to generate a fundamental frequency; rather the higher-order harmonics are required for that purpose. So heterojunction bipolar transistor (HBT)–based push–push technology may be the solution to this kind of problems. An oscillator is characterized by required frequency of oscillation (f_{osc}), wide frequency tuning range, low phase noise, high frequency stability over temperature, low gain, and low output power. Being a nonlinear element oscillator, linearity may be presented either as a selective, positive feedback amplifier or by a resonant tank with a shunt negative resistance circuit. The radio frequency integrated circuits (RFICs) for wireless communication are mostly based on the mixed-mode technology and so it is very difficult and costly for the testing of the same. So the different analog, RF, and digital circuits can be tested separately. The specifications of VCO and so of IC such as gain, phase noise, and free running frequency can be affected by the parametric defects and process variations causing the IC to be inoperative. So it is very essential to have a self-test circuit to avoid the faults in VCO and phase-locked loop (PLL).

Earlier, it was very expensive for the testing of an RFIC due to the setup and maintenance cost of the automatic test equipment. So, to reduce the cost, reconfigurable test circuits were integrated inside the IC itself working in the loopback principle. Built-in self-test (BIST) is the most preferred testing technique, but it requires a specific die area and very difficult to be integrated with the RFIC due to its large area consumption, mainly due to the passive inductive components. Thus, the use of active inductors (AIs) in the RFIC leaves some head room for the testing circuits for integration.

18.1.1 INDUCTOR–CAPACITOR VOLTAGE-CONTROLLED OSCILLATOR

A resonator-based VCO uses an inductor–capacitor (LC) pair tank circuit as the resonator in shunt with a core circuit, which provides the necessary feedback for sustained oscillation. Whenever a pure sinusoid is required, designers opt for this type of VCO. As it behaves as a bandpass filter, the out-of-band noise cannot affect the signal here.

18.1.1.1 Linear Feedback Approach

This model uses the concept of negative feedback to achieve positive feedback for realizing an oscillator. The block diagram of the linear feedback model of VCO is shown in Figure 18.1. A and β are the gain of the amplifier and feedback network transfer function, respectively, and both are functions of frequency.

$$V_o = \frac{A(V_{osc}, \omega)}{1 - A(V_{osc}, \omega)\beta(V_{osc}, \omega)} V_{in} \qquad (18.1)$$

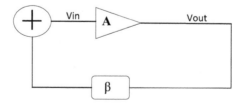

FIGURE 18.1 Linear feedback model.

The close loop gain should be infinite to get an output signal out of nothing. The broadband noise from the active and passive components of the circuit upon biasing causes the oscillation to start. Initially, $V_{in}=0$, and the gain is independent of the voltage, but gradually V_{in} increases to V_{osc} where the gain, as well as a function of input voltage, becomes linear. The conditions for sustained oscillation are as follows: the loop gain phase shift should be 180°, and magnitude should be greater than 1. Hartley oscillator, Armstrong oscillator [1], Colpitts oscillator [2], and Clapp oscillator [3] fall under this category.

Figure 18.2 shows a low-power Colpitts VCO [2]. It employs the current-reused series–shunt feedback topology, which takes care of the power requirement of the VCO. The Q-factor of the circuit is improved by the inversion metal oxide semiconductor (MOS) varactor (enhanced varactor) along with the metal–insulator–metal (MIM) capacitor and so the improvement of noise performance is there.

18.1.1.2 Cross-Coupled Approach

The most popular VCO topology is the cross-coupled VCO topology supported by all bipolar junction transistors, HBTs, and MOS transistors. Its popularity is because one can achieve high oscillation frequency even by the help of a small g_m, which increases with the biasing current. The requirement of buffer is eliminated here, simplifying the circuit design and minimizing the area requirement. But there is a problem in MOS version, as the MOS transistor enters to the triode region during the oscillation affecting both frequency and noise.

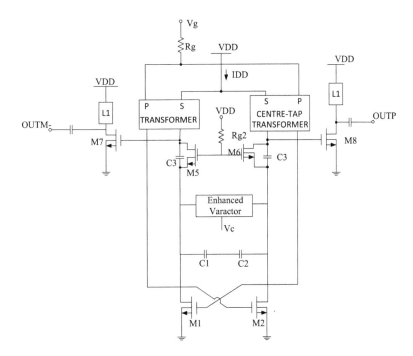

FIGURE 18.2 Colpitts oscillator with enhanced varactor configuration.

18.2   RING VOLTAGE-CONTROLLED OSCILLATOR

Ring oscillators are generally used as waveform-based oscillators. As they are based on the principle of switching, they are always modeled as a nonlinear oscillators. The basic cell of a ring oscillator is an inverter/delay cell and may be single ended or a differential one. Figure 18.3 shows an N-stage ring VCO made of differential inverter; N may be odd or even.

The total period of the output is $T = 2Nt_d$, where t_d is the delay of individual stage. Then the frequency of oscillation becomes $f_o = 1/T$, and the gain of VCO is $K_{vco} = 2\pi(df_o/dV_{tune})$. The minimum number of stages (N) depends on the t_d and ω_o of a given technology, and phase shift of each stage is 180°/N. The delay cell may be either a fast slewing saturated delay cell (Schmidt trigger and buffer based), slow slewing saturated delay cell (source-coupled pair based), or nonsaturated delay cell (voltage inverter based). The delay cells may use an active load (current mirror or AI) or a passive load (resistor or inductor). The phase noise of the ring VCO is defined as the average square of phase deviation of frequency for a range from $f - \left(\dfrac{\Delta f}{2}\right)$ to $f + \left(\dfrac{\Delta f}{2}\right)$, $\overline{\theta^2} = S_\theta(f)\Delta f$, where $S_\theta(f)$ is the power spectral density of the frequency f. For the saturated delay cell–based VCO, the noise process is nonstationary as the transistors turned off in each cycle, and delay is affected by the noise quite after the crossing of V_{out} and $V_{threshold}$. But in nonsaturated delay cell–based VCO, the noise is stationary as all the transistors never turned off, and here the noise can be calculated by the linear modeling of the transistors. Thus, the former has better phase noise performance.

18.3   ACTIVE INDUCTOR

The AI is based on a gyrator-C topology [4], which exploits the intrinsic capacitance of the transistors. AIs may be lossless or lossy type. For a lossless gyrator, both input and output impedances are infinite, and the transconductances of the transconductor are constant as shown in Figure 18.4a, whereas in lossy gyrator network, any one of the input impedances or the output impedance is finite as shown in Figure 18.4b. Generally, all practical gyrator circuits are lossy.

Again, the gyrator-C AI may be single-ended or floating type. For single-ended AI, one of the terminals is connected to either power supply or ground, whereas for

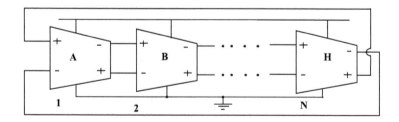

FIGURE 18.3   Block diagram of ring VCO. VCO, voltage-controlled oscillator.

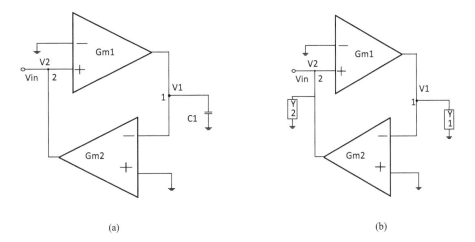

(a)   (b)

FIGURE 18.4   (a) Lossless AI and (b) lossy AI. AI, active inductor.

floating AI, none of the terminals are connected to either power supply or ground. Looking into port 2, the admittance is

$$Y = \frac{I_{in}}{V_{in}} = \frac{1}{S\left(\dfrac{C1}{G_{m1}G_{m2}}\right)} \tag{18.2}$$

which behaves as an inductor having inductance value $L = \left(\dfrac{C1}{G_{m1}G_{m2}}\right)$, i.e., the inductance is directly proportional to the load capacitance and inversely proportional to the product of transconductances. This gyrator-C AI is inductive over all the frequency spectra. The equivalent circuit of the lossy single-ended gyrator-C AI is shown in Figure 18.5. The AI has many advantages over the passive counterpart, such as tunability, high quality factor, and small in size. But it has also the linearity issue along with noise-producing factor, which makes the designer to design carefully to overcome these issues.

18.4   VOLTAGE-CONTROLLED OSCILLATOR USING ACTIVE INDUCTOR

As the current technology demands for low power, low area, low phase noise, and highly tuned VCOs, researchers have successfully embedded the AIs as a part of the VCO for different applications. In such VCOs, the AIs are used for the coarse frequency tuning, and the varactors are used for the fine frequency tuning.

The author [5] has proposed a reconfigurable VCO for multiband RF applications as well as in Internet-of-things applications as shown in Figure 18.6. It uses an AI having a monitoring interface and n-bit controlled inverter module to control the transconductance of the biasing of transistor M2 as shown in Figure 18.7a. The figure of merit proposed for this VCO is

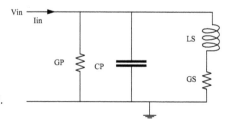

FIGURE 18.5 Equivalent circuit of lossy AI. AI, active inductor.

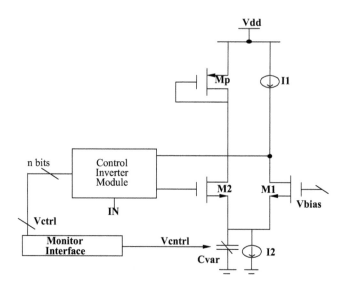

FIGURE 18.6 Reconfigurable VCO with AI. AI, active inductor; VCO, voltage-controlled oscillator.

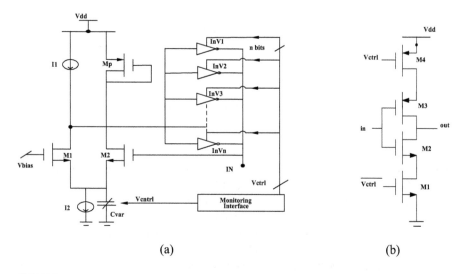

(a)   (b)

FIGURE 18.7 (a) Active inductor structure and (b) inverter structure.

$$FOM = L(\Delta f) + 10\log(P_{dis}(mW)) - 20\log\left(\frac{f_{osc}}{f_{offset}}\right) \qquad (18.3)$$

The inverter is designed by domino logic as per Figure 18.7b. The tuning range of this VCO is from 1.22 to 2.6 GHz with a very low power consumption of 4 mW and die area of .0031 μm². The VCO shown in Figure 18.8 is based on a PMOS cross-coupled pair of transistor core, connected to the supply through a current source to avoid the supply noise, minimizing the phase noise [6]. The two inductors L1 and L2 are realized by the AI. The tail current sources for the AI are implemented by two transistors. The AI has a resistive feedback between the two transconductors to increase the quality factor of the AI, thus reducing the phase noise of the VCO. Here with this design, there is a 94% of tuning range from 120 MHz to 2 GHz with a phase noise variation of −80 to −90 dbc Hz at a frequency offset of 1 MHz. The power consumption of the VCO is 7 mW with a 2.2 mW for the AI from a supply of 1.8 V. The VCO with same concept is adopted in Ref. [7]. The only difference is that here the AI uses an extra additive capacitor C_a at node 1 and the transistors operate in subthreshold region instead of saturation, thus reducing the power consumption. The figure of merit proposed in this work is

$$FOM_T = L(\Delta f) - 20\log\left(\frac{f_0}{\Delta f}\frac{F_T}{10}\right) + 10 \qquad (18.4)$$

where $L(\Delta f)$ is the phase noise, F_T is the frequency tuning range, and P_{dis} is the power dissipation. This VCO can be used in low-power, multiband, and wide-tuning applications.

An ultrawideband CMOS (complementary metal oxide semiconductor) Hartley VCO is designed as per Figure 18.9a, consisting both common source stages as negative transconductor and pi-feedback through two parallel AIs in series with a capacitor [8]. The AI in Figure 18.9b uses the active resistor in the feedback path. The tuning range is 3.8–7.4 GHz with a high-quality factor of 90 and a little bit high

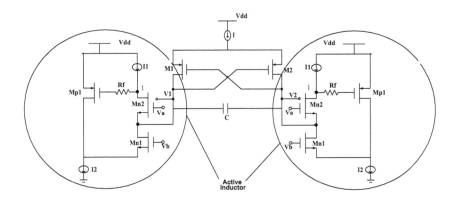

FIGURE 18.8 Active inductor–based VCO. VCO, voltage-controlled oscillator.

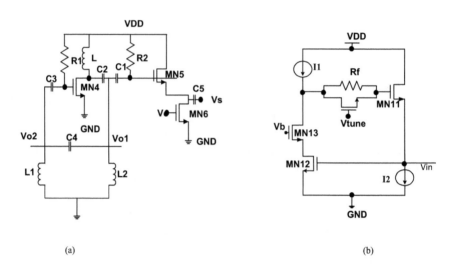

FIGURE 18.9   (a) Hartley VCO and (b) active inductor. VCO, voltage-controlled oscillator.

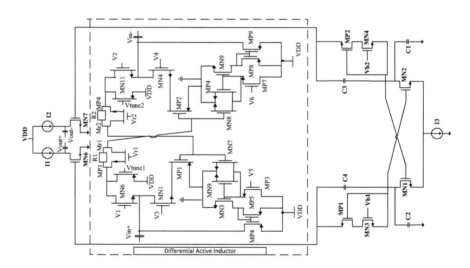

FIGURE 18.10   VCO with differential AI. AI, active inductor; VCO, voltage-controlled oscillator.

power dissipation of 29.1 mW. The phase noise offered by this VCO is −92.05 dBc/ Hz. Better phase noise and VCO gain can be obtained by varying the biasing condition and by the use of active load. A LC VCO with differential AI [9] with a balance between the phase noise and the power dissipation is designed in TSMC 180-nm process as in Figure 18.10. The differential AI is designed with a current-mirror feedback networks and a cascode structure as shown in the dotted box. The VCO hunts a 13.6 mW power for a supply of 1.8 V. The frequency tuning ranges from 1.126 to 2.713 GHz, with a phase noise of −117.2 dBc/Hz at 1-MHz offset frequency. So far, we have discussed the AI-based LC VCO that includes the LC and cross-coupled

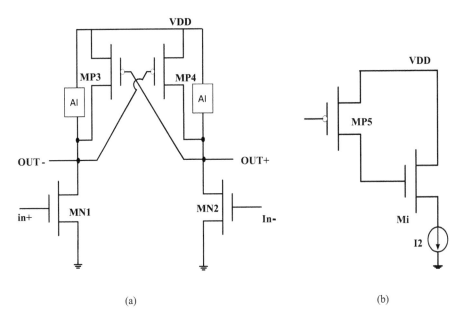

FIGURE 18.11   (a) Cross-coupled delay cell and (b) active Inductor.

VCOs. Now the circuit appeared in Figure 18.11a is a cross-coupled delay cell with AI, as shown in Figure 18.11b, as load which is used to design a fully differential VCO [10]. The VCO has eight stages and tested with resistive and current source load and found smaller delay as compared with its passive counterpart. The AI is a self-biased one, and the VCO provides full rail-to-rail voltage swing without any loss. But there is a reduction in frequency tuning range of the VCOs to 0–1.1 GHz from 1.5 to 2.2 GHz with AI as load instead of PMOS transistor load.

18.5 DISCRETE FOURIER TRANSFORM FOR VOLTAGE-CONTROLLED OSCILLATOR

Almost all the RFICs have the facilities of wafer-level testing in the present scenario to reduce the testing overhead. Most of them have a test pattern generation circuit for VCO or PLL. There are two kinds of test strategies for RF front end. The first one has separate test of the individual blocks of the transceiver; each behaves as a device under test (DUT). They mostly lead to a high-test overhead but also a high-test coverage. The researchers followed the end-to-end test methodology that includes boundary scan (BS) for digital circuit testing and DSP (digital signal processor)-based BS for the RF circuit testing, which includes a separate system controller to generate different test pattern for each DUT.

The second type has the transceiver whole as the DUT, and the blocks are connected in a loop and known as loopback technique [11–16] as shown in Figure 18.12. Many scientists have tried to implement the different loopback techniques for the testing. The loopback circuit consists of an attenuator, offset VCO, band-pass filter,

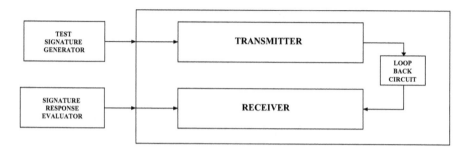

FIGURE 18.12 Loopback technique for testing RFIC. RFIC, radio frequency integrated circuit.

and the baseband analog-to-digital converter. The test pattern is generated by a DSP, and Gaussian minimum shift keying is used for the testing of GSM transceiver. A BIST PLL technique is presented for the testing of VCO in Ref. [17] where no additional circuitry is required for the test pattern generation, but the evaluation circuitry consists of the XOR networks, DRAM, and OR gate which monitor the unintended frequency of oscillation due to internal faults rather the voltage of VCO. Ten numbers of catastrophic fault models for the VCO were generated and tested as

- gate-to-source short
- gate-to-drain short
- drain-to-source short
- gate open
- drain open
- source open
- resistor open
- resistor short
- capacitance short
- capacitance open

A discrete Fourier transform for LC VCO based on a single test signal is presented in Ref. [18]. The fault model may be hard, i.e., due to shorts and opens in the circuit, or soft, i.e., due to parametric fault and corner states. Here, sense transistors are introduced along with the VCO to sense the fault. Under fault-free case, the current value of the sense transistor was measured, and in the presence of the fault, it is observed that there is a drop in the current value in the sensing transistor. This fault is reflected as a compression in the VCO output waveform as well as a single-bit digital output signal.

18.6 SUMMARY

This chapter emphasis on the different types of VOCs and their use in various fields. The ring VCOs and digitally controlled oscillator are good for the optical fiber communication and for the high-speed data rate applications, where there is a very high

TABLE 18.1
Comparison of AI-Based VCOs

	Unit	[19]	[20]	[21]	[22] With AC	[22] Without AC	[8]	[5]	[6]
Tech.	µm	0.18 CMOS	0.18 SiGe BiCMOS	0.18 SiGe BiCMOS	0.18 CMOS	0.18 CMOS	0.18 CMOS	0.13 CMOS	0.18 CMOS
V_{DD}	V	1.8	2.5	3.3	1.8	1.8	1.8	1.1	1.8
Tuning range	GHz	0.5–3.0	2–15	8–17	1.13–2.67	1.3–3.24	3.80–7.40	1.22–2.6	0.1–2
	%	143	–	72	–	–	–	72	94
K_{vco}	MHz V	2500	–	–	–	–	2400	–	2350
DC power	mW	6–2.8	88–120	75–115	2.2–13	2.2–13	29.1	3.6–4.3	7
P_0	dBm	–14 to –22	–8 to –6	–15 to 12	–	–	–	–	–4.7 to 1.5
$L (\Delta f)$	dBc/Hz	–101 to 118	–112 (10 MHz)	–	–82.8 to –92.2	–74.2 to –88.6	–75.42 to –92.05	–82 to –87	–80 to –90
Active area	mm²	0.15×0.3	0.67×0.25	0.65	0.171×0.171	0.171×0.171	–	0.0031	–
FOMT	dBc Hz	–	–	–	–168	–166.3	–	–151	–

AC, additive capacitor; AI, active inductor; FOM, figure of merit; VCO, voltage-controlled oscillator; CMOS, complementary metal oxide semiconductor; BiCMOS, bipolar CMOS.

tuning range requirement and the phase noise can be tolerated to some extent. On the other hand, the LC VCOs and the cross-coupled VCOs are on high demand for the wireless communication systems such as Bluetooth, ZigBee, Wi-Fi, and so on. As we discussed that the LC VCOs' tuning range can be enhanced by the implication of the AI with proper design, maintaining the phase noise least effected, such designs are now adopted in industries with technology compatibilities. A comparison of performance parameters of the various AI-based VCOs is presented in Table 18.1, and the results are up to the industry standard. Also, a clear picture of the testing of the wireless transceiver ICs is presented here, the DUT being VCO and PLL, as they are the key modules for the signal generation in any RFIC, mostly using the loopback techniques.

REFERENCES

1. C. C. Hsiao, C. W. Kuo, C. C. Ho, and Y. J. Chan, "Improved quality-factor of 0.18-μm CMOS active inductor by a feedback resistance design," *IEEE Microw. Wirel. Components Lett.*, vol. 12, no. 12, pp. 467–469, 2002.
2. T. P. Wang, "A K-band low-power colpitts VCO with voltage-to-current positive-feed-back network in 0.18 μm CMOS," *IEEE Microw. Wirel. Components Lett.*, vol. 21, no. 4, pp. 218–220, 2011.
3. N. Pohl, H. M. Rein, T. Musch, K. Aufinger, and J. Hausner, An 80 GHz SiGe bipolar VCO with wide tuning range using two simultaneously tuned varactor pairs. *2008 IEEE CSIC Symp. GaAs ICs Celebr. 30 Years Monterey, Tech. Dig. 2008*, pp. 1–4, 2008.
4. F. Yuan, CMOS *active inductors and transformers*, Switzerland. 2008.
5. F. Haddad, I. Ghorbel, and W. Rahajandraibe, "Design of reconfigurable inductorless RF VCO in 130nm CMOS," *Bionanoscience*, vol. 9, no. 2, pp. 285–295, 2019.
6. H. B. Kia and A. K. A'ain, "A wide tuning range voltage controlled oscillator with a high tunable active inductor," *Wirel. Pers. Commun.*, vol. 79, no. 1, pp. 31–41, 2014.
7. Y.-J. Jeong, Y.-M. Kim, H.-J. Chang, and T.-Y. Yun, "Low-power CMOS VCO with a low-current, high-Q active inductor," *IET Microwaves, Antennas Propag.*, vol. 6, no. 7, p. 788, 2012.
8. M. Mehrabian, A. Nabavi, and N. Rashidi, A 4~7GHz ultra wideband VCO with tunable active inductor. In *Proceedings of The 2008 IEEE International Conference on Ultra-Wideband, ICUWB 2008*, vol. 2, pp. 21–24, 2008.
9. Y. Zhang *et al.*, A Novel LC VCO with high output power and low phase noise using differential active inductor. In *2018 IEEE 3rd International Conference on Integrated Circuits and Microsystems, ICICM 2018*, pp. 90–93, 2018.
10. F. Yuan, A fully differential VCO cell with active inductor load for GBPS serial links. *3rd Int. IEEE Northeast Work. Circuits Syst. Conf. NEWCAS 2005*, vol. 2005, pp. 183–186, 2005.
11. D. Lupea, U. Pursche, and H. J. Jentschel, RF-BIST: Loopback spectral signature analysis. in *Proceedings -Design, Automation and Test in Europe*, pp. 478–483, 2003.
12. M. Jarwala, D. Le, and M. S. Heutmaker, End-to-end test strategy for wireless systems. *IEEE Int. Test Conf.*, pp. 940–946, 1995.
13. A. Halder, S. Bhattacharya, G. Srinivasan, and A. Chatterjee, A system-level alternate test approach for specification test of RF transceivers in loopback mode. In *Proceedings of the IEEE International Conference on VLSI Design*, no. 2003, pp. 289–294, 2005.

14. G. Srinivasan, F. Taenzler, and A. Chatterjee, "Loopback DFT for low-cost test of single-VCO-based wireless transceivers," *IEEE Des. Test Comput.*, vol. 25, no. 2, pp. 150–159, 2008.

15. S. Ozev and C. Olgaard, Wafer-level RF test and DfT for VCO modulating transceiver architecures. *Proc. IEEE VLSI Test Symp.*, no. Vts, pp. 217–222, 2004.

16. G. Srinivasan, A. Chatterjee, and F. Taenzler, Alternate loop-back diagnostic tests for wafer-level diagnosis of modern wireless transceivers using spectral signatures. In *Proceedings of the IEEE VLSI Test Symposium*, vol. 2006, pp. 222–227, 2006.

17. I. Toihria, R. Ayadi, and M. Masmoudi, "High performance BIST PLL approach for VCO testing," 2014, International Conference on Advanced Technologies for Signal and Image Processing (ATSIP), pp. 517–522.

18. L. Dermentzoglou, Y. Tsiatouhas, and A. Arapoyanni, "A design for testability scheme for CMOS LC-tank voltage controlled oscillators," *J. Electron. Test. Theory Appl.*, vol. 20, no. 2, pp. 133–142, 2004.

19. L. H. Lu, H. H. Hsieh, and Y. Te Liao, "A wide tuning-range CMOS VCO with a differential tunable active inductor," *IEEE Trans. Microw. Theory Tech.*, vol. 54, no. 9, pp. 3462–3468, 2006.

20. T. Kanar and G. M. Rebeiz, "A 2–15 GHz VCO with harmonic cancellation for wideband systems," *IEEE Microw. Wirel. Components Lett.*, vol. 26, no. 11, pp. 933–935, 2016.

21. S. Shankar, S. Horst, and J. D. Cressler, Frequency- and amplitude-tunable X-to-Ku band SiGe ring oscillators for multiband BIST applications. In *Proceedings of the IEEE Bipolar/BiCMOS Circuits and Technology Meeting*, pp. 9–12, 2010.

22. J. Shim and J. Jeong, "A band-selective low-noise amplifier using an improved tunable active inductor for 3–5 GHz UWB receivers," *Microelectron. J.*, vol. 65, no. January, pp. 78–83, 2017.

19 Fault Simulation Algorithms: Verilog Implementation

Sobhit Saxena
Lovely Professional University

CONTENTS

19.1 Introduction 325
19.2 Logic Simulation 325
19.3 Fault Simulation 327
19.4 Verilog Coding for Simulation 330
References 338

19.1 INTRODUCTION

Simulation is a technique normally used for functionality verification during circuit designing before fabrication. Fault is the effect of any defect created in the hardware of the circuit during fabrication, which can cause the circuit malfunction.

Every hardware unit of the circuit needs to be tested in the production house, and segregation between faulty and fault-free units is done before it is out for customer use. Testing of hardware units can be done by applying all possible inputs, and responses are recorded and matched with the expected results. This process requires a lot of time. In order to reduce the time and also to identify the type of fault with the location of the fault in the circuit, instead of all possible inputs, test vectors/patterns need to be applied for testing. To identify these test vectors/patterns, simulation exercises are required for every possible input to check their effectiveness in testing and identification of the possible faults in the circuit. This type of simulation is known as fault simulation.

Therefore, the simulation is classified into two broad categories: logic simulation and fault simulation as depicted from the block diagram shown in Figure 19.1.

19.2 LOGIC SIMULATION

Logic simulation is the simulation of a fault-free circuit. In logic simulation, the available digital circuit is given a set of test stimuli, and the output response is compared with expected response in terms of waveforms.

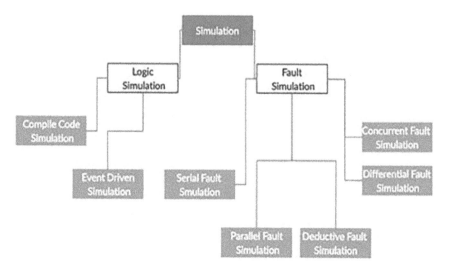

FIGURE 19.1   Types of simulation.

a. Compiled Code Simulation:

In compiled code simulation, the digital circuits is converted into a set of machine instructions called codes that model the functionality of individual gates with the interconnecting wires. Before the generation of codes, logic optimization and logic levelization are performed. Different types of code generation techniques [1] are used depending upon requirements, namely, high level programming language source code, native machine code, and interpreted code. The code is executed with the input pattern written in the test bench in the host machine at each clock cycle.

The compiled code simulation technique is valid only for zero delay circuits. Its models do not include gate and wire delays; that is why, it glitches, and race around conditions cannot be handled. Apart from this, low simulation efficiency is another limitation.

b. Event-Driven Simulation:

In event-driven simulation, the gate evaluation is performed only when event takes place at the input of that gate. Event means any change in the logic value. Since all the gates are not evaluated for every clock cycle, the simulation efficiency increases much compared with compiled code simulation. Event-driven simulator includes a scheduler that schedules the necessary gate evaluation according to the occurrence of the events. The events list includes which gate need to be evaluated and when it is evaluated according to the time stamps when gate and wire delays are considered.

To increase the efficiency of the scheduler, the timing wheel concept is used [2]. Event-driven simulation is a well-suited approach for models and includes delays, and there is a possibility of detecting hazards.

19.3   FAULT SIMULATION

Testing of a digital circuit includes the propagation of available inputs through various logic gates to the output and analyzing the output response with the expected response of the same digital circuit. Inputs applied to the circuit are known as test vectors; the circuit that is used for testing is known as circuit under test; and comparison of output responses and the expected responses is done by output response analyzer as shown in Figure 19.2.

Fault simulation can be performed in many ways:

a. Serial fault simulation
b. Parallel fault simulation
c. Deductive fault simulation
d. Concurrent fault simulation
e. Differential fault simulation

Among these, deductive fault simulation technique can be faster than the parallel fault simulation but is slower than the concurrent fault simulation because concurrent fault simulation simulates only the fault-affected parts of the circuits. However, differential fault simulation is expected to be 12 times faster than the concurrent fault simulation. The memory requirement of the concurrent fault simulation is quite high than the deductive fault simulation because of the I/O values of every bad gates available in the target circuit. The advantage of concurrent fault simulation is that it can be used for functional models and it can perform the sequential fault simulation with ease.

a. Serial Fault Simulation

This is the simplest fault simulation technique. First, the fault-free circuit is simulated for a particular input pattern, and the response is stored for comparison purposes later. Then, one stuck at fault is injected in the circuit, and simulation is performed for the same input pattern and the response is stored. Similar process is repeated for every fault individually. Those faults for which response is different from fault-free circuit can be detected by that particular input pattern and become the test vector for those faults. The same process is repeated for different input patterns till all the faults found their test vectors.

b. Parallel Fault Simulation

Parallel fault simulation is similar to parallel logic simulation. Fault simulation can take the permanent advantage of bitwise parallelism in computer to

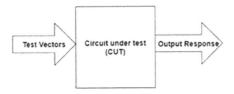

FIGURE 19.2   Basic testing process.

reduce the fault simulation time. To realize this bitwise parallelism, there are two ways: one is parallel fault simulation and another one is parallel pattern simulation. Assuming that binary logic is used, one bit is sufficient to store the logic value of the signal. It means that the computer is using n-bit wide data words; each signal is connected with a data word of which $n-1$ bits are issued for $n-1$ faulty circuits and the remaining bit is reserved for fault-free circuit. According to this, $n-1$ faulty circuits and one fault-free circuit are processed in parallel using bitwise logic operation that corresponds to a speedup factor of approximately $n-1$ compared with serial fault simulation. A fault is detected if the bit value is different from the fault-free circuit at any way of the outputs.

c. Deductive Fault Simulation:

Deductive fault simulation is very different type of simulation in which no fault injection is required. In this simulation, for a particular input pattern, the possible excited faults at primary inputs are identified. The propagation of these faults depends upon the passing gate. Faults that can be excited at the output of that gate will be added to the list. The overall faults that are propagated at the output can be detected by that input pattern and become the test vector for the propagated faults. The same process is repeated for different input patterns till all possible faults got propagated at the output of the circuit. This deductive fault simulation can be very fast because only fault-free simulations have to be performed [3].

d. Concurrent Fault Simulation:

Out of the available fault simulation techniques, the concurrent fault simulation technique is best suited for the combinational and the sequential circuits. Concurrent fault simulation is an event-driven simulation in which both the fault-free and the faulty circuits are simulated together. Concurrent fault simulation works by simulating the differential parts of the circuit that are affected by the fault (Figure 19.3).

In concurrent fault simulation, every gate has a set of bad gates, and these bad gates are basically the gates with I/O values in the presence of the respective faults. In concurrent fault simulation, the bad gate and the good gate are simulated concurrently, and then it is checked whether the fault is propagated to the output or not. If the fault is visible at the output, then it can be observed that the fault is propagated to the output for the given test vector, which is called a good event. For some test vectors, the fault is not visible at the output, which is called a bad event [4].

Advantages

1. Concurrent fault simulation is faster because it simulates the differential parts of the circuit.
2. Concurrent fault simulation can deal with multivalued simulations.
3. It can perform sequential simulation with ease.
4. It is suited for both the combinational and the sequential circuits.

The flow of the concurrent fault simulation is given in flowchart in which there is a collapsed fault list, which consists of list of the faults and the test

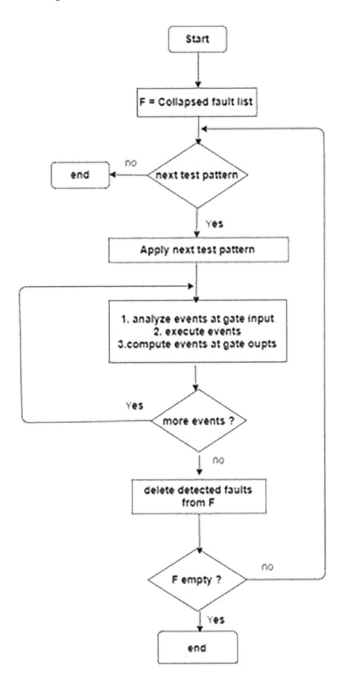

FIGURE 19.3 Flowchart for concurrent fault simulation [5].

vector set that is given to simulate the fault-free and the faulty circuits. These fault-free and the faulty gates are evaluated, and then the output response of the circuit is compared with the fault-free circuit. If any fault is detected, then that is deleted from the collapsed fault list. This process will be continued until all the faults are covered in the collapsed fault list.

e. Differential Fault Simulation:

Differential fault simulation is the combination of concurrent fault simulation and the single fault simulation technique for sequential fault simulation [6]. Initially, for first input pattern, fault-free circuit simulation is done, and the output is stored; then, first fault is injected in the circuit, and the simulation is occurred. The difference between the states (of memory elements, e.g., flip-flops) of fault-free and faulty circuit is stored. Then, first fault is removed and second fault is inserted. and the difference between the states of faulty circuit with first fault and faulty circuit with second fault is stored. This process is repeated till for all faults get simulated for first input pattern. For the next input pattern, the state of the fault-free circuit is first restored, and then pattern is applied. After simulation, the output is stored. The state of faulty circuit with first fault is restored by inserting the difference between the states of fault-free and faulty circuit with first input pattern. After that, first fault is injected as an event, and the process continues as done for first input pattern. The state of every circuit is restored from the last simulation. Those faults that are detected are dropped from the list.

19.4   VERILOG CODING FOR SIMULATION

For understanding how fault simulation is being done, the Verilog code is written for parallel and concurrent fault simulation techniques by taking an example of very commonly used circuit in the digital design, i.e., multiplexer. The circuit and functionality of multiplexer is explained for understanding, and then coding and simulation is performed.

Multiplexer is a data selector that allows a single signal out of all the available analog or digital signals. The multiplexer propagates the inputs to the output by the selection lines. The size of a multiplexer is $2n \times 1$ where n is the number of selection lines, 2n inputs and a single output. For a simple 2×1 multiplexer, there are two inputs and a single selection line. The gate-level schematic of a 2×1 multiplexer with the inputs a, b, and c and the output y is given (Figure 19.4).

The explanation of the above 2×1 is better explained through the below truth table, which depicts the detailed flow of each set of inputs. Here, b is the selection line; when b=0, then the value present at terminal a is propagated to the output, and when b=1, the value present at terminal c is propagated to the output (Table 19.1).

Multiplexers can be used for various applications such as communication systems, computer memory, telephone network, transmission from the computers to satellites, and so on.

Parallel Fault Simulation:

Assuming that width of the computer word is 3 bits, first bit stores fault-free circuit output response, and the remaining second and third bits are storing

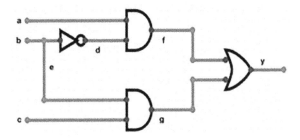

FIGURE 19.4 Circuit diagram of 2×1 multiplexer.

TABLE 19.1

Truth Table of 2×1 Multiplexer

a	b	c	y
0	0	0	0
0	0	1	0
0	1	0	0
0	1	1	1
1	0	0	1
1	0	1	1
1	1	0	0
1	1	1	1

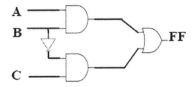

FIGURE 19.5 Fault-free circuit.

the value of x-faulty and y-faulty circuits' output responses (Figure 19.5). x and y are two faults, one is "a" stuck at "1" and another one is "g" stuck at "0." To make a stuck at "1," insert extra OR gate with one signal input "1." Similarly, for stuck at "0," insert extra AND gate with one signal input "0" (Figure 19.6).

While simulation, one fault is considered for one time. It means while propagating the x-fault, y-fault is not considered. Similarly, according to that, while propagating y-fault, x-fault is not considered. While propagating the x-faulty (F_x) circuit, output response is different from the fault-free (FF) circuit's output response. It means that x-fault is detected and "x" has become high at output response. Similarly, while propagating the y-faulty (F_y) circuit, output response is different from the fault-free circuit's output

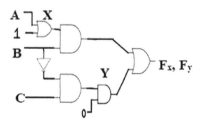

FIGURE 19.6   Circuit with injected faults.

TABLE 19.2

Truth Table of Fault-Free and Faulty Circuits

Inputs			Outputs				
A	B	C	FF	F_x	F_y	x	y
0	0	0	0	0	0	0	0
0	0	1	1	1	0	0	1
0	1	0	0	1	0	1	0
0	1	1	0	1	0	1	0
1	0	0	0	0	0	0	0
1	0	1	1	1	0	0	1
1	1	0	1	1	1	0	0
1	1	1	1	1	1	0	0

response. It means that the y-fault is detected and "y" has become high at output response, which has been shown in Table 19.2.

In Table 19.2, a, b, and c are inputs to the circuit. FF, F_x, and F_y are the fault-free, x-fault, and y-fault output responses. If F_x and FF output responses are different, then "x" is high at output response. Similarly, if F_y and FF output responses are different, then "y" is high at output response.

Verilog Code

```
module prall(a,b,c,f,x,y,fx,fy);
input a,b,c;
output f,x,y,fx,fy;
wire [8:0]w;
not (w[0],b);
and af1(w[1],a,b);
and af2(w[2],c,w[0]);
or af3(w[3],w[1],w[2]);
// fault-free circuit Programming //

or ax0(w[4],1,a);
and ax1(w[5],b,w[4]);
or ox2(w[6],w[2],w[5]);
// x-faulty circuit Programming //
```

```
and oy0(w[7],0,w[2]);
or oy1(w[8],w[7],w[1]);
// y-faulty circuit programming //

assign fx=w[6];
assign fy=w[8];
assign f=w[3];
xor (x,w[6],w[3]);
xor (y,w[8],w[3]);
endmodule
```

Test Bench

```
module prall_tb();
reg a,b,c;
wire f,x,y;
wire fx,fy;
 prall u(a,b,c,f,x,y,fx,fy);
initial
begin
a=0;b=0;c=0;
#5 a=0;b=0;c=1;
#5 a=0;b=1;c=0;
#5 a=0;b=1;c=1;
#5 a=1;b=0;c=0;
#5 a=1;b=0;c=1;
#5 a=1;b=1;c=0;
#5 a=1;b=1;c=1;
#10 $stop;
end
endmodule
```

Simulation Results:

Figure 19.7 shows that y-fault is detected. When [a, b, c] are equal to [0, 0, 1],"FF" and "Fy" output responses are different. So that the y output response is high (Figures 19.7 and 19.8).

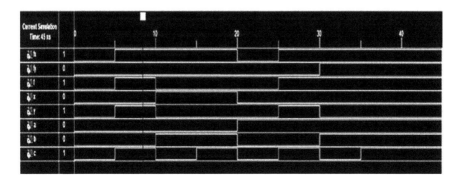

FIGURE 19.7   y-fault is detected at output response.

[waveform simulation image]

FIGURE 19.8   x-fault is detected at output response.

Figure 19.8 shows that x-fault is detected. When [a, b, c] are equal to [0, 1, 0], "FF" and "Fx" output responses are different. So that the x output response is high.

Concurrent Fault Simulation:

To analyze the concurrent fault simulation, we have considered a 2×1 multiplexer as we have discussed earlier. This has inputs a, b, and c and the output y. The d wire is output of the not gate whose input is b. f is the output of and gate whose inputs are a and d. g is the output of the and gate whose inputs are b and c. To understand the concurrent fault simulation simpler, we have considered two single stuck at faults. One is the a stuck at 0 and other is the g stuck at 1.

Let us first consider the case of a stuck at 0. At this point, the and gate with inputs a and d will have the fault. The fault-free and gate and the faulty and gate are simulated together, and this fault is further propagated to the output of the circuit, and it is verified with the expected response. If the fault is detectable, then that event is called a good event or else the event is called bad event. The 2×1 multiplexer with the a stuck at 0 is given in Figure 19.9 where ff is the output of the faulty gate output. If the a stuck at 0, then output of the and gate is ff=0. The output yfa (faulty circuit output with the fault a stuck at 0) now depends only on the g which is equal to b and c.

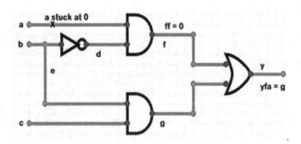

FIGURE 19.9   2×1 multiplexer with a stuck at 0.

In the concurrent fault simulation, the fault list has single fault (a stuck at 0); then the test vectors are applied until the fault is detected and propagated to the output. The detailed propagation of values of the faulty circuit and fault-free circuit is given in Table 19.3.

In Table 19.3, a, b, and c are the inputs, f is the fault-free output of the and gate, y is the output of the fault-free circuit, ff is the faulty and gate output, and yfa is the faulty circuit output. The fdet signal is used for detecting the fault of that gate. The ydet signal is used for checking whether the fault is propagated to the output or not. We can observe that the fault (a stuck at 0) is observed at only the test vectors (abc = 100,101) at the and gate output (ff), so the ff = 1 at those test vectors, which are good events. Now, we check if the fault due to the a stuck at 0 is propagated to the output. We observe the ydeta = 1 only at the test vectors (abc = 100,101). The simulated waveform for the 2×1 multiplexer with a stuck at fault is given.

Verilog Code

```verilog
module ckt(a,b,c,k,y,yfa,yfg,fdet,ydeta,ydetg);
input a,b,c,k;
output y,yfa,yfg;
output reg fdet,ydeta,ydetg;
wire d,f,g;
wire ff; // faulty and gate
supply1 vdd;
supply0 gnd;
not n1(d,b);
and a1(f,a,d);// fault free circuit
and a2(g,b,c);
or o1(y,f,g);
and a3(ff,gnd,d); // faulty and gate (a|0)
or o2(yfa,ff,g);
or o3(yfg,f,vdd); // faulty and gate
always@(f,ff,y,yfa,yfg,fdet,ydeta,ydetg)
begin
case(k)
```

TABLE 19.3
Faulty Circuit Truth Table for a Stuck at 0

a	b	c	f	Y	Ff	yfa	fdet	ydeta
0	0	0	0	0	0	0	0	0
0	0	1	0	0	0	0	0	0
0	1	0	0	0	0	0	0	0
0	1	1	0	1	0	1	0	0
0	0	0	1	1	0	0	1	1
0	0	1	1	1	0	0	1	1
0	1	0	0	0	0	0	0	0
0	1	1	0	1	0	1	0	0

```
1'b0 : begin
if(f!=ff)
                begin
                        assign fdet= vdd;
                    if(y==yfa)
                            assign ydeta = gnd;
                        else
                            assign ydeta = vdd;
                    end
            else
                begin
                        assign fdet= gnd;
                        assign ydeta= gnd;
                    end
            end
1'b1 : begin
    if(y==yfg)
                assign ydetg = gnd;
                else
                    assign ydetg = vdd;
    end
endcase
end
endmodule
```

Test Bench

```
module ckt_tb();
reg a,b,c,k;
wire y,yfa,yfg,fdet,ydeta,ydetg;
ckt u(a,b,c,k,y,yfa,yfg,fdet,ydeta,ydetg);
initial
begin
k=1;a=0; b=0; c=0;
#5 k=1; a=0; b=0; c=1;
#5 k=1; a=0; b=1; c=0;
#5 k=1; a=0; b=1; c=1;
#5 k=1; a=1; b=0; c=0;
#5 k=1; a=1; b=0; c=1;
#5 k=1; a=1; b=1; c=0;
#5 k=1; a=1; b=1; c=1;
#5 $stop;
end
endmodule
```

Simulation Results:

In the above simulation results, the signal k is given 0 because we have used case statement in the Verilog program (Figure 19.10). So, if we give k=0, then the a stuck at 0 fault simulation is carried out, and if we give k=1, then g stuck at 1 is carried out.

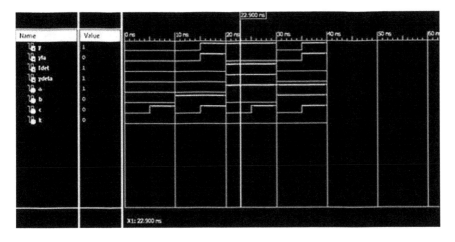

FIGURE 19.10   Simulation of 2×1 multiplexer with a stuck at 0.

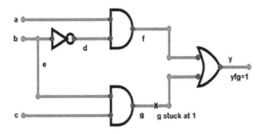

FIGURE 19.11   2×1 multiplexer with g stuck at 1.

Now, we consider the g stuck at 1 fault; then the collapsed fault list now contains the g stuck at 1. Since the output of the multiplexer is yfg=f+g, the output will be yfg=1 (Figure 19.11).

Since g stuck at 1, there are no gates to be evaluated, and only the output gate has to be evaluated. The faulty output is yfg that is compared with output faulty free circuit, which is y. The circuit is simulated with test vectors set and is best understood by Table 19.4. The gf signal is the output of the and gate which is stuck at 1, and yfg is the faulty output signal which is also stuck at 1 due to g stuck at 1.

The fault of g stuck at 1 can be observed at the test vectors (abc=000,001,010,110), and these are called good events. The simulated results for the g stuck at 1 are given (Figure 19.12).

Here to simulate the g stuck at 1, the k is given, so the signals fdet and ydeta are in don't care state.

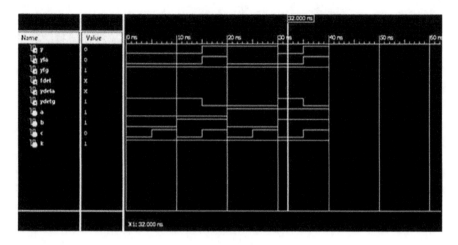

FIGURE 19.12   Simulation of 2×1 with g stuck at 1.

TABLE 19.4
Faulty Circuit Truth Table for g Stuck 1

a	b	c	G	gf	y	yfg
0	0	0	0	1	0	1
0	0	1	0	1	0	1
0	1	0	0	1	0	1
0	1	1	1	1	1	1
1	0	0	0	1	1	1
1	0	1	0	1	1	1
1	1	0	0	1	0	1
1	1	1	1	1	1	1

REFERENCES

1. L.-T. Wang, N. E. Hoover, E. H. Porter, and J. J. Zasio, SSIM: A software levelized compiled-code simulator, in *Proc. Des. Automat. Conf.*, June 1987, pp. 2–8.
2. E. G. Ulrich, Exclusive simulation of activity in digital networks, *Commun. ACM*, 12(2), 102–110, 1969.
3. D. B. Armstrong, A deductive method for simulating faults in logic circuits, *IEEE Trans. Comput.*, C-21(5), 464–471, 1972.
4. E. G. Ulrich and T. Baker, Concurrent simulation of nearly identical digital networks, *IEEE Trans. Comput.*, 7(4), 39–44, 1974.
5. L.-T. Wang, C.-W. Wu and X. Wen, *VLSI Test Principles and Architectures*, Morgan Kaufmann Publishers, Elsevier, San Francisco, 2006.
6. W. T. Cheng and M. L. Yu, Differential fault simulation: A fast method using minimal memory, in *Proc. Des. Automat. Conf.*, June 1989, pp. 424–428.

20 Hardware Protection through Logic Obfuscation

Jyotirmoy Pathak and Suman Lata Tripathi
Lovely Professional University

CONTENTS

20.1 Introduction 339
20.2 Logic Working 341
20.3 Comparisons 345
20.4 Future Works 346
20.5 Conclusions 347
References 347

20.1 INTRODUCTION

With the recent burst in technology, the use of integrated circuits (ICs) in various fields has seen a tremendous increase. Due to this high demand, certain steps in the process of fabrication are done outside of the foundry, which makes it prone to several attacks. There are several processes employed for securing ICs against malicious attacks. One such method is known as hardware obfuscation. Hardware obfuscation basically means hiding the IC's structure and function, which makes it much more difficult to reverse-engineer by the adversaries. The adversaries generally make use of reverse engineering to decipher the IC. Reverse engineering basically stands for the technology that is used to describe be the structure, functionality, and design of an IC. Once there is adequate knowledge of the inner structure of the circuit, it is quite easy to tamper it or use it illegally. Thereby, obfuscation provides a means of making the circuit structurally and functionally difficult to comprehend, which increases the cost and time required to reverse-engineer it, providing security.

Hardware obfuscation provides a security not only against reverse engineering but also against IC piracy, overbuilding of chips, duplication, etc. According to Moore's law, the number of transistors in a circuit is increasing twice every 18 months. This is made possible because the size of the transistor is reducing to a larger extent; currently, work is being done in 7-nm technology. With the shrinkage in the size of the transistors, it was believed that reverse engineering of the design will get difficult, but it is not so. Apple's iPhone 6 and iPhone 6s were reverse-engineered successfully, and the processing units present inside it were identified.

Reverse engineering has made IC theft and espionage a major threat to IC industry. There has been a tremendous increase in cases involving hardware theft [1]. This makes it even more urgent to concentrate on the hardware security perspective. Research has been going on obfuscation at many levels. The basic idea of obfuscation is premanufacture safety, which includes prevention from the addition of hardware Trojans while designing a circuit. This Trojan can easily leak the confidential information of the IC or can make the IC work in an unfamiliar manner. There have been many researches that have been going on to detect the presence of Trojan and also prevent the addition of Trojan. Another area of concern is postmanufacture in which an attacker can use the manufacturer IC in ways that incur a loss to the customer or manufacturer. There have been researches on gate-level obfuscation [2] in which researchers make use of AOI gates instead of simple AND–OR logic gates and, after that, reverse the obtained output in order for obtaining a correct value. At layout-level obfuscation, they make use of special camouflaged gates instead of simpler library gates, which prevent the hacker to identify its correct function even through an SEM [3]. They also add dummy contacts to the layout, making it much more difficult to see the connectivity of the circuit, thereby making it a time-consuming task to reverse-engineer the design [4].

HDL (hardware description language)-level obfuscation includes obfuscation of the design by modification of HDL codes. Here, the codes that are written in Verilog or VHDL are made much more difficult to understand. This is achieved by removing comment lines in coding [5] or by using an encryption technique using the AES algorithm. Later, this encrypted code can be decrypted at the user end with the help of a decryption key [6].

Illegal and unauthorized use of IC can be prevented by various other techniques such as watermarking [7, 8] and cryptography [9, 10].

Obfuscation can be done using two basic ways:

- Structurally
- Functionally

In the case of structural obfuscation, we complicate the structure of the given design, making it a tedious task to find its functionality [11]. In Ref. [12], we have seen that by making use of high-level transformation (HLT) techniques, circuit structures were modified to look alike but perform a different functionality. So the circuit obtained looks structurally the same but are functionally different.

In functional obfuscation, we make use of a key. Until and unless the entered key is correct, the design does not give the desired output. This key can be stored in the form of fuses in the circuit [13]. However, saving the key in the nonvolatile memory of the circuits in the form of fuses makes it insecure, as it can be obtained by the adversaries. So there was a development of a technique called physical unclonable function (PUF) that provides with the key to the circuitry [14]. The key can be provided either using an FSM or a PUF. On entering the correct key, the circuit operated in a normal mode, and on entering a wrong key, the circuit operated in an obfuscated mode. In Ref. [15], to make it much more difficult to reverse-engineer, the author devised a technique to use dynamic obfuscation in which if they entered incorrect key, the circuit goes into obfuscated mode but does not always give wrong output. Based on a random number, sometimes, the output obtained is correct and other times incorrect, increasing the complexity of design and making it much more secure.

20.2 LOGIC WORKING

Basically, logic obfuscation is a technique which modifies the circuit design in such a way that it becomes tedious to understand the working of the circuitry while reverse-engineering it, which makes it much more secure by increasing the complexity of the circuit, which results in more time and cost requirement for deciphering it. In order to achieve maximum security against reverse-engineering design, we must ensure the following: (i) it is hard to differ the original design from the obfuscated design; (ii) the output obtained in obfuscated mode is different than the one obtained in normal mode; and (iii) the obfuscation does not affect the functionality of the chip in any way.

With respect to technique, logic obfuscation is divided into structural obfuscation and functional obfuscation as shown in Figure 20.1. Structural obfuscation involves modifying the structure of the given circuit, making sure that it does not affect its functionality. This is achieved by using HLT and various other techniques. Functional obfuscation can also be termed as an encryption technique in which for the correct operation of the circuit, we need to enter the correct key, failing to which results in a different output.

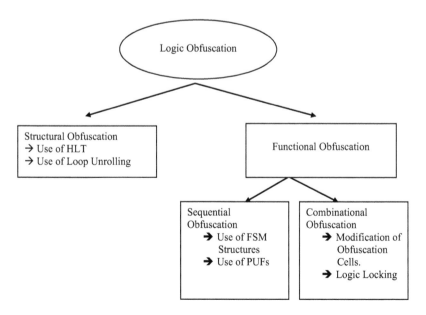

FIGURE 20.1 Taxonomy for logic obfuscation. FSM, finite-state machine; HLT, high-level transformation; PUF, physical unclonable function.

In Ref. [16], Chakorborty provided protection to the hardware through net-list obfuscation. Modification of gate-level net list in this paper results in structural and functional obfuscation. They modified the state transition table and the internal logic for structural obfuscation. They introduce a new gate with one input as the enable input. If the enable input is 0, the circuit operated as the normal operational circuit, but if the enable input is 1, the circuit results in the inverted output. This is expressed by the equation as:

$$f_{mod} = f.\overline{en} + \overline{f}.g.en \tag{20.1}$$

Besides this, they also introduce a small FSM structure. At initialization, it is in obfuscated mode. On entering the correct sequence of inputs, it enters the normal mode. They also made use of a proper MKG function with an increased number of inputs for providing better security. They performed the experiments using XOR gates and OR gates and found that the technique used in this paper increases the nodes, failing considerably with the increases in the number of modified nodes. Thereby, this technique provided a solution for piracy and tampering of IC with a low area overhead, acceptable power overhead, and low complexity.

In Ref. [17], Chakorborty devised a technique for providing security at the register transfer level (RTL). He converted the RTL core to control and dataflow graph and then introduced an FSM for key-based obfuscation. The design operates in two modes. If the entered key is correct, it operates in normal mode, but if the entered key is incorrect, it operates in obfuscated mode, thereby giving an output that is totally different from the output obtained under a normal mode of operation. He showed that the number of verification failures is increased with the area overhead less than 10%. Besides this security, another major concern regarding hardware security is the introduction of malicious hardware in the circuit at any time during fabrication, which can be later activated and used to leak confidential information regarding the hardware concerns. This is another serious concern that has been researched by many researchers. In Ref. [18], Farinaz did a survey on hardware Trojans insertion in a circuit and how to detect the location of those Trojans. She described in detail various timing and power-based analysis used to detect the presence of malicious hardware as well as the prevention technique from the most used attacks, i.e., side channel attacks. The Trojan inserted in the circuitry gets activated at a particular situation, thus making it difficult to detect its presence before it gets activated. In Ref. [19], a technique was developed to provide protection against Trojan that gets triggered at rare circuit conditions. They make use of extra state elements and make the unreachable states that are present in normal mode as reachable states in obfuscated mode, thereby making it difficult for the adversary to add a well-hidden Trojan to the circuit. Basically for a Trojan to be effective, the adversary has to determine the rare event as the trigger node for which the adversary had to perform numerous simulations but the adversary has no way of knowing if the simulations performed are in normal mode or obfuscated mode. The results showed better protection with a low overhead of area. In Ref. [20], the technique was modified to that present in Ref. [36]; here, key gates are inserted in the design with one of the inputs as the key at various random site locations. If the provided key is incorrect, the output obtained is fixed and incorrect. The key is stored in a tamper-free memory inside the chip. The greater the key size, the longer the time required for the attacker to decipher the key. The area overhead, in this case, is quite high even when only 5% of gates are inserted as key gates compared with the total gates in the design. There has been a constant development in the area of hardware security recently. Numerous techniques and area are researched under this. Classification on the hardware security on the different area has been described in Ref. [21] for a better understanding of the people researching in this area. In Ref. [22], they developed a novel technique of obfuscating a digital signal–processing circuit by use of HLT. HLT was mainly used for area, delay, and power trade-offs, but it was observed that, apart from this, HLT also provides structural and functional obfuscation. They also introduced multiple modes that are meaningful but functionally incorrect, thereby providing a higher level of security. When the incorrect key is applied from the FSM, the reconfiguration can activate either a meaningful or nonmeaningful mode, which is functionally incorrect. This technique provided an advantage of lower area overhead by providing a higher level of security. As we can observe from the papers mentioned earlier the area overhead for these techniques is a concern, thereby in Ref. [23], there is a mechanism for providing security to the circuit without affecting the area overhead of the design. In this paper, the original net list of the IC is modified to obtain obfuscated net list. This obfuscated cell (OC) results in a correct output if the entered key is correct. The advantage of OC is that the adversary cannot identify it on reverse engineering by image processing techniques. This makes only the authorized chips to work properly. The previous techniques of using XOR/XNOR gates resulted in a high area overhead, which is not in this case. It resulted in an area overhead of 0.63% and a power overhead of 2.6% in the circuit.

In Ref. [24], they have classified hardware obfuscation at different levels. They have researched on various techniques employed to reverse-engineer a design, checked the obfuscation techniques at various level, and then focused on the obfuscation at device level by manipulating the interconnects and the transistors in the circuit. They have varied the doping concentration and introduce dummy contacts for increasing the effects of reverse-engineering a design. They also showed that changing the logic family from CMOS (complementary metal oxide semiconductor) to pass transistor logic or transmission gates also increases the efforts of the adversary to reverse-engineer a design. For a more detailed understanding of the layout- and net list–level obfuscation, it must be noted that the survey paper in Ref. [25] has very precisely classified all the possible techniques for various layout-level obfuscation along with their demerits as well as the net list–level obfuscation techniques. In Ref. [26], there is a slight modification of the technique that was presented in Ref. [22]. This paper provides a structural obfuscation by using five basic techniques such as loop unrolling, loop invariant code motion, tree height reduction, logic transformation, and removal of redundant logics. After application of these modifies HLT techniques, particle swarm optimization (PSO) is performed for the optimization of the design. This results in a low cost of obfuscation. Along with structural obfuscation using HLT, there have been researches going on various logic locking and camouflaging techniques. These techniques were found to be imminent in using as the logic obfuscation technique in Ref. [27], but it is observed that these techniques are not resilient to SAT attacks. To overcome these solutions, various anti-SAT and anti-tree SAT techniques were developed [28]. In Ref. [29], they have done a survey to show how successful these anti-SAT techniques are to prevent the misuse of an IC, and it is observed that the protection provided by these techniques is quite bleak. They have discussed the security challenges and the removal attack of the anti-SAT logic in detailed explanation. To solve these issues of rarely being able to successfully hide the functionality of a design when encountered with SAT attacks, in Ref. [30], they have introduced the concept of the cyclic topology of the circuit in which they add a minimum number of dummy contacts to make a cyclic loop, which becomes difficult to remove. It results in an SAT-resilient circuit with 40% less area overhead and a delay constraint of less than 20%. For camouflaging technique of obfuscation, it is observed that the clock signal used for all the blocks inside the circuit is the same, so in Ref. [31], they varied the clock period and added false paths and wave pipelining paths in the circuit, which allows two waves to travel in a path at the same time with careful time constraints for each one of them to prevent mixing of them. It increases the security at a negligible cost, but care must be taken that the adversary is not aware that this technique has been used in the circuitry to provide security; else it is very easy to find out the paths where wave pipelining has been utilized and reverse-engineering it would become a lot easier. In Ref. [32], they have made a slight modification than all other previous functional logic obfuscation techniques discussed earlier. They have made use of different obfuscation cells rather than a single common obfuscation cell, which increases the probability of structural mismatch, making reverse-engineering a bit more difficult. The key is used for the circuit to perform in a normal or obfuscated mode like in previous papers, which are then forwarded to different OCs. It is observed that the performance overhead of the circuit using different OCs is low compared with the one using similar OCs. By reading many papers on functional logic obfuscation, it is observed that the basic working of the circuit is the same in which the circuit operates in two modes, the normal mode and the obfuscated mode. The key is provided to the circuit, and depending on the correctness of the entered key, the circuit either gives a correct output or an incorrect output. In Ref. [22], it was further modified to give a meaningful incorrect output to make it much more difficult for the adversary to understand the correct functioning of the IC, and in Ref. [15], this concept has been further modified by introducing the concept of dynamic obfuscation. It results in the circuit behaving in an unpredictable manner in which it performs in logically correct way if the key is a correct, but it may or may not perform correctly in case of entering incorrect key. This random correctness and incorrectness of the output further complicate the circuit. To make the circuit behave in much randomized manner, an extra circuitry called a trigger circuit is used which triggers the circuit in a random manner whenever an incorrect key is entered. The key length does not need to be long for providing better security as in the previous related work. With the development of various techniques for obfuscation, one of the most widely used techniques is that of logic locking. In Ref. [33], it is observed that the circuit has introduced an XOR/AND/OR gate or buffer or mux for locking of the logic in case a wrong key is entered. This technique is not very secure as it is very easy to know that logic is locked to 1 if any one of the inputs of the OR gate is 1. So to make it much more difficult to understand, a modification is done in which the use of XNOR/NAND/NOR gate is done for securing the logical functioning, but this raises another issue of a large area overhead in the circuit. In Ref. [34], they have used the circuitry already present inside the IC for logic-locking procedure. Nowadays, testing of an IC is equally important, and to do a thorough test, extra circuitry is included in the design known as test points. This paper makes use of those test points as the logic-locking gates. It introduces the PUF and scrambler in the circuit. PUF is used to generate unclonable keys, and scrambler makes the circuit operated in either test mode/functional mode or obfuscated mode.

20.3 COMPARISONS

In Ref. [35], protection is provided to IP cores by embedding a digital signature in the IP. This consumes a very less area overhead but provides security against fake ownership and does not affect the design from overbuilding and reverse-engineering attacks. To provide protection against IP piracy, Rajendran (2012) [36] explored the concept of EPIC in which they locked the design using combinational locking mechanism using a key. This rendered the design useless if the entered key is wrong. This includes the area overheads of the wires and gates for additional locking, TRNG, and so on but provides better security compared with digital signatures. Based on the length of the key entered, the security becomes better. For a 64-bit key, it requires a year to break it as there are more than 2^{20} combinations available. The protection mainly focuses on IP vendors and the end users to modify it. In Ref. [16], they described a means of providing security at manufacturers, IP vendors, chip designers, and end users by obfuscation of net list. This is done by using XOR and OR gates and randomly placing internal wires at a different location. This is performed at different area constraints, and it is observed that the larger the area constraint, the larger the failing vectors. The timing constraints and the power overhead were under the average of 22%, which is well within the manageable limit. This was modified in Ref. [17] to provide protection in RTL by using key, and it was observed that the number of failing vectors, in this case, is very high and the area and power overhead is under 10%. These circuits fail to provide protection against Trojan attacks, which triggers at a rare event, so in Ref. [19], they devised an approach to make the present Trojans benign and also prevent the rare triggering condition to occur. This technique provides a high Trojan coverage to provide protection at an overall area and power overhead of 11% on an average. Ref. [36] was modified in Ref. [37] by inserting the extra gates at a random location, weighted and unweighted locations. It results in a high area overhead and a power–delay product of 25% for random placement and 21% on unweighted placement. In Ref. [38], they have devised another clever method of obfuscation by using dynamic traversal of paths by locking the mode using code lock during state transition time with a nominal area overhead of 7.8%. Regardless of the above safety measures, if an attacker inserts the hardware Trojan in the chip with low controllability, it becomes very difficult to know the affected area, so in Ref. [39], they developed a technique for encryption of the circuit that prevents the introduction of the Trojans in the circuit areas with low controllability. In Ref. [22], the one-level protection is modified to two-level protection, i.e., both structural and functional protection. One of the major benefits of this process is the meaningful output nodes that are functionally correct but still the output is wrong, which makes it hard for the adversary to know the proper working of the design. In Ref. [40], they have utilized the layout-level obfuscation by generating special gates called as obfuscates, which are stored in the obfuscated library and used, which required the adversary to provide extra effort in reverse engineering it. It gives an area overhead of 5.67× on average, but it can be further reduced by using a less number of inputs in obfuscates. The area and power overhead along with proper security is becoming an issue when we talk about the obfuscation, so in Ref. [23], a scheme is developed making use of inverter and mux to form an OC, which helps in providing a better security even if the gate-level net list is obtained by hiding the functionality with an average area overhead of 0.63% and a power overhead of 2.6%. To further modify the structural obfuscation presented in Ref. [22], Yasin (2017) [26] developed a few higher transformation techniques as well as a low cost–driven design by using PSO. It provides an obfuscation of 22% within reduction in the cost of design by 55%. In the aforementioned technique, we use the same OCs for obfuscation of the design; it is easier for the adversary to reverse-engineer all of these once the adversary reverse-engineers one, so Yasin et al. [32] developed a technique of using different OCs in design, making it difficult to reverse-engineer. The area–power–delay constraint in this technique is negligible. For further increasing the security of the circuit, Chakraborty et al. [15] made use of dynamic obfuscation technique in which when the entered key is wrong, the output of the design is inconsistent. The inconsistency can be further randomized, and this prevents the circuits for SAT attacks. The area and power overhead, in this case, is 1%, and security is increased effectively. Another technique developed to make the circuit resilient to SAT-based attack is discussed in Ref. [30]. They have made use of cyclic obfuscation by creating dummy paths in which way this produces an area overhead of 40% less than the use of XOR gate in camouflaging and a delay of 20% less than the delay of use of XOR gate in camouflaging. The technique mentioned before obfuscated the design by adding an additional gate in case of logic-locking technique, but in Ref. [34], they made use of test points to function as the logic-locking architecture to function in testing or normal operation as well as to lock the design in case of wrongly entered key. This increases the design for testability along with providing necessary protection to the circuit.

20.4 FUTURE WORKS

With the rapid increase in the chip density, conventional techniques for IC fabrication have reached its limit. We are shifting to newer technologies such as FinFET, CNTs, DG-MOSFET, and so on. One such technology is the 3D IC. This 3D IC consists of more than two ICs piled vertically on top of each other and connected through TSVs (Through Silicon-Via). Thereby, this can be much more secure. In Ref. [41], split manufacturing technology is being used in 2.5D IC in which the stack of IC is manufactured in foundries, but the connection between them is made at a secure place. This technique is called as split manufacturing technology; the same technique is used in case of 3D IC manufacturing as done in Ref. [42]. This technique provides a better means of security than the conventional technique while having a side advantage of low power and Moore's law utilization.

Another technique is the use of a modified device in place of conventional ones. One such device is the spin-orbit torque device. In Ref. [43], they have made use of spin-orbit torque device that works on the principle of magnetization. The functionality of the device is determined by the magnetization state providing a much more secure design. Another technique used is the hybrid design model of technology. In Ref. [44], they have used a hybrid design by mixing memristor and CMOS structure, which is found to be immune against machine learning attacks. These hybrid models are used for the construction of a PUF circuit in which the provided key is a lot safer against the attack. Another technique that has been actively researched upon for providing obfuscation is cryptography technique [9, 10].

20.5   CONCLUSIONS

With the increase in the use of IC chips in numerous areas, the reverse engineering of these chips has seen a tremendous increase in the past decade. This technique can lead to IP piracy, overbuilding, leakage of secure information, Trojan introduction, malfunctioning of IC, and so on. Securing the IC has become more important than ever. To ensure the safety of the device, many techniques have been researched upon, among which logic obfuscation is most widely used. This provides security to design both structurally and functionally. It makes two functionally different designs look structurally the same, thereby making it difficult for the adversary to reverse-engineer. It also includes the use of camouflaging, dummy gates, and the introduction of different logic blocks. It makes use of a key to validate a circuit; in case of entering a wrong key, the circuit does not operate as intended. To provide more security, the key length was focused upon, which came with a cost of area overhead. This leads to the research in various other alternatives to make the design much more secure and increases the complexity of the circuit keeping the area, power, and delay overhead constraints within permissible limits. These include making use of PUF, hybrid PUF, dynamic obfuscation, different OCs, and so on. As per the need of each design with a specific area and delay constraint, these techniques are utilized. Future works for security on different technologies such as 3D ICs can be a major turn in providing a much higher level of security against reverse engineering.

REFERENCES

1. Foreign infringement of intellectual property rights implications on selected U.S. industries. http://www.usitc.gov/publications/332/working_papers/id_14_100505.pdf.
2. Becker, Georg T., Francesco Regazzoni, Christof Paar, and Wayne P., Burleson. "Stealthy dopant-level hardware trojans: Extended version." *Journal of Cryptographic Engineering* 4, no. 1 (2014): 19–31.
3. Rajendran, Jeyavijayan, Michael Sam, Ozgur Sinanoglu, and Ramesh Karri. Security analysis of integrated circuit camouflaging. In *Proceedings of the 2013 ACM SIGSAC Conference on Computer & Communications Security*, pp. 709–720 (2013).

4. Brzozowski, Maciej, and Vyacheslav N. Yarmolik. Obfuscation as intellectual rights protection in VHDL language. In *2007 CISIM'07. 6th International Conference on Computer Information Systems and Industrial Management Applications*, pp. 337–340 (2007).

5. Elbirt, Adam J., Wei Yip, Brendon Chetwynd, and Christof Paar. "An FPGA-based performance evaluation of the AES block cipher candidate algorithm finalists." *IEEE Transactions on Very Large Scale Integration (VLSI) Systems* 9, no. 4 (2001): 545–557.

6. Nagra, Jasvir, and Christian Collberg. *Surreptitious Software: Obfuscation, Watermarking, and Tamperproofing for Software Protection: Obfuscation, Watermarking, and Tamperproofing for Software Protection*. Pearson Education, USA, 2009.

7. Zhu, William Feng. Concepts and Techniques in Software Watermarking and Obfuscation. *PhD Dissertation*, ResearchSpace@ Auckland, 2007.

8. Ghosh, Santosh, Debdeep Mukhopadhyay, and Dipanwita Roychowdhury. "Secure dual-core cryptoprocessor for pairings over barreto-naehrig curves on FPGA platform." *IEEE Transactions on Very Large Scale Integration (VLSI) Systems* 21, no. 3 (2013): 434–442.

9. Mazumdar, Bodhisatwa, Debdeep Mukhopadhyay, and Indranil Sengupta. "Construction of RSBFs with improved cryptographic properties to resist differential fault attack on grain family of stream ciphers." *Cryptography and Communications* 7, no. 1 (2015): 35–69.

10. Li, Li, and Hai Zhou. Structural transformation for best-possible obfuscation of sequential circuits. In *IEEE International Symposium on Hardware-Oriented Security and Trust (HOST)*, 2013, pp. 55–60 (2013).

11. Lao, Yingjie, and Keshab K. Parhi. Protecting DSP circuits through obfuscation. In *2014 Circuits and Systems (ISCAS)*, pp. 798–801 (2014).

12. Bitansky, Nir, and Vinod Vaikuntanathan. Indistinguishability obfuscation from functional encryption. In *2015 IEEE 56th Annual Symposium on Foundations of Computer Science (FOCS)*, pp. 171–190 (2015).

13. Suh, G. Edward, and Srinivas Devadas. Physically unclonable functions for device authentication and secret key generation. In *Design Automation Conference, 2007. DAC'07. 44th ACM/IEEE*, pp. 9–14 (2007).

14. Koteshwara, Sandhya, Chris H. Kim, and Keshab K. Parhi. "Key-based dynamic functional obfuscation of integrated circuits using sequentially triggered mode-based design." *IEEE Transactions on Information Forensics and Security* 13, no. 1 (2018): 79–93.

15. Chakraborty, Rajat Subhra, and Swarup Bhunia. "HARPOON: An obfuscation-based SoC design methodology for hardware protection." *IEEE Transactions on Computer-Aided Design of Integrated Circuits and Systems* 28, no. 10 (2009): 1493–1502.

16. Chakraborty, Rajat Subhra, and Swarup Bhunia. RTL hardware IP protection using key-based control and data flow obfuscation. In *2010 23rd International Conference on VLSI Design, 2010. VLSID'10*, pp. 405–410 (2010).

17. Tehranipoor, Mohammad, and Farinaz Koushanfar. "A survey of hardware trojan taxonomy and detection." *IEEE Design & Test of Computers* 27, no. 1 (2010): 10–25.

18. Chakraborty, Rajat Subhra, and Swarup Bhunia. "Security against hardware Trojan attacks using key-based design obfuscation." *Journal of Electronic Testing* 27, no. 6 (2011): 767–785.

19. Rajendran, Jeyavijayan, Youngok Pino, Ozgur Sinanoglu, and Ramesh Karri. Security analysis of logic obfuscation. In *Proceedings of the 49th Annual Design Automation Conference*, pp. 83–89 (2012).

20. Jin, Year. "Introduction to hardware security." *Electronics* 4, no. 4 (2015): 763–784.

21. Lao, Yingjie, and Keshab K. Parhi. "Obfuscating DSP circuits via high-level transformations." *IEEE Transactions on Very Large Scale Integration (VLSI) Systems* 23, no. 5 (2015): 819–830.
22. Zhang, Jiliang. "A practical logic obfuscation technique for hardware security." *IEEE Transactions on VLSI Systems* 24, no. 3 (2016): 1193–1197.
23. Vijayakumar, Arunkumar, Vinay C. Patil, Daniel E. Holcomb, Christof Paar, and Sandip Kundu. "Physical design obfuscation of hardware: A comprehensive investigation of the device and logic-level techniques." *IEEE Transactions on Information Forensics and Security* 12, no. 1 (2017): 64–77.
24. Becker, Georg T., Marc Fyrbiak, and Christian Kison. Hardware obfuscation: Techniques and open challenges. In *Foundations of Hardware IP Protection*, pp. 105–123. Springer, 2017.
25. Sengupta, Anirban, Dipanjan Roy, Saraju P. Mohanty, and Peter Corcoran. "DSP design protection in CE through algorithmic transformation based structural obfuscation." *IEEE Transactions on Consumer Electronics* 63, no. 4 (2017): 467–476.
26. Yasin, Muhammad, Bodhisatwa Mazumdar, Ozgur Sinanoglu, and Jeyavijayan Rajendran. "Removal attacks on logic locking and camouflaging techniques." *IEEE Transactions on Emerging Topics in Computing* 1 (2017): 1–1.
27. Yasin, Muhammad, Bodhisatwa Mazumdar, Ozgur Sinanoglu, and Jeyavijayan Rajendran. Security analysis of anti-sat. In *Design Automation Conference (ASP-DAC), 2017 22nd Asia and South Pacific*, pp. 342–347 (2017).
28. Yasin, Muhammad, Bodhisatwa Mazumdar, Ozgur Sinanoglu, and Jeyavijayan Rajendran. "Removal attacks on logic locking and camouflaging techniques." *IEEE Transactions on Emerging Topics in Computing* 1 (2017): 1.
29. Shamsi, Kaveh, Meng Li, Travis Meade, Zheng Zhao, David Z. Pan, and Yier Jin. Cyclic obfuscation for creating sat-unresolvable circuits. In *Proceedings of the on Great Lakes Symposium on VLSI 2017*, pp. 173–178 (2017).
30. Zhang, Grace Li, Bing Li, Bei Yu, David Z. Pan, and Ulf Schlichtmann. TimingCamouflage: Improving circuit security against counterfeiting by unconventional timing. In *2018 Design, Automation & Test in Europe Conference & Exhibition (DATE)*, pp. 91–96 (2018).
31. Sumathi, G., L. Srivani, D. Thirugnana Murthy, Anish Kumar, and K. Madhusoodanan. Hardware obfuscation using different obfuscation cell structures for PLDs. In SG-CRC, pp. 143–157 (2017).
32. Yasin, Muhammad, Jeyavijayan, JV Rajendran, Ozgur Sinanoglu, and Ramesh Karri. "On improving the security of logic locking." *IEEE Transactions on Computer-Aided Design of Integrated Circuits and Systems* 35, no. 9 (2016): 1411–1424.
33. Chen, Michael, Elham Moghaddam, Nilanjan Mukherjee, Janusz Rajski, Jerzy Tyszer, and Justyna Zawada. "Hardware protection via logic locking test points." *IEEE Transactions on Computer-Aided Design of Integrated Circuits and Systems*, 37, no. 12 (2018): 3020–3030.
34. Castillo, Encarnacin, Uwe Meyer-Baese, Antonio García, Luis Parrilla, and Antonio Lloris. "IPP@ HDL: Efficient intellectual property protection scheme for IP cores." *IEEE Transactions on Very Large Scale Integration (VLSI) Systems* 15, no. 5 (2007): 578–591.
35. Roy, Jarrod A., Farinaz Koushanfar, and Igor L. Markov. EPIC: Ending piracy of integrated circuits. In *Proceedings of the Conference on Design, Automation and Test in Europe*, pp. 1069–1074 (2008).
36. Rajendran, Jeyavijayan, Youngok Pino, Ozgur Sinanoglu, and Ramesh Karri. Security analysis of logic obfuscation. In *Proceedings of the 49th Annual Design Automation Conference*, pp. 83–89 (2012).

37. Desai, Avinash R., Michael S. Hsiao, Chao Wang, Leyla Nazhandali, and Simin Hall. Interlocking obfuscation for anti-tamper hardware. In *Proceedings of the Eighth Annual Cyber Security and Information Intelligence Research Workshop*, p. 8 (2013).
38. Dupuis, Sophie, Papa-Sidi Ba, Giorgio Di Natale, Marie-Lise Flottes, and Bruno Rouzeyre. A novel hardware logic encryption technique for thwarting illegal overproduction and hardware trojans. In *On-Line Testing Symposium (IOLTS), 2014 IEEE 20th International*, pp. 49–54 (2014).
39. Malik, Shweta, Georg T. Becker, Christof Paar, and Wayne P. Burleson. Development of a layout-level hardware obfuscation tool. In *2015 IEEE Computer Society Annual Symposium on VLSI (ISVLSI)*, pp. 204–209 (2015).
40. Imeson, Frank, Ariq Emtenan, Siddharth Garg, and Mahesh V. Tripunitara. Securing computer hardware using 3D integrated circuit (IC) technology and split manufacturing for obfuscation. In *USENIX Security Symposium*, pp. 495–510 (2013).
41. Xie, Yang, Chongxi Bao, Yuntao Liu, and Ankur Srivastava. 2.5 D/3D integration technologies for circuit obfuscation. In *2016 17th International Workshop on Microprocessor and SOC Test and Verification (MTV)*, pp. 39–44 (2016).
42. Yang, Jianlei, Xueyan Wang, Qiang Zhou, Zhaohao Wang, Hai Li, Yiran Chen, and Weisheng Zhao. "Exploiting spin-orbit torque devices as reconfigurable logic for circuit obfuscation." *IEEE Transactions on Computer-Aided Design of Integrated Circuits and Systems* (2018).
43. Mathew, Jimson, Rajat Subhra Chakraborty, Durga Prasad Sahoo, Yuanfan Yang, and Dhiraj K. Pradhan. "A novel memristor-based hardware security primitive." *ACM Transactions on Embedded Computing Systems (TECS)* 14, no. 3 (2015): 60.

Index

A

Active inductor (AI), 314–315
  equivalent circuit of lossy, 316
  gyrator-C, 314–315
  reconfigurable VCO with, 316
  VCO with differential, 318
  voltage-controlled oscillator using, 315–319
Active inductor–based VCO
  active inductor, 314–315
  discrete Fourier transform for VCO,
    319–320
  overview, 311–313
  ring voltage-controlled oscillator, 314
  voltage-controlled oscillator using active
    inductor, 315–319
  for wireless communication, 311–322
Adaptive antenna array, 294
Adder and subtractor circuit, 25–27
Addition of ceramic fillers, 228–229
AHPL (A Hardware Programming
    Language), 95
AlGaN/GaN heterostructure field-effect
    transistor (HFET)
  model description, 216
  overview, 215–216
  results and discussions, 216–221
Algorithmic modeling, 54–55
Analog design
  described, 18
  vs. digital design, 18
Analog-to-digital conversion, 297–298
Analog-to-digital converters, 297–305
  basics of analog-to-digital conversion,
    297–298
  delta-sigma, 305
  flash, 303–304
  pipelined, 304–305
  quantization, 299–300
  sampling, 298–299
Analytes, 148
APL (artificial language), 94–95
Architecture, 72
  behavioral modeling, 73
  dataflow modeling, 72
  structural modeling, 73–74
Armchairedge GNR (AGNR), 138–139
Artificial neural network (ANN), 259–261, 263
Asynchronous counters, 33–34
Atmel, 9
@ (sensitivity_list), 49

B

Beetle/chip FET, 182, 184
Behavioral modeling, 54–55
  Verilog, 107–108
  VHDL, 73, 101
Bell, John, 306
Binary number systems, 19
Biochemical-based biosensor, 170
Bio-FETs
  applications, 170, 178
  cell-based, 180–183
  classification and advances in, 176–179
  different improvement techniques, 177–178
  DNA-modified FET or GenFET, 178–179
  dual-gate ISFET (DG-ISFET), 172–173
  enzyme FET (EnFET), 176
  FET, a Bio-FET's basic structure, 170–171
  ISFET, a pH Sensor, 171–172
  ISFET bioapplications, 173
  major benchmarks in Bio-FET device, 177
  types of, 184
  working, 176–177
Bioreceptors, 148
Biosensors
  advantages, 148–149
  biochemical-based, 170
  components, 148
  defined, 169–170
  FET-based, 150–165
  introduction, 147–148
  low-power FET-based (*see* Bio-FETs)
  penicillin-sensitive biosensor, 176
  silicon nanowire, 173–174
  types of, 149–150
Blocking and nonblocking statements, 55–56
Boolean equation, 22–24
Bulk MOSFET, 115–116
Butler matrices, 294
Byte, defined, 2

C

Calorimetric biosensor, 150
Cantatore, Eugenio, 302
Carbon nanotube biosensor, 153–154
Carbon nanotube field-effect transistors
    (CNTFETs), 137
Cell-based Bio-FETs, 180–183
  application, 182–183
  beetle/chip FET, 182
  working, 181–182

351

Ceramic fillers, addition, 228–229
Chandrakasan, Anantha P., 303
Channel length (L_{CH}), 113
Charge pump (CP-PLLs), 273
Charge pump circuits, in PLL for reference spur
    reduction, 277–281
Choi, Seunghwar, 304
CNT FET, 139
Code writing
    with Verilog, 108–109
    with VHDL, 101–103
Combinational logic circuits
    adder and subtractor, 25–27
    analysis and design, 24–25
    Boolean equation, 22–24
    decoder, 27
    demultiplexer, 28–29
    encoder, 27, 28
    introduction, 24
    multiplexer, 27–29
    vs. sequential circuits, 31
Comment lines, 41–42, 74
Compiled code simulation, 326
Complementary metal oxide semiconductor
    (CMOS), 22, 23, 112
    challenges of scaling down, 132–136
    deep submicron, 210
    existing low-power techniques, 235–239
    nanodevices beyond, 136–142
    proposed low-power adiabatic logic
        techniques, 239–241
    scaling down, 130
    scaling down solutions with FET geometries,
        132–136
    SRAM cell architectures based on,
        204–206
Complex programmable array logic (CPLD),
    5, 8–9
Concurrent fault simulation, 326, 327,
    328–330
Configurable logic blocks, 10–12
Constant value, Verilog HDL, 45
Continuous assignments, 76
Conventional complementary metal oxide
    semiconductor, 235–236
Conventional positive-feedback adiabatic logic,
    240–241
CoolRunner, 9
Copolymerization, 228
Cortes, Corinna, 261
"Coulomb blockade" phenomenon, 141
Counters
    asynchronous, 33–34
    digital circuits, 33–34
    registers, 34, 35
    synchronous, 34, 35
Cypress Semiconductor, 9

D

Dataflow model/modeling, 56–57
    Verilog, 107
    VHDL, 72, 100
DDL, 95
Debugging, hardware description
    languages, 96
Decimator, 305–306
Decoder, 27
Deductive fault simulation, 326, 327, 328
De Jonghe, Dimitri, 258, 259
Delta–sigma analog-to-digital converter, 305
Delta time, 70
Demultiplexer, 28–29
Depleted lean channel transistor (DELTA), 117
Design simulation, hardware description
    languages, 96
Design under Test (DUT), 79
Devices for biomedical signals, 14
D flip-flops, 31, 32
Dielectrically modulated field-effect transistor
    (FET)-based biosensors, 158–165
    electric field, 160–162
    modeling of, 158–165
    sensitivity, 163–164
    surface potential, 158–160
    threshold voltage, 162–163
Differential fault simulation, 326, 327, 330
Differential pass-transistor logic style, 237
Digital circuits
    counters, 33–34
    memory, 34–36
    storage elements, 30–33
Digital design
    vs. analog design, 18
    hardware description language for, 71
    introduction, 18–19
    Verilog HDL for, 41–54
Digital integrated circuit, 19–20
Digital logic-circuit, 19–20
Digital-to-analog converters, 297–306
    decimator, 305–306
    feedback digital-to-analog converter, 305
Diode–transistor logic (DTL), 20, 21
Discrete Fourier transform for voltage-controlled
    oscillator, 319–320
DNA-modified FET or GenFET, 178–179
Dong, Ning, 259
Dong-Shin Jo, 305
Double-gate MOSFET (DG-MOSFET), 117,
    191, 276
Double-gate silicon-on-insulator (DGSOI)
    MOSFET, 117
Driver, System Verilog, 80–81
Dual-gate ISFET (DG-ISFET), 172–173
Dynamic RAM (DRAM), 36

E

E²ROM, *see* Electrically erasable programmable read-only memory (EEPROM)
8T SRAM cell, 205–206
Electrically erasable programmable read-only memory (EEPROM), 5
Electric field, of FET-based biosensor, 160–162
Electrochemical biosensor, 149–150
Electrolytes
  gel electrolytes, 226, 229
  liquid electrolytes, 226, 229
  solid electrolytes, 226, 227
Emitter-coupled logic (ECL), 20
Encoder, 27, 28
Entity declaration, 72
Enzyme FET (EnFET), 176
  applications, 178
  different improvement techniques, 177–178
  DNA-modified FET or GenFET, 178–179
  working, 176–177
Erasable programmable read-only memory (EPROM), 4–5
Event-driven simulation, 326
Event processing, 70
Extreme learning machine (ELM), 262–263

F

Factory-programmable devices
  EEPROM, 5
  EPROM, 4–5
  PROM, 4
  ROM, 2–4
Fall delay, 69
Farooq, Muhammad Umer, 259
Fault simulation, 327–330
  concurrent, 327, 328–330
  deductive, 326–328
  differential, 326, 327, 330
  parallel, 326, 327–328
  serial, 326, 327
Fault simulation algorithms
  fault simulation, 327–330
  logic simulation, 325–326
  overview, 325
  Verilog coding for simulation, 330–338
Feedback digital-to-analog converter, 305
  field effect transistor (FET), 170–171; *see also* Bio-FETs
  based biosensors, 150–165
  CNT, 139
  graphene, 139
  nanowire, 139
Field-effect transistor (FET)-based biosensors, 150–165
  modeling, 156–165

modeling of dielectrically modulated, 158–165
  types of, 152–156
  working of, 151–152
Field programmable devices
  CPLD, 5, 8–9
  FPGA, 5, 10–14
  GAL, 5, 8
  PAL, 5, 6–7
  PLA, 5, 7–8
Field-programmable gate array (FPGA)
  applications in medical imaging, 14
  configurable logic blocks, 10–12, 11
  design flow of, 12–14, 13
  internal architecture, 10–12
  IOBs, 12
  overview, 10
  programmable interconnect, 12
FinFET, 117, 134–136
  SRAM cell architectures based on, 206–209
Flash analog-to-digital converter, 303–304
Flash memory, 36
Flip-flops
  D, 31, 32
  JK, 31, 32, 33
  SR, 31, 32
  storage elements, 31–33
  T, 32, 33
Flynn, Michael P., 304, 306
4×1 multiplexer
  using conventional complementary metal oxide semiconductor, 241, 242
  using conventional positive-feedback adiabatic logic, 243
  using differential pass-transistor logic style, 242, 243
  using gate diffusion input logic style, 242
  using pass-transistor logic style, 241
  using transmission gate logic style, 242
  using two-phase adiabatic static clocked logic, 245
Functionality, defined, 131
Functions, VHDL, 98–99

G

Galton, Ian, 306
Gate-all-around (GAA) MOSFET
  brief review on sentaurus TCAD, 193–195
  device design and simulation, 195
  overview, 189–193
  results and discussions, 195–200
Gate delays, 69
Gate diffusion input logic style, 238–239
Gate leakage, 210
Gate-level model, 57–58
Gate primitives, 50

Gateway Design Automation, 40
Geim, A., 123
Gel electrolytes, 226, 229
Generate statement, 50
Generator, System Verilog, 80
Generic array logic (GAL) devices, 5, 8, 117
Geometric programming, 256, 263
Gielen, Georges, 258, 259
Graphene FET (GFET), 139
   biosensors, 175–176
Graphene nanoribbon (GNR) transistors, 137–139
Graphene transistors, 123
Gray, Paul R., 304
Gu, Chenjie, 258

H

Hamilton, Tara Julia, 306
Hardware description languages (HDL)
   basic principles of, 68–69
   concepts of, 69–70
   debugging with, 96
   design flow, 97
   design flow expense, 96
   design simulation with, 96
   design tool suites, 70
   design using, 71
   for digital design, 71
   hardware simulation process, 70
   history, 94–95
   introduction to, 40–41, 68, 94
   motivation of developing, 94
   and programming language, 95
   structure of, 94
   simulation of circuits, 94
   timing and concurrency, 69–70
   types of, 70–71
   verification with, 96
   Verilog, 40–41
   very high-scale integrated circuits,
      71–74
   VHSIC, 40
Hardware obfuscation
   comparisons, 345–346
   future works, 346–347
   logic working, 341–345
   overview, 339–341
Hardware security, 340, 342, 343
Hardware simulation process, 70
Harpe, Pieter, 302
Hayashi, Y., 117, 191
Hexadecimal number systems, 19
Hieu, Bui Van, 304
High-electron mobility transistor (HEMT),
      123–124
High-k gate dielectric, 119
Hyung-Jong Ko, 305

I

Identifiers, 42, 74
IEEE International Solid-State Circuits
      Conference, 303
IEEE Journal of Solid-State Circuits, 301,
      302–303, 304, 305, 306
ImmunoFET, 179–180
   applications, 180
Inductor–capacitor voltage-controlled oscillator
      (VCO)
   cross-coupled approach, 313
   linear feedback approach, 312–313
Integrated circuit (IC), 19–20
Intel (Altera), 9, 10
Interface, System Verilog, 80
International Conference on Green
      Technologies, 304
*International Journal of Soft Computing and
      Engineering (IJSCE)*, 304
International Technology Roadmap for
      Semiconductors (ITRS), 114, 129,
      131–132
International Telecommunication Union
      (ITU), 292
International Union of Pure and Applied
      Chemistry (IUPAC), 169
I/O blocks (IOBs), 12
Ion-sensitive field-effect transistor biosensor
      (ISFET), 152, 153
Irfansyah, Astria Nur, 306
ISFET, 176–183
   bioapplications, 173
   a pH sensor, 171–172
Ishikuro, Hiroki, 302
ISPS, 95
ITRS (International Technology Roadmaps for
      Semiconductors), 190

J

Jambunathan, Sreenivas, 306
Jenkins, Julian, 306
JK flip-flops, 31–33
Junctionless field-effect transistor biosensor
      (JLT), 155–156
Junctionless MOSFET, 119–120
Junction tunneling leakage, 211

K

Keywords, Verilog HDL, 41, 75
Klein, Leonhard, 306
Kolhe, Amit, 304
Kumar, Pradeep, 304
Kuroda, Tadahiro, 302
Kyoung-Jun Moon, 305

L

Latches, 30–31
Lattice Semiconductor, 8, 9, 10
Leakage reduction techniques, 211–213
    lower self-controllable voltage level, 211
    self-controllable voltage level technique, 211
    upper self-controllable voltage level, 211–213
Lehmann, Torsten, 306
Lewis, Stephen H., 304
Lexical tokens, 74–75
Libraries, VHDL, 98–99
Liquid electrolytes, 226, 229
Logic blocks, 9
    configurable, 10–12
Logic families
    complementary metal oxide semiconductor,
        22, 23
    digital integrated circuit, 19–20
    diode–transistor logic, 20, 21
    emitter-coupled logic, 20
    resistor–transistor logic, 20, 21
    transistor–transistor logic, 22
Logic gate delays, 69
Logic simulation, 325–326
    compiled code simulation, 326
    event-driven simulation, 326
Logic working, 341–345
Loop filter or low-pass filter, 274
Lower self-controllable voltage level, 211
Low-noise fast PLLs using multi-PFD
        architectures, 281–283
Low-power adiabatic logic techniques
    conventional positive-feedback adiabatic
        logic, 240–241
    proposed, 239–241
    two-phase adiabatic static clocked
        logic, 241
Low-power FET-based biosensors
    cell-based Bio-FETs, 180–183
    classification and advances in, 176–179
    immunofet, 179–180
    organic field-effect transistor, 174–176
    overview, 169–170
    principle of operation, 170–173
    silicon nanowire biosensor, 173–174
Low-power techniques
    comparative analysis, 245–247
    conventional complementary metal oxide
        semiconductor, 235–236
    differential pass-transistor logic style, 237
    existing, 235–239
    existing design, 241–243
    gate diffusion input logic style, 238–239
    pass-transistor logic style, 236–237
    proposed design, 243–245
    transmission gate logic style, 237–238

M

Macromodeling
    nonparametric, 259–263
    overview, 251–252
    parametric-based, 252–259
Magnetic storage, 36
Mahapatra, K.K., 304
McCulloch, Warren, 259
Memory
    described, 34
    digital circuits, 34–36
    flash, 36
    magnetic storage, 36
    optical storage, 36
    RAM, 36
    ROM, 35
Memristor, 122–123
Metal gate electrodes, 119
Metal oxide semiconductor field-effect transistor
        (MOSFET), 112
    bulk, 115–116
    double-gate silicon-on-insulator, 117
    junctionless, 119–120
    PDSOI, 116
    scaling down, 129–130
    SOI, 116
    surrounding-gate silicon-on-insulator, 118
    topologies, 119–124
    triple-gate silicon-on-insulator, 118
Michael Choi, 305
Microsemi (Actel), 10
Mixed model, 59
Model order reduction, 256–259, 263
Monitor, System Verilog, 81
Moore, Gordon, 113
Moore's law, 113, 129, 131–132
More than Moore (MtM), 113
Multibridge channel MOSFET (MBCFET), 118
Multiplexer, 27–29, 330–338

N

Nanobeam stacked channels, 118
Nanoelectronic devices, *see* Nanotechnology
Nanofillers, 230
Nanotechnology
    applications, 143
    beyond CMOS, 136–142
    emerging improvements, 115–124
    limitations, 143
    material technology, 118–119
    overview, 112–113
    technical challenges and solutions, 142–144
Nanowire (NW) FETs, 136, 139
Nanowire field-effect transistor biosensor, 153
Nanowire (NW) transistors, 136–137

Nauta, Bram, 306
Nicholson, Andrew, 306
Nikas, Antonios, 306
90-nm CMOS technology, 233–248
Nonparametric macromodeling, 259–263
    artificial neural network (ANN),
        259–261, 263
    extreme learning machine (ELM), 262–263
    support vector machine (SVM),
        261–262, 263
Number systems
    binary, 19
    hexadecimal, 19
    octal, 19
Number value, Verilog HDL, 45

O

Octal number systems, 19
Operators, 52–54, 75
    Verilog, 105, 106
Optical biosensor, 150
Optical storage, 36
Organic field-effect transistors (OFETs),
        139–140, 174–176
    applications, 175
    graphene FET (GFET) biosensors, 175–176
Ortmanns, Maurits, 306

P

Packages, VHDL, 98–99
Parallel fault simulation, 327–328
Parameters, Verilog HDL, 45
Parametric-based macromodeling, 252–259
    geometric programming, 256, 263
    model order reduction, 256–259, 263
    posynomial templates, 256, 263
    symbolic modeling, 253–255, 263
Pass-transistor logic style, 236–237
Pavan, Shanthi, 306
PDSOI MOSFET, 116
Penicillin-sensitive biosensor, 176
Performance-linked (PLL) architectures
    charge pump circuits in PLL for reference
        spur reduction, 277–281
    charge pump current mismatch, 284
    dead zone, 284–285
    delay in feedback loop, 285
    loop filter leakage current, 284
    low-noise fast PLLs using multi-PFD
        architectures, 281–283
    PFD architectures in PLL to reduce dead
        zone, 274–277
    PLL sampling effect, 285
    wide-lock range PLLs using AFC techniques,
        283–284

Performance-linked phase-locked loop
        architectures
    design challenges, 284–285
    overview, 271–272
    performance-linked phase-locked loop
        components, 272–274
    recent performance-linked
        architectures of phase-locked
        loop blocks, 274–284
Performance-linked phase-locked loop
        components
    charge pump (CP-PLLs), 273
    loop filter or low-pass filter, 274
    phase frequency detector (PFD), 272–273
    voltage-controlled oscillator (VCOs),
        273–274
Perl, 71
PFD architectures in PLL to reduce dead zone,
        274–277
Pham, Long, 306
Phase frequency detector (PFD), 272–273
Piezoelectric (PE) biosensor, 150
Pipelined analog-to-digital converter, 304–305
Pitts, Walter, 259
Plasticization, 229–230
Polymer-based composites, 226–227
    addition of ceramic fillers, 228–229
    copolymerization, 228
    methods of synthesis, 228–230
    nanofillers, 230
    overview, 225–226
    plasticization, 229–230
    sol–gel process, 229
    solution casting method, 228
Ports, 52
Posynomial templates, 256, 263
Power reduction techniques, 211
Procedural assignment, 76–77
Programmable array logic (PAL), 5, 6–7
Programmable logic array (PLA), 5, 7–8
    advantages, 8
    problems with, 8
Programmable logic devices (PLDs)
    hierarchical structure, 3
    overview, 1–2
    types of, 2
Programmable read-only memory (PROM), 4
Programming languages, 95
Program structure
    Verilog, 104–105
    VHDL, 97–98
PROM programmer, 4
Proposed low-power adiabatic logic techniques
    conventional positive-feedback adiabatic
        logic, 240–241
    two-phase adiabatic static clocked
        logic, 241

Q

Quantization, 299–300

R

Ramachandran, R., 304
Random access memory (RAM), 5, 36
Read-only memory (ROM), 2–4, 35
Receiver, in SAs, 294–296
Registers, 34, 35
Register transfer-level model, 56–57
Resistor–transistor logic (RTL), 20, 21
Ring voltage-controlled oscillator (VCO), 314
Rise delay, 69
Roermund, Arthur van, 302
Rosenblatt, Frank, 259
Roychowdhury, Jaijeet, 259

S

Sampling, 298–299
Scoreboard, System Verilog, 81–82
Scripting language, 71
Sekigawa, T., 117, 191
Sekimoto, Ryota, 302
Self-controllable voltage level technique, 211
Sensitivity, of FET-based biosensor, 163–164
Sentaurus Structure Editor, 193–194
Sentaurus TCAD
    brief review on, 193–195
    codes, 194–195
Sequential circuits
    block diagram of, 29
    vs. combinational circuits, 31
    designing, 29–30
    introduction, 29
    types of, 30
Sequential statements
    begin … end construct, 45
    case, 46–47
    continuous assignment statement, 48–49
    if … else statement, 46
    loops, 47–48
    Verilog HDL, 45–49
Serial fault simulation, 326, 327
Serial protocol, 5
Seung-Tak Ryu, 305
7T CMOS SRAM cell, power consumed, 209
7T SRAM cell, 205
    architectures based on CMOS technology,
        204–206
    architectures based on FINFET technology,
        206–209
    different types of leakage current in SRAM,
        210–213
    hold process, 209

operations of, 208–209
overview, 203–204
proposed design, 207–208
read process, 208
result analysis, 209
write process, 208
Shikata, Akira, 302
Shockley–Red–Hall (SRH) generation, 194
Short-channel effects (SCEs), 113–115, 134
Silicon nanowire (SiNW) biosensor, 153, 173–174
    applications, 174
    fabrication process, 174
Simulations
    compiled code, 326
    event-driven, 326
    fault, 327–330
    logic, 325–326
    time, 70
    Verilog coding for, 109, 330–338
    VHDL, 103–104
Single-electron transistors (SETs), 140–142
6T SRAM cell, 204–205
Smart antennas (SAs)
    adaptive antenna array, 294
    architecture of, 294–297
    Butler matrices and, 294
    receiver, 294–296
    switched beam (SB) antennas, 294
    transmitter, 296–297
SOI MOSFET, 116
    multigate, 116–117
    schematic view of multigate, 117
Sol–gel process, 229
Solid electrolytes, 226, 227
Solution casting method, 228
Spin transistors, 140
Spintronics, 121–122
SRAM (static random access memory), 36
    different leakage reduction techniques,
        211–213
    different types of leakage current in, 210–213
    gate leakage, 210
    junction tunneling leakage, 211
    lower self-controllable voltage level, 211
    self-controllable voltage level technique, 211
    subthreshold leakage current, 210
    upper self-controllable voltage level, 211–213
SRAM cell architectures
    based on CMOS technology, 204–206
    based on FinFET technology, 206–209
    8T SRAM cell, 205–206
    operations of, 7T SRAM, 208–209
    proposed design, 7T SRAM cell, 207–208
    7T SRAM cell, 205
    6T SRAM cell, 204–205
    10T SRAM cell, 206
    12 T SRAM cell, 206

SR flip-flops, 31, 32
Storage elements
    digital circuits, 30–33
    flip-flops, 31–33
    latches, 30–31
Strained silicon, 119
Structural modeling/model, 57–58
    Verilog, 106–107
    VHDL, 73–74, 99–100
Subprograms, 98–99
Subthreshold leakage current, 210
Successive approximation register (SAR) ADC,
        300–303
Sukumaran, Amrith, 306
Support vector machine (SVM), 261–262, 263
Surface potential, 158–160
Switched beam (SB) antennas, 294
Switch-level model, 58–59
Switch-level primitives, 51
Symbolic modeling, 253–255, 263
Synchronous counters, 34, 35
Synthesis
    Verilog, 109
    VHDL, 103–104
SystemC models, 95
System Verilog, 41
    described, 78
    Design under Test (DUT), 79
    driver, 80–81
    environment, 82–83
    generator, 80
    interface, 80
    miniproject, 85–91
    monitor, 81
    scoreboard, 81–82
    test bench structure for, 78–84
    top module, 83–84
    transaction, 79–80

T

Taylor, Gerry, 306
T channel FET (ITFET), 118
10T SRAM cell, 206
Test bench
    simulator directives, 60–62
    System Verilog, 78–84
    Verilog HDL, 60–62, 77–78
T flip-flops, 32, 33
Threshold voltage (V_{th}), 162–163
Timescale compiler directive, 75
Time value (#), 49
Timing, and concurrency, 69–70
Top module, System Verilog, 83–84
Transducers, 148
Transistors, *see specific types*
Transistor scalability, 113

Transistor–transistor logic (TTL), 22
Transmission gate logic style, 237–238
Transmitter, in SAs, 296–297
Trigate, 118
Tristate gates, 50–51
Tunnel field-effect transistor (TFET), 121
    biosensor, 155
Turnoff delay, 69
12 T SRAM cell, 206
Twin-Si nanowire (TSNWFET), 118
Two-phase adiabatic static clocked logic, 241

U

Upper self-controllable voltage level, 211–213
U.S. Department of Defense, 95
User-defined primitives (UDPs)
    described, 59
    Verilog HDL, 59–60

V

Vapnik, Vladimir, 261
Varghese, George Tom, 304
Variables
    net data, 42–43
    register, 43–44
    Verilog, 105
    Verilog HDL, 42–44
    VHDL, 97–98
Vector data, 44–45
Verification, hardware description languages, 96
Verilog, 10, 95, 96
    design elements, 106–109
    introduction, 104–109
    module declarations, 105
    operators, 105, 106
    program structure, 104–105
    simulation and synthesis, 109
    variables, 105
Verilog-A, 63–65
Verilog-AMS, 62–63, 84–85
Verilog coding for simulation, 330–338
    multiplexer, 330–338
Verilog HDL (VHDL), 10, 40–41
    always and initial blocks, 77
    for analog design, 62–63
    @ (sensitivity_list), 49
    code writing, 101–103
    comment lines, 41–42, 74
    continuous assignments, 76
    data types, 75
    design elements, 99–103
    design module, 51–52
    for digital design, 41–54
    functions, 98–99
    gate primitives, 50

generate statement, 50
identifiers, 42, 74
introduction, 96–104
keywords of, 41, 75
lexical tokens, 74–75
libraries, 98–99
miniproject, 85–91
modeling types, 54–59
number or constant value, 45
operator types, 52–54, 75
packages, 98–99
parameters, 45
procedural assignment, 76–77
sequential statements, 45–49
simulations and synthesis, 103–104
switch-level primitives, 51
test bench, 60–62, 77–78
timescale, 75
time value (#), 49
tristate gates, 50–51
user-defined primitives, 59–60
variables, 42–44, 98
vector data, 44–45
whitespace, 42, 74
Verilog IEEE standard, 41
VHSIC (very highspeed integrated circuits), 40, 95
VLSI technology, 234
Voelker, Matthias, 306
Voltage-controlled oscillator (VCOs), 273–274
comparison of AI-based, 321
discrete Fourier transform for, 319–320
ring, 314
used for frequency translation, 274
using active inductor, 315–319

W

Wan Kim, 305
Whitespace, 42, 74
Wide-lock range PLLs using AFC techniques, 283–284
Wireless communication
active inductor, 314–315
active inductor–based VCO for, 311–322
deployment of SAs for, 292–293
discrete Fourier transform, 319–320
inductor–capacitor voltage-controlled oscillator, 312–313
ring voltage-controlled oscillator, 314
voltage-controlled oscillator using active inductor, 315–319

X

Xilinx, 9, 10
Xueyi Liu, 257

Y

Yip, Marcus, 303
Yong Chen, 258

Z

Zeus, 95
Zigzag-edge GNR (ZGNR), 138–139

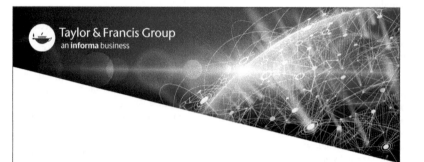